Springer Proceedings in Mathematics & Statistics

Volume 336

Springer Proceedings in Mathematics & Statistics

This book series features volumes composed of selected contributions from workshops and conferences in all areas of current research in mathematics and statistics, including operation research and optimization. In addition to an overall evaluation of the interest, scientific quality, and timeliness of each proposal at the hands of the publisher, individual contributions are all refereed to the high quality standards of leading journals in the field. Thus, this series provides the research community with well-edited, authoritative reports on developments in the most exciting areas of mathematical and statistical research today.

More information about this series at http://www.springer.com/series/10533

Gregory E. Fasshauer • Marian Neamtu
Larry L. Schumaker
Editors

Approximation Theory XVI

Nashville, TN, USA, May 19–22, 2019

Springer

Editors
Gregory E. Fasshauer
Department of Applied Mathematics
and Statistics
Colorado School of Mines
Golden, CO, USA

Marian Neamtu
Department of Mathematics
Vanderbilt University
Nashville, TN, USA

Larry L. Schumaker
Department of Mathematics
Vanderbilt University
Nashville, TN, USA

ISSN 2194-1009 ISSN 2194-1017 (electronic)
Springer Proceedings in Mathematics & Statistics
ISBN 978-3-030-57463-5 ISBN 978-3-030-57464-2 (eBook)
https://doi.org/10.1007/978-3-030-57464-2

This Springer imprint is published by the registered company Springer Nature Switzerland AG.
The registered company address is: Gewerbestrasse 11, 6330 Cham, Switzerland

Preface

These proceedings are based on papers presented at the international conference on *Approximation Theory XVI*, which was held May 19–22, 2019 at Vanderbilt University in Nashville, Tennessee. The conference was the sixteenth in a series of meetings in Approximation Theory held at various locations in the USA. The previous conferences in the series were held in Austin (1973, 1976, 1980, and 1992), College Station (1983, 1986, 1989, and 1995), Nashville (1998), St. Louis (2001), Gatlinburg (2004), and San Antonio (2007, 2010, 2013, and 2016).

The conference was attended by 134 participants from 20 countries and included 8 plenary lectures, 73 minisymposium talks, and 36 contributed talks. We would like to thank all who attended, and in particular our plenary speakers Costanza Conti (Università degli Studi di Firenze), John A. Evans (University of Colorado at Boulder), Frances Kuo (University of New South Wales), Doug Hardin (Vanderbilt University), Deanna Needell (University of California at Los Angeles), Rodrigo B. Platte (Arizona State University), Gerlind Plonka-Hoch (University of Göttingen), and Michael Unser (Swiss Federal Institute of Technology Lausanne).

Our thanks are due to the Department of Mathematics at Vanderbilt for providing logistical support and to our reviewers who helped select papers for this volume and for providing suggestions to the authors on ways to improve their papers.

Golden, CO, USA — Gregory E. Fasshauer

Nashville, TN, USA — Marian Neamtu

Nashville, TN, USA — Larry L. Schumaker

Contents

Time-Variant System Approximation via Later-Time Samples

Roza Aceska and Yeon Hyang Kim

Abstract We develop a mathematical framework and efficient computational schemes to obtain an approximate solution of partial differential equations (PDEs) via sampled data. Recently, DeVore and Zuazua revisited the classical problem of inverse heat conduction, and they investigated how to recover the initial temperature distribution of a finite body from temperature measurements made at a fixed number of later times. In this paper, we consider a Laplace equation and a variable coefficient wave equation. We show that only one sensor employed at a crucial location at multiple time instances leads to a sequence of approximate solutions, which converges to the exact solution of these PDEs. This framework can be viewed as an extension of the novel, dynamical sampling techniques.

Keywords Dynamical system · Evolutionary systems representations · Near-best approximation · Initial datum

1 Introduction

Efficient data processing is essential in large data applications, whether the phenomenon of interest is sound, heat, electrostatics, electrodynamics, fluid dynamics, elasticity, or quantum mechanics. The spatial/time distribution of these aspects can be described similarly in terms of PDEs. When solving a PDE of interest, we need to know the initial conditions, described by some function [5, 6, 8]. However, in real-life applications, full knowledge of the initial conditions is often impossible due to unavailability of a large number of sensors [1, 7]. The way to overcome this impairing is to exploit the evolutionary nature of the sampling environment, while

R. Aceska (✉)
Ball State University, Muncie, IN, USA
e-mail: raceska@bsu.edu

Y. H. Kim
Central Michigan University, Mt Pleasant, MI, USA
e-mail: kim4y@cmich.edu

G. E. Fasshauer et al. (eds.), *Approximation Theory XVI*, Springer Proceedings in Mathematics & Statistics 336, https://doi.org/10.1007/978-3-030-57464-2_1

working with a reduced number of sensors, i.e., employ the concept of dynamical sampling [2–4].

The concept of dynamical sampling is beneficial in setups where the available sensing devices are limited due to some access constraints. In such an under sampled case, we use the coarse system of sensors multiple times to compensate for the lack of samples at a single time instance. Our focus is on developing methods which efficiently approximate solutions of important PDEs by engaging the dynamical nature of the setup dictated by the initial conditions. We develop the theory and algorithms for a new sampling and approximation framework. This framework combines spatial samples of various states of approximations and eventually provides an exact reconstruction of the solution. We assume that the initial state of the solution is in a selected Sobolev class.

Recent results [7] show that only one sensor employed at a crucial location at multiple time instances leads to a sequence of approximate solutions, which converges to the exact solution of the heat equation:

$$u_t = u_{xx},$$

$$u(0,t) = u(\pi,t) = 0,$$

$$u(x,0) = f(x),$$

under the assumption that the initial condition function f is in a compact class of Sobolev type. As a result, the sine basis decomposition coefficients of the initial function have controlled decay. We apply this approach to solve other PDEs, while using one spatial sensor multiple times for data collection: We use an appropriate basis decomposition, and work under the assumption that the basis decomposition coefficients of the initial state function have controlled decay. In other words, we assume that the initial state of the solution is in a selected Sobolev class.

2 Laplace Equation

We study the problem of solving an initial value problem (IVP) from discrete measurements made at appropriate instances/locations; thus, the initial conditions are not known in full detail. We aim to show that with a carefully selected placement and activation of the sensing devices, the unknown initial conditions can be completely determined by the discrete set of measurements; thus, the general solution to the IVP of interest is derived.

Under some initial and boundary conditions, the Laplace equation

$$u_{xx} + u_{yy} = 0, \quad x \in [0,1], \quad y \geq 0 \tag{1}$$

$$u_x(0,y) = u_x(1,y) = 0, \quad \lim_{y\to\infty} u(x,y) = 0 \quad u(x,0) = f(x),$$

has a general solution

$$u(x, y) = \sum_{k=0}^{\infty} a_k \cos(k\pi x)\, e^{-k\pi y}, \qquad \text{where } a_k = 2\int_0^1 f(x)\cos(k\pi x)\, dx. \tag{2}$$

The solution to (1) is the steady state temperature $u(x, y)$ in the semi-infinite plate $0 \leq x \leq 1$, $y \geq 0$, with the assumption that the left and right sides are insulated and assume that the solution is bounded. The temperature along the bottom side is assumed to be a known function $f(x)$.

In case the values $f(x)$ are not fully known at all $x \in [0, 1]$, we propose to take samples $u_k := u(x_0, y_k)$, $k \geq 0$, at an array of space-time locations (x_0, y_k), such that $|\cos(k\pi x_0)| \geq d_0 k^{-1}$ for some $d_0 > 0$ and for all k integers, $k \neq 0$. For the condition $|\cos(k\pi x_0)| \geq d_0 k^{-1}$ for some $d_0 > 0$ and for all k integers, we choose $\alpha \in (0, 3/2)$ so that

$$dist\left(\alpha, \left\{\frac{1}{2k}, \frac{3}{2k}, \ldots, \frac{2k+1}{2k}\right\}\right) \geq \frac{c_0}{k^2}, \; k = 1, 2, \ldots,$$

with c_0 an absolute constant. Then we have

$$dist\left(\alpha k\pi, \left\{\frac{\pi}{2}, \frac{3\pi}{2}, \ldots, \frac{(2k+1)\pi}{2}\right\}\right) \geq \frac{c_0\pi}{k}, \; k = 1, 2, \ldots,$$

We then take $x_0 = \alpha$. We further assume that $y_1 < y_2 < \ldots$. We work with $(c_k)_{k\geq 0}$ such that for some $r > 0$,

$$\sum c_k^2 k^{2r} \leq 1. \tag{3}$$

The function

$$F_0(z) := \sum_{k=0}^{\infty} c_k z^{-k} \tag{4}$$

is an analytic function in the unit disk $D = \{z \in C : |z| < 1\}$, which is uniquely determined by the set of coefficients $(c_k)_{k\geq 0}$. Furthermore, for the choice of $z = e^{-\pi y}$ and $c_k = a_k \cos(k\pi x_0)$, $k \geq 0$, we have: $F_0(e^{-\pi y}) = u(x_0, y)$.

Note that the evaluations

$$F_0(z_k) = u_k, \; k \geq 0, \tag{5}$$

where $z_k = e^{-\pi y_k}$, fully determine the function F_0. In case there was another analytic function on the open disc G_0, which satisfied $G_0(z_k) = u_k$, $k \geq 0$, then we'd have an analytic function $F_0 - G_0$ with countably many zeroes in D (since

$(F_0 - G_0)(z_k) = 0, k \geq 0)$; thus, $F_0 - G_0$ must be the zero function. This implies that $\{u_k | k = 0, 1, 2, ..\}$ uniquely determines (2).

Next, we sample $u(x, y)$ at locations (x_0, y_k), $k \geq 0$, where

$$y_0 > 0, \; y_n = \rho^n y_0, \; n \geq 1,$$

for some $\rho > 2$. The samples have an expansion

$$u_j = \sum_{k=0}^{\infty} c_k e^{-k\pi y_j} = \sum_{k=0}^{\infty} c_k e^{-k\pi \rho^j y_0}, \quad j = 1, 2, \ldots . \tag{6}$$

Notice that by (6) it holds

$$c_0 = u_n - \sum_{k=1}^{\infty} c_k e^{-k\pi \rho^n y_0},$$

$$c_1 = u_{n-1} e^{\pi \rho^{n-1} y_0} - c_0 e^{\pi \rho^{n-1} y_0} - \sum_{k=2}^{\infty} c_k e^{-k\pi \rho^{n-1} y_0},$$

$$c_2 = u_{n-2} e^{2\pi \rho^{n-2} y_0} - c_0 e^{2\pi \rho^{n-2} y_0} - c_1 e^{\pi \rho^{n-2} y_0} - \sum_{j=3}^{\infty} c_j e^{-(j-2)\pi \rho^{n-2} y_0},$$

$$\ldots$$

$$c_n = u_n e^{n\pi y_0} - c_0 e^{n\pi y_0} - c_1 e^{(n-1)\pi y_0} - \ldots - \sum_{j=n+1}^{\infty} c_j e^{-(j-n)\pi y_0}.$$

We take $n+1$ samples, and aim at approximating the initial value f, and respectively the solution (2). We define

$$\bar{c}_0 := u_n,$$

$$\bar{c}_1 := u_{n-1} e^{\pi \rho^{n-1} y_0} - \bar{c}_0 e^{\pi \rho^{n-1} y_0},$$

$$\bar{c}_2 := u_{n-2} e^{2\pi \rho^{n-2} y_0} - \bar{c}_0 e^{2\pi \rho^{n-2} y_0} - \bar{c}_1 e^{\pi \rho^{n-2} y_0},$$

$$\ldots$$

$$\bar{c}_n := u_n e^{n\pi y_0} - \bar{c}_0 e^{n\pi y_0} - \bar{c}_1 e^{(n-1)\pi y_0} - \ldots - \bar{c}_{n-1} e^{n\pi y_0} e^{-(n-1)\pi y_0}.$$

For each $j = 1, \ldots, n$, we denote the error in recovering c_j by $E_j := |\bar{c}_j - c_j|$. Since $\rho > 2$, $|c_j| \leq j^{-r} \leq k^{-r}$ for $j > k$, and $\frac{1}{1-e^{-\pi \rho^n y_0}} \leq \frac{1}{1-e^{-\pi y_0}}$, we estimate

$$E_0 \le \sum_{j=1}^{\infty} |c_j| e^{-j\pi\rho^n y_0} \le \sum_{j=1}^{\infty} e^{-j\pi\rho^n y_0} = \frac{e^{-\pi\rho^n y_0}}{1 - e^{-j\pi\rho^n y_0}} \le \frac{e^{-\pi\rho^n y_0}}{1 - e^{-\pi y_0}}.$$

Lemma 1 *For each $j \ge 0$, we have*

$$E_j \le 2^j \frac{e^{-\pi\rho^{n-j} y_0}}{1 - e^{-\pi y_0}}.$$

Proof We use mathematical induction. The claim is verified for $j = 0, 1$. Suppose the claim holds true for all $j \le k - 1$ for some $k \ge 1$. Then

$$\begin{aligned}
&E_k \\
&\le E_0 e^{\pi\rho^{n-k} y_0 k} + E_1 e^{\pi\rho^{n-k} y_0 (k-1)} + \ldots + E_{k-1} e^{\pi\rho^{n-k} y_0} + \left(\frac{1}{k+1}\right)^r \frac{e^{-\pi\rho^{n-k} y_0}}{1 - e^{-\pi y_0}} \\
&\le \sum_{j=0}^{k-1} 2^j \frac{e^{-\pi\rho^{n-j} y_0}}{1 - e^{-\pi y_0}} e^{\pi\rho^{n-k} y_0 (k-j)} + \left(\frac{1}{k+1}\right)^r \frac{e^{-\pi\rho^{n-k} y_0}}{1 - e^{-\pi y_0}} \\
&\le \frac{e^{-\pi\rho^{n-k} y_0}}{1 - e^{-\pi y_0}} \left[\sum_{j=0}^{k-1} 2^j e^{-\pi\rho^{n-j} y_0 - \pi\rho^{k-j} y_0 - \pi\rho^{n-k} y_0} + \left(\frac{1}{k+1}\right)^r\right]
\end{aligned}$$

Since $\rho > 2$,

$$e^{-\pi\rho^{n-j} y_0 - \pi\rho^{k-j} y_0 - \pi\rho^{n-k} y_0} \le 1, \tag{7}$$

which implies that

$$\sum_{j=0}^{k-1} 2^j e^{-\pi\rho^{n-j} y_0 - \pi\rho^{k-j} y_0 - \pi\rho^{n-k} y_0} + \left(\frac{1}{k+1}\right)^r \le 2^k.$$

For the inequality (7)

$$e^{-\pi\rho^{n-j} y_0 - \pi\rho^{k-j} y_0 - \pi\rho^{n-k} y_0} \le 1, \tag{8}$$

since

$$-\pi\rho^{n-j} y_0 - \pi\rho^{k-j} y_0 - \pi\rho^{n-k} y_0 = -\pi t_0(\rho^{k-j} - (k - j + 1)),$$

we need to have, for $0 \le j < k$,

$$\rho^{k-j} > k - j + 1.$$

This implies $\rho^k > k+1$ for $k \geq 1$. When $k = 1$, we have $\rho > 2$. By the mathematical induction, we have $\rho^k > k+1$ for $k \geq 1$ if $\rho > 2$.

We define an approximation $F_n(x)$ to $f(x)$ as

$$F_n(x) := \sum_{j=0}^{m} \frac{\bar{c}_j}{\cos j\pi x_0} \cos j\pi x, \quad m := \lceil \frac{n}{2} \rceil.$$

Theorem 1 *Given any fixed choice of* $y_1 > 0,\ \rho > 2$, *let* $y_k := \rho^k y_0,\ k \geq 1$. *Then for* $f \in \{\sum a_k \cos k\pi x \in L_2([0,1])\} : \sum_{k=1}^{\infty} k^{2r} |a_k|^2 \leq 1\}$, *whenever*

$$e^{-\pi y_k} \leq 2^{-k} k^{-r-1},$$

we have

$$\lim_{n\to\infty} \|f - F_n\| = 0.$$

Proof By Lemma 1, we have

$$\begin{aligned}
\|f - F_n\|^2 &\leq \sum_{j=0}^{m} \frac{E_j^2}{|\cos j\pi x_0|^2} + \sum_{j=m+1}^{\infty} |a_j|^2 \\
&\leq \sum_{j=0}^{m} \left(\frac{j}{d_0}\right)^2 \left(2^j \frac{e^{-\pi \rho^{n-j} y_0}}{1 - e^{-\pi y_0}}\right)^2 + m^{-2r} \\
&\leq \left(\frac{m}{d_0}\right)^2 \left(\frac{e^{-\pi \rho^m y_0}}{1 - e^{-\pi y_0}}\right)^2 \sum_{j=0}^{m} 2^{2j} + m^{-2r} \\
&= \left(\frac{m}{d_0}\right)^2 \left(\frac{e^{-\pi y_m}}{1 - e^{-\pi y_0}}\right)^2 \sum_{j=0}^{m} 2^{2j} + m^{-2r} \\
&\leq \left(\frac{3}{4d_0^2(1 - e^{-\pi y_0})^2} + 1\right) m^{-2r} \to 0,
\end{aligned}$$

as $n \to \infty$.

3 Variable Coefficient Wave Equation

In this section, we consider the following generalization of the wave equation:

$$u_{xx} + (1+t)^2 u_{tt} + \frac{1}{1+t} u_t = 0, \ t \geq 0, \tag{9}$$

where $x \in [0, \pi]$ and $t \geq 0$. A simple calculation shows that the solution of this equation is

$$u(x,t) = \sum_{k \geq 1} a_k \sin(kx) \frac{1}{(1+t)^k},$$

where $(a_k)_{k=1}^{\infty}$ are the Fourier sine coefficients of $f(x) = u(x, 0)$. Thus if the initial function is $f(x) = u(x, 0)$ is given, then we can obtain $u = u(x, t)$. In case $f(x)$ is not known at all $x \in [0, \pi]$, we use later time samples, which are available at one fixed location x_0 and at time instances

$$t_1 < t_2 < \ldots < t_s < \ldots$$

to recover the initial datum f, and consequently u. To do this, we first choose x_0 using a similar argument as in Sect. 2 so that we have $|\sin(kx_0)| \geq d_0 k^{-1}$ for some $d_0 > 0$ and for all $k \geq 1$.

We note that the samples satisfy

$$u_s := u(x_0, t_s) = \sum_{k \geq 1} a_k \sin(kx_0) \frac{1}{(1+t_s)^k} = \sum_{k \geq 1} c_k \frac{1}{(1+t_s)^k}, \tag{10}$$

where $c_k := a_k \sin(kx_0)$. We further assume that we have $\sum_k c_k^2 k^{2r} \leq 1$. We will impose conditions on the time instances employed so we can construct an approximation of the initial datum and thus recover $u(x, t)$. As we will see, the choice for $t_1 = \rho > \sqrt{2}$, $t_k \geq \rho^{2^{k-1}} - 1$ when $k \geq 2$, will provide good convergence rate. We set the algorithm as follows:

$$\bar{c}_1 = u_n(1 + t_n),$$

and for $2 \leq k \leq n$ we set

$$\bar{c}_k = u_{n-k}(1 + t_{n-k+1})^k - \sum_{j=1}^{k-1} \bar{c}_j \frac{(1 + t_{n-k+1})^k}{(1 + t_{n-k+1})^j}.$$

Lemma 2 *For every $n \geq 1$ and $1 \leq k \leq n$, we have*

$$E_k := |c_k - \bar{c}_k| \leq 2^{k-1} A_0 \frac{1}{1 + t_{n-k+1}}, \tag{11}$$

where $A_0 = 2^{-r} \frac{1}{1-(1+t_1)^{-1}}$.

Proof First, we note that for the choice of t_k it holds:

$$\frac{1}{1+t_{n-j+1}}\frac{(1+t_{n-k})^{k+1}}{(1+t_{n-k})^j} \leq \frac{1}{1+t_{n-k}} \quad \text{when } j \leq k. \tag{12}$$

Then

$$E_1 \leq \sum_{k>1} |c_j| \frac{1+t_n}{(1+t_n)^j} \leq 2^{-r} \sum_{k>1} (1+t_n)^{-(j-1)}$$

$$= 2^{-r}(1+t_n)^{-1}\frac{1}{1-(1+t_n)^{-1}} \leq 2^{-r}(1+t_n)^{-1}\frac{1}{1-(1+t_1)^{-1}}.$$

Suppose for every $j \leq k$ it holds $E_j \leq 2^{j-1}A_0\frac{1}{1+t_{n-j+1}}$. Then

$$E_{k+1} \leq \sum_{j<k+1} E_j \frac{(1+t_{n-k})^{k+1}}{(1+t_{n-k})^j} + \sum_{j>k+1} |c_j| \frac{(1+t_{n-k})^{k+1}}{(1+t_{n-k})^j}$$

$$\leq \sum_{j<k+1} 2^{j-1}A_0 \frac{1}{1+t_{n-j+1}}\frac{(1+t_{n-k})^{k+1}}{(1+t_{n-k})^j} + \sum_{j>1} |c_j| \frac{1}{(1+t_{n-k})^j}.$$

By (12), it holds

$$E_{k+1} \leq \sum_{j<k+1} 2^{j-1}A_0\frac{1}{1+t_{n-k}} + \frac{1}{(k+1)^r}\frac{1}{1+t_{n-k}}\frac{1}{1-(1+t_{n-k+1})^{-1}}$$

$$\leq \sum_{j<k+1} 2^{j-1}A_0\frac{1}{1+t_{n-k}} + \frac{1}{(k+1)^r}\frac{1}{1+t_{n-k}}A_0 \leq 2^k A_0 \frac{1}{1+t_{n-k}}.$$

To simplify our calculations, we assume we always take $n = 2m$ samples, and define

$$F_n := \sum_{k=1}^{m} \bar{c}_k f_k. \tag{13}$$

Theorem 2 *Let $t_1 = \rho > \sqrt{2}$ and $t_k \geq \rho^{2^{k-1}} - 1$ when $k \geq 2$. Then, whenever $f \in \{\sum a_k \sin kx \in L_2(\mathbb{R}) : \sum_{k=1}^{\infty} k^{2r}|a_k|^2 \leq 1\}$, we have*

$$\lim_{n\to\infty} \|f - F_n\| = 0.$$

Proof By the decay assumption on $(c_k)_{k\geq 1}$, we obtain

$$\|f - F_n\|_2^2 \leq \sum_{k=1}^{m} |c_k - \bar{c}_k|^2 + \sum_{k>m} |c_k|^2 \leq \sum_{k=1}^{m} 2^{k-1} A_0 \frac{1}{1 + t_{n-k+1}} + \frac{1}{m^{2r}}.$$

Since $t_{n-k+1} - 1 = \rho^{2^{n-k}}$, for $k = 1, 2, \ldots, m$ we have

$$\frac{2^{k-1}}{1 + t_{n-k+1}} = \frac{2^{k-1}}{\rho^{2^{n-k}}} \leq \frac{2^{k-1}}{\rho^{2^m}}.$$

Thus

$$\|f - F_n\|_2 \leq A_0 \frac{1}{\rho^{2^m}} \sum_{k=1}^{m} 2^{k-1} + \frac{1}{m^{2r}} = \frac{2^m A_0}{\rho^{2^m}} + \frac{1}{m^{2r}} \leq \frac{C}{m^{2r}}.$$

Acknowledgments This material is based upon work supported by the National Security Agency under Grant No. H98230-18-0144 while the authors were in residence at the Mathematical Sciences Research Institute in Berkeley, California, during Summer 2018. The authors thank the referees for their valuable comments.

References

1. Aceska, A., Arsie, A., Karki, R.: On near-optimal time samplings for initial data best approximation. Le Matematiche **74**(1), 173–190 (2019)
2. Aldroubi, A., Petrosyan, A.: Dynamical sampling and systems from iterative actions of operators. In: Pesenson, I., et al. (eds.) Frames and Other Bases in Abstract and Function Spaces: Novel Methods in Harmonic Analysis, vol. 1, pp. 15–26. Springer International Publishing, Cham (2017)
3. Aldroubi, A., Davis, J., Krishtal, I.: Dynamical sampling: time space trade-off. Appl. Comput. Harmon. Anal. **34**(3), 495–503 (2013)
4. Aldroubi, A., Cabrelli, C., Molter, U., Tang, S.: Dynamical sampling. Appl. Comput. Harmon. Anal. **42**(3), 378–401 (2017)
5. Back, I.V., Blackwell, B., St. Claire, C.R.: Inverse Heat Conduction. Wiley, New York (1985)
6. Choulli, M.: Une Introduction aux Problèmes Inverses Elliptiques et Paraboliques. Mathématiques & Applications, vol. 65. Springer, Berlin (2009)
7. DeVore, R., Zuazua, E.: Recover of an initial temperature from discrete sampling. Math. Models Methods Appl. Sci. **24**(12), 2487–2501 (2014)
8. DeVore, R., Howard, R., Micchelli, C.: Optimal nonlinear approximation. Manuscr. Math. **63**, 469–478 (1989)

C^1-Quartic Butterfly-Spline Interpolation on Type-1 Triangulations

Domingo Barrera, Costanza Conti, Catterina Dagnino, María José Ibáñez, and Sara Remogna

Abstract In this paper, we construct and analyse C^1 quartic interpolating splines on type-1 triangulations, approximating regularly distributed data. This is achieved by defining the associated Bernstein-Bézier coefficients from point values of the function to be approximated in such a way that C^1 regularity is obtained for enough regular functions as well as the optimal order of approximation. We construct such interpolating splines by combining a quasi-interpolating spline with one step of an interpolatory subdivision scheme. Numerical tests confirming the theoretical results are provided.

Keywords Spline approximation · Bernstein-Bézier form · Type-1 triangulation

1 Introduction

The use of spline interpolation and quasi-interpolation for the approximation of functions and data is widely developed in the literature and many approaches have been proposed. Schemes based on the construction of finite elements, macro-elements and local stable minimal determining sets for general (refined or not) triangulations of a polygonal domain have been proposed (see e.g. [16, 17] and references therein), as well as the definition of such approximating splines in the space spanned by a family of compactly supported functions (see e.g. [8] and [20]). In the uniform case, box splines have been also extensively used (see e.g. [5, 10, 24]

D. Barrera · M. J. Ibáñez
Department of Applied Mathematics, University of Granada, Granada, Spain
e-mail: dbarrera@ugr.es; mibanez@ugr.es

C. Conti
Dipartimento di Ingegneria Industriale, University of Florence, Florence, Italy
e-mail: costanza.conti@unifi.it

C. Dagnino · S. Remogna (✉)
Dipartimento di Matematica, Università di Torino, Torino, Italy
e-mail: catterina.dagnino@unito.it; sara.remogna@unito.it

G. E. Fasshauer et al. (eds.), *Approximation Theory XVI*, Springer Proceedings in Mathematics & Statistics 336, https://doi.org/10.1007/978-3-030-57464-2_2

and references therein, and [3, 4, 18, 19]), and a new procedure was introduced based on the definition of the Bernstein-Bézier (BB-) coefficients of the splines on each triangle in the partition by using only point values in a neighbourhood of the triangle. The BB-coefficients are properly defined to produce globally C^1 splines and to achieve the required polynomial reproduction (see e.g. [1, 14, 21, 23] and [22] for the 3D case).

The C^1 quartic scheme exact on cubic polynomials introduced in [23] is a particular case of a general family derived in [2] that depends on some free parameters. The BB-coefficients with respect to any triangle of the quasi-interpolating splines are defined from the values at a large number of points lying in a neighbourhood of the triangle, so it is quite natural to think about reducing the number of evaluations needed to compute the BB-coefficients. This issue is dealt with in [2], where it is proved that only evaluation at vertices and midpoints of edges of triangles are needed.

In this paper, we follow this approach and combine a quasi-interpolating spline with one step of the so called modified Butterfly interpolatory subdivision scheme to construct C^1 quartic interpolating splines on regular type-1 triangulations, whose BB-coefficients are defined uniquely from the values at the vertices of the triangulation.

The organization of the paper is as follows. In Sect. 2, some results on the representation of C^1-quartic splines on three-directional triangulations are recalled, as well as the notations to be used in the paper. In Sect. 3 we recall the family of quasi-interpolation operators studied in [2]. In particular, all of them depend on some free parameters, so we propose some specific choices for them. In Sect. 4 we present the construction of a family of interpolating splines, obtained by combining the quasi-interpolating splines of Sect. 3, with one step of the modified Butterfly interpolatory subdivision scheme. We also discuss the approximation properties of the corresponding operators. Finally, in Sect. 5, we propose some numerical tests to confirm the theoretical results established in the previous sections.

2 Notations and Preliminaries

We consider the type-1 triangulation Δ defined by the directions $d_1 := (h, h)$, $d_2 := (h, -h)$ and $d_3 := d_1 + d_2$, with $h > 0$. Its vertices $v_{i,j}$ are linear combinations of directions d_1 and d_2 with integer coefficients, i.e. $v_{i,j} := id_1 + jd_2$, $i, j \in \mathbb{Z}$. The two-dimensional lattice $\mathcal{V} := \{v_{i,j} : i, j \in \mathbb{Z}\}$ decomposes the real plane into parallelograms $P_{i,j}$ with vertices $v_{i,j}$, $v_{i,j+1}$, $v_{i+1,j+1}$ and $v_{i+1,j}$ (see Fig. 1(left)), each of which is subdivided into two triangles $T_{i,j}$ and $\widetilde{T}_{i,j}$ obtained by connecting the vertices $v_{i,j}$ and $v_{i+1,j+1}$, so that

$$\Delta := \bigcup_{i,j\in\mathbb{Z}} \left(T_{i,j} \cup \widetilde{T}_{i,j}\right).$$

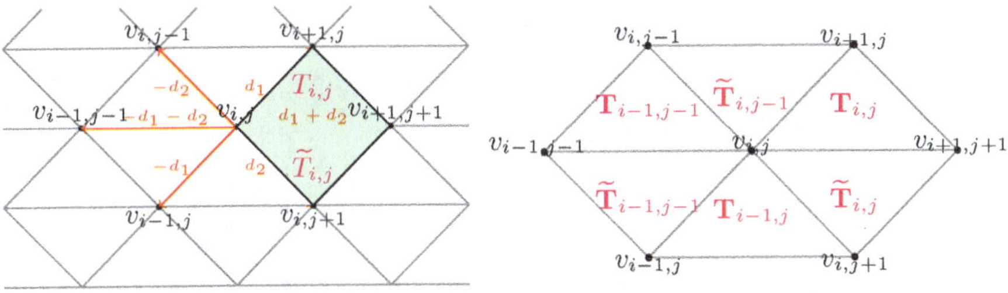

Fig. 1 The triangulation Δ (left) and the hexagon $H_{i,j}$ (right)

The triangles sharing a vertex $v_{i,j}$ determine an hexagon, denoted by $H_{i,j}$ (see Fig. 1(right)).

The approximating splines will be constructed in the space

$$\mathcal{S}_4^1(\Delta) := \left\{ s \in C^1\left(\mathbb{R}^2\right) : s_{|T} \in \mathbb{P}_4, \text{ for all } T \in \Delta \right\},$$

where $\mathbb{P}_4$ stands for the space of bivariate quartic polynomials. Such splines will be defined by directly setting their BB-coefficients on the triangles of Δ (see e.g. [16]). The restriction to a triangle $T \in \Delta$ with vertices v_0, v_1 and v_2 of a spline $s \in \mathcal{S}_4^1(\Delta)$ can be expressed as

$$s_{|T} = \sum_{i+j+k=4} c_{i,j,k}^T B_{i,j,k}^T,$$

where $B_{i,j,k}^T := \frac{4!}{i!j!k!} b_0^i b_1^j b_2^k$, $i, j, k \geq 0$, $i + j + k = 4$, are the Bernstein polynomials of degree 4 associated with T and the barycentric coordinates (b_0, b_1, b_2) w.r.t. T satisfy the equalities $(x, y) = b_0 v_0 + b_1 v_1 + b_2 v_2$, $b_0 + b_1 + b_2 = 1$ for $(x, y) \in T$. To alleviate the notation, no reference is made to the triangle with respect to which the barycentric coordinates are determined.

Each BB-coefficient $c_{i,j,k}^T$ of the quartic polynomial $s_{|T}$ is associated to the domain point $\xi_{i,j,k}^4 := (iv_0 + jv_1 + kv_2)/4$ in T. Let $\mathcal{D}_4$ be the subset of the domain points arising when all triangles in Δ are run. Each vertex gives rise to a single point in $\mathcal{D}_4$. The same is applicable for any domain point on an common edge to two triangles (see Fig. 2). If quadratic splines are considered instead of quartic splines, their BB-representations give rise to coefficients associated with the domain points $\xi_{i,j,k}^2 := (iv_0 + jv_1 + kv_2)/2$, from which the subset $\mathcal{D}_2$ is defined. Finally, the subset $\mathcal{D}_1$ is the collection of all vertices of the triangulation Δ (see Fig. 3).

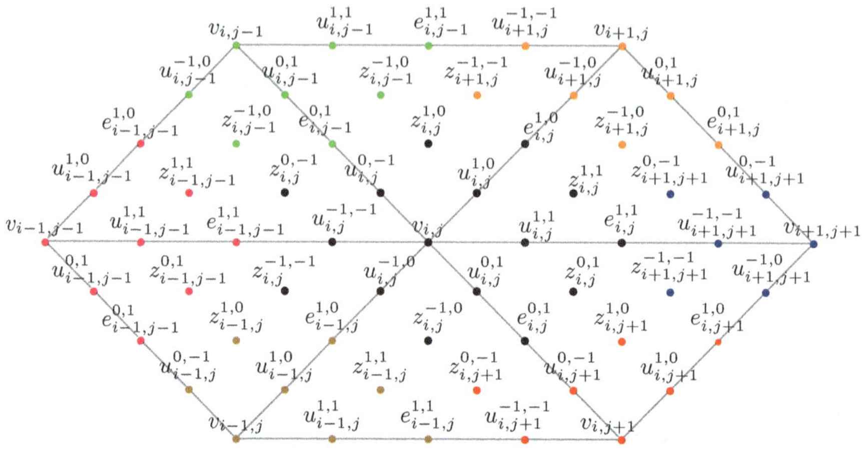

Fig. 2 The points of $\mathcal{D}_4$ relative to $H_{i,j}$

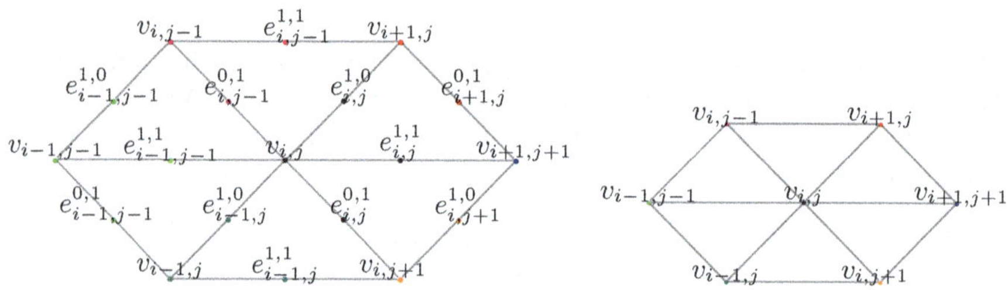

Fig. 3 The points of $\mathcal{D}_2$ (left) and $\mathcal{D}_1$ (right) relative to $H_{i,j}$

Moreover, we define $\mathcal{D}_\ell := \bigcup_{i,j} \mathcal{D}_\ell^{i,j}$, $\ell = 2, 4$ with

- $\mathcal{D}_4^{i,j} := \left\{v_{i,j}\right\} \cup \left\{e_{i,j}^{k,m}, k, m \in \{0,1\}, k+m \neq 0\right\}$
  $\cup \left\{u_{i,j}^{k,m}, z_{i,j}^{k,m}, k, m \in \{-1,0,1\}, k+m \neq 0\right\}$, where
  - $e_{i,j}^{k,m}$ is the midpoint of $\left[v_{i,j}, v_{i+k,j+m}\right]$,
  - $u_{i,j}^{k,m} := \frac{1}{4}\left(3v_{i,j} + v_{i+k,j+m}\right)$,
  - $z_{i,j}^{k,m} := \frac{1}{4}\left(2v_{i,j} + v_{i+k,j+m} + v_{r,s}\right)$, with $v_{r,s}$ the third vertex of $\left[v_{i,j}, v_{i+k,j+m}, v_{r,s}\right] \in \Delta$ counting counterclockwise;
- $\mathcal{D}_2^{i,j} := \left\{v_{i,j}, e_{i,j}^{1,0}, e_{i,j}^{0,1}, e_{i,j}^{1,1}\right\}$.

3 C^1 Quartic Quasi-interpolating Splines on $\mathcal{D}_2$

In order to make the paper self-contained, here we briefly recall how the quasi-interpolating spline $Q_{4,2}f \in \mathcal{S}_4^1(\Delta)$ in [2] is defined. Such a spline is constructed using the values of f on $\mathcal{D}_2$.

We take advantage of the fact that Δ is a uniform triangulation to define the BB-coefficients of the restriction of the quasi-interpolating $Q_{4,2}f$ to each triangle. For instance, we write the restriction of $Q_{4,2}f$ to the triangle $T_{i,j}$ as

$$\begin{aligned}
Q_{4,2}f_{|T_{i,j}} =& c\left(v_{i,j}\right) B_{4,0,0}^{T_{i,j}} + c\left(u_{i,j}^{1,1}\right) B_{3,1,0}^{T_{i,j}} + c\left(u_{i,j}^{1,0}\right) B_{3,0,1}^{T_{i,j}} \\
&+ c\left(e_{i,j}^{1,1}\right) B_{2,2,0}^{T_{i,j}} + c\left(z_{i,j}^{1,1}\right) B_{2,1,1}^{T_{i,j}} + c\left(e_{i,j}^{1,0}\right) B_{2,0,2}^{T_{i,j}} \\
&+ c\left(u_{i+1,j+1}^{-1,-1}\right) B_{1,3,0}^{T_{i,j}} + c\left(z_{i+1,j+1j}^{0,-1}\right) B_{1,2,1}^{T_{i,j}} + c\left(z_{i+1,j}^{-1,0}\right) B_{1,1,2}^{T_{i,j}} \\
&+ c\left(u_{i+1,j}^{-1,0}\right) B_{1,0,3}^{T_{i,j}} + c\left(v_{i+1,j+1}\right) B_{0,4,0}^{T_{i,j}} + c\left(u_{i+1,j+1}^{0,-1}\right) B_{0,3,1}^{T_{i,j}} \\
&+ c\left(e_{i+1,j}^{0,1}\right) B_{0,2,2}^{T_{i,j}} + c\left(u_{i+1,j}^{0,1}\right) B_{0,1,3}^{T_{i,j}} + c\left(v_{i+1,j}\right) B_{0,0,4}^{T_{i,j}},
\end{aligned}$$

where the notation $c(p)$ is used for the BB-coefficient relative to the domain point $p \in \mathcal{D}_4$.

Moreover, once defined the BB-coefficients relative to $T_{i,j}$, those corresponding to the other five triangles around the vertex $v_{i,j}$ are defined by translation and/or rotation.

In order to obtain an interpolatory spline at the vertices, we define $c\left(v_{i,j}\right) := f\left(v_{i,j}\right)$.

The domain points $p \in \mathcal{D}_4$ have been labelled as u, e and z-points. Their BB-coefficients will be defined as linear combinations of the values of f at 19 points in $\mathcal{D}_2$ (see Fig. 3). As an example,

$$\begin{aligned}
c\left(u_{ij}^{1,1}\right) =& \gamma_0 f\left(v_{ij}\right) + \gamma_1 f\left(e_{i,j}^{1,1}\right) + \gamma_2 f\left(e_{i,j}^{1,0}\right) + \gamma_3 f\left(e_{i,j-1}^{0,1}\right) + \gamma_4 f\left(e_{i-1,j-1}^{1,1}\right) \\
&+ \gamma_5 f\left(e_{i-1,j}^{1,0}\right) + \gamma_6 f\left(e_{i,j}^{0,1}\right) + \gamma_7 f\left(v_{i+1,j+1}\right) + \gamma_8 f\left(e_{i+1,j}^{0,1}\right) \\
&+ \gamma_9 f\left(v_{i+1,j}\right) + \gamma_{10} f\left(e_{i,j-1}^{1,1}\right) + \gamma_{11} f\left(v_{i,j-1}\right) + \gamma_{12} f\left(e_{i-1,j-1}^{1,0}\right) \\
&+ \gamma_{13} f\left(v_{i-1,j-1}\right) + \gamma_{14} f\left(e_{i-1,j-1}^{0,1}\right) + \gamma_{15} f\left(v_{i-1,j}\right) + \gamma_{16} f\left(e_{i-1,j}^{1,1}\right) \\
&+ \gamma_{17} f\left(v_{i,j+1}\right) + \gamma_{18} f\left(e_{i,j+1}^{1,0}\right).
\end{aligned}$$

The coefficients used to define the above linear combination form the *mask* $\boldsymbol{\gamma} \in \mathbb{R}^{19}$, and $c\left(u_{i,j}^{1,1}\right) = \mathbf{f}^{i,j} \cdot \boldsymbol{\gamma}$, where the vector $\mathbf{f}^{i,j} \in \mathbb{R}^{19}$ contains the values of $f(p)$,

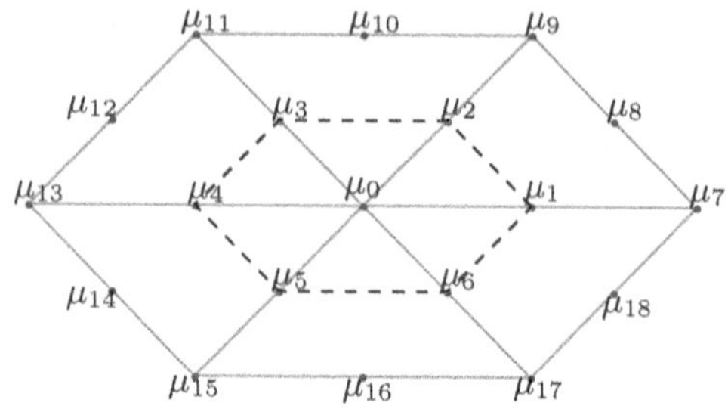

Fig. 4 Order for enumerate $\mathbf{f}^{i,j}, \boldsymbol{\alpha}, \boldsymbol{\beta}, \boldsymbol{\gamma}$

$p \in \mathcal{D}_4 \cap H_{i,j}$, enumerated as indicated in Fig. 4, as well as $\boldsymbol{\gamma}$. The BB-coefficients associated with the other u-points ($u_{i,j}^{1,0}, u_{i,j}^{0,-1}, u_{i,j}^{-1,-1}, u_{i,j}^{-1,0}$, and $u_{i,j}^{0,1}$) are defined analogously from the rotated versions of $\boldsymbol{\gamma}$.

In the same way, $c\left(e_{i,j}^{1,1}\right)$ and $c\left(z_{i,j}^{1,1}\right)$ are defined by considering the masks $\boldsymbol{\alpha}$ and $\boldsymbol{\beta}$, respectively. It means $c\left(e_{i,j}^{1,1}\right) = \mathbf{f}^{i,j} \cdot \boldsymbol{\alpha}$ and $c\left(z_{i,j}^{1,1}\right) = \mathbf{f}^{i,j} \cdot \boldsymbol{\beta}$. The BB-coefficients $c(e)$ and $c(z)$ relative to an e-point and a z-point, respectively, are defined from the rotated versions of $\boldsymbol{\alpha}$ and $\boldsymbol{\beta}$.

It is known that a quasi-interpolation operator in $\mathcal{S}_4^1(\Delta)$ can reproduce the space of cubic polynomials (see e.g. [10, 15]). Therefore, the masks $\boldsymbol{\alpha}$, $\boldsymbol{\beta}$ and $\boldsymbol{\gamma}$ must be defined to obtain C^1-regularity and the exactness on $\mathbb{P}_3$ of the quasi-interpolation operator. In other words, we require that the following constrains are satisfied:

$$Q_{4,2}f \in C^1\left(\mathbb{R}^2\right) \quad \text{and} \quad Q_{4,2}f = f \text{ for all } f \in \mathbb{P}_3. \tag{1}$$

In [2] the following result is established.

Proposition 1 *The imposition of (1) results in infinitely many solutions depending on the first three elements $\beta_1, \beta_2, \beta_3$ of the mask $\boldsymbol{\beta}$. The mask $\boldsymbol{\alpha}$ is uniquely determined from the following values (see Fig. 5):*

$$\alpha_0 = \alpha_7 = -\tfrac{1}{3}, \alpha_1 = \tfrac{2}{3}, \alpha_2 = \alpha_6 = \alpha_8 = \alpha_{18} = \tfrac{1}{3}, \alpha_9 = \alpha_{17} = -\tfrac{1}{6},$$
$$\alpha_j = 0, \ j \in \{3, 4, 5, 10, 11, 12, 13, 14, 15, 16\}$$

The values of the masks $\boldsymbol{\beta}$ and $\boldsymbol{\gamma}$ are given in Figs. 6 and 7.

Concerning the error estimates, the following classical result (see e.g. [10, 16]) holds.

Theorem 1 *For an arbitrary triangle T in Δ and for a given $f \in C^{m+1}\left(\mathbb{R}^2\right)$, $0 \le m \le 3$,*

$$\left\| D^\nu \left(f - Q_{4,2}f\right)\right\|_{\infty,T} \le K_{|\nu|} h^{m+1-|\nu|} \left\| D^{m+1} f \right\|_{\infty,\Omega_T},$$

Fig. 5 The BB-coefficient $c\left(e_{i,j}^{1,1}\right)$ is evaluated from the mask $\boldsymbol{\alpha}$

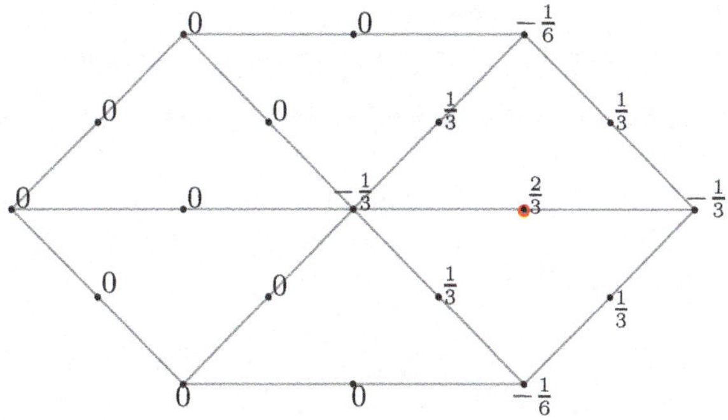

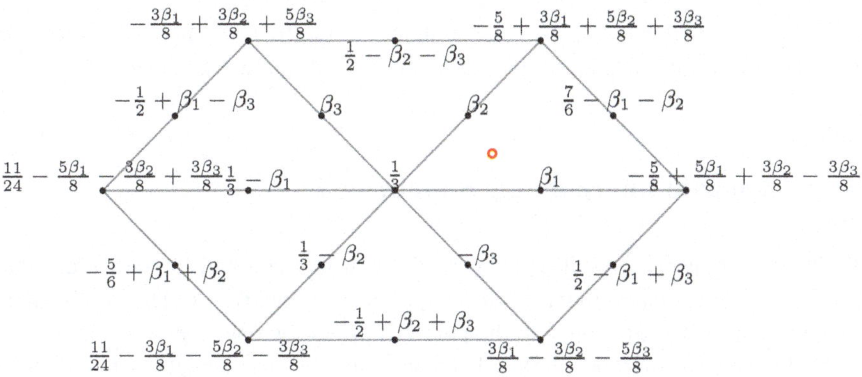

Fig. 6 The evaluation of the BB-coefficient $c\left(z_{i,j}^{1,1}\right)$ requires the mask $\boldsymbol{\beta}$, whose values depend on three free parameters

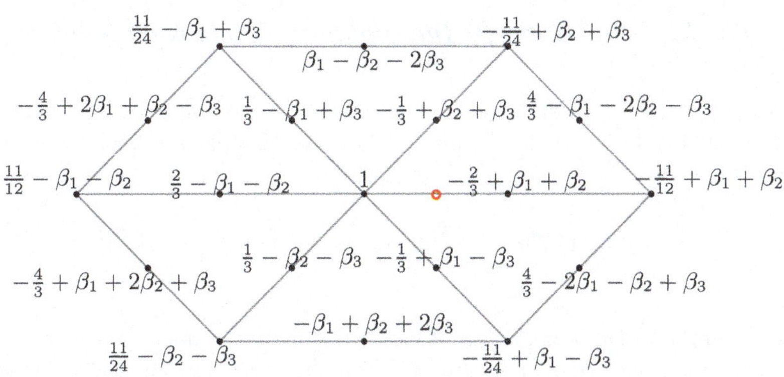

Fig. 7 The mask $\boldsymbol{\gamma}$ needed to evaluate $c\left(u_{i,j}^{1,1}\right)$ also depends on three free parameters

for all $0 \leq |\nu| \leq m$, $\nu := (\nu_1, \nu_2)$, *where* $K_{|\nu|}$ *are constants independent on* h *and* Ω_T *denotes the union of all triangles* $T \in \Delta$ *that intersect* T.

Masks $\boldsymbol{\beta}$ and $\boldsymbol{\gamma}$ depend on three parameters, so a strategy is needed to choose them.

The first strategy is reduced to assigning zero values to these parameters. Another possibility is to set $\beta_1 = \beta_2$ and $\beta_3 = 0$, so that the resulting masks have certain symmetries. Since

$$\left\| Q_{4,2} \right\|_\infty \leq \max \left\{ \|\boldsymbol{\alpha}\|_1, \|\boldsymbol{\beta}\|_1, \|\boldsymbol{\gamma}\|_1 \right\}$$

and $\|\boldsymbol{\alpha}\|_1 = 3$, it is easy to minimize the upper bound $U(\beta_1) := \max \left\{3, \|\boldsymbol{\beta}\|_1, \|\boldsymbol{\gamma}\|_1\right\}$ to obtain that $\left\| Q_{4,2} \right\|_\infty$ is bounded by 3 if $\beta_1 \in \left[\frac{13}{36}, \frac{41}{84}\right]$.

In the following we will use such a choice for the free parameters, obtaining a family of quasi-interpolating splines depending on β_1 and we denote it by $Q_{4,2}^{\beta_1} f$.

4 C^1 Quartic Interpolating Splines on $\mathcal{D}_1$

In this section, we discuss the construction of new interpolating splines by application of a 'preprocessing' to the quasi-interpolating splines $Q_{4,2}^{\beta_1} f$. The idea is, first, to approximate the function f at the points of type e by one step of a subdivision algorithm suitable for type-1 triangulated data, and then use the quasi-interpolating operator $Q_{4,2}^{\beta_1}$. The result is a spline interpolating at the points of $\mathcal{D}_1$ since the quasi-interpolant $Q_{4,2}^{\beta_1}$ has this property.

4.1 The Modified Butterfly Interpolatory Subdivision Scheme

We recall that a bivariate subdivision scheme is an iterative algorithm for refining a set of points $\mathbf{f} = \{f_j,\ j \in \mathbb{Z}^2\}$ by repeatedly applying a linear refinement operator $S_{\mathbf{a}}$ of type

$$(S_{\mathbf{a}}\mathbf{f})_i = \sum_{j \in \mathbb{Z}^2} \mathrm{a}_{i-2j} f_j, \qquad i \in \mathbb{Z}^2. \tag{2}$$

From (2) we see that, at each step of the recursion, the 'refined' points are linear combinations of the 'coarse' points with real coefficients being the subdivision mask $\mathbf{a} = \left\{\mathrm{a}_i,\ i \in \mathbb{Z}^2\right\}$ (for more details, see [6, 7, 11], and reference therein). From (2) we also see that one step of a subdivision scheme transforms a set of data points attached to $\mathbb{Z}^2$ into a set of data points attached to $\frac{1}{2}\mathbb{Z}^2$.

Even though subdivision schemes usually keep refining data till convergence to a continuous limit, the idea here is to use just one step of the so called Modified Butterfly Interpolatory Subdivision Scheme (MBISS) for data on type-1 triangulations. The MBISS is an interpolatory scheme (see [13]) meaning that at each step the coarse set of points is included into the refined one and new points are inserted. This translates into the refinement rule ('duplication' rule)

$$p_{2i}^{[k+1]} = p_i^{[k]}, \quad i \in \mathbb{Z}^2.$$

Each of the three insertion rules of the MBISS, transforming a sequence from level k to level $k+1$, is involving 10 points lying around the point to be inserted and are exactly the same for all three possible directions of insertion. They are as follows:

– for the 'horizontal' insertion

$$\begin{aligned} p_{2i+(1,0)}^{[k+1]} = (\tfrac{1}{2} - \omega)\left(p_i^{[k]} + p_{i+(0,1)}^{[k]}\right) + (\tfrac{1}{8} + 2\omega)\left(p_{i-(1,0)}^{[k]} + p_{i+(1,1)}^{[k]}\right) \\ +\omega\left(p_{i-(0,1)}^{[k]} + p_{i+(0,1)}^{[k]}\right) \\ +(-\tfrac{1}{16} - \omega)\left(p_{i+(1,0)}^{[k]} + p_{i+(1,2)}^{[k]} + p_{i-(1,1)}^{[k]} + p_{i-(1,-1)}^{[k]}\right), \end{aligned} \tag{3}$$

– for the 'vertical' insertion

$$\begin{aligned} p_{2i+(0,1)}^{[k+1]} = (\tfrac{1}{2} - \omega)\left(p_i^{[k]} + p_{i+(1,0)}^{[k]}\right) + (\tfrac{1}{8} + 2\omega)\left(p_{i-(0,1)}^{[k]} + p_{i+(1,1)}^{[k]}\right) \\ +\omega\left(p_{i-(1,0)}^{[k]} + p_{i+(2,0)}^{[k]}\right) \\ +(-\tfrac{1}{16} - \omega)\left(p_{i+(0,1)}^{[k]} + p_{i-(1,1)}^{[k]} + p_{i+(1,-1)}^{[k]} + p_{i+(2,1)}^{[k]}\right), \end{aligned} \tag{4}$$

– for the 'diagonal' insertion

$$\begin{aligned} p_{2i+(1,1)}^{[k+1]} = (\tfrac{1}{2} - \omega)\left(p_i^{[k]} + p_{i+(1,1)}^{[k]}\right) + (\tfrac{1}{8} + 2\omega)\left(p_{i+(1,0)}^{[k]} + p_{i+(0,1)}^{[k]}\right) \\ +\omega\left(p_{i+(2,2)}^{[k]} + p_{i-(1,1)}^{[k]}\right) \\ +(-\tfrac{1}{16} - \omega)\left(p_{i-(0,1)}^{[k]} + p_{i+(2,1)}^{[k]} + p_{i+(1,2)}^{[k]} + p_{i-(1,0)}^{[k]}\right). \end{aligned} \tag{5}$$

For the MBISS we mention the following properties that are relevant in our discussion (see [13] for all details).

Proposition 2 *If* $w \in \left(-\frac{1}{32}, \frac{1}{32}\right)$, *then the MBISS is convergent. Moreover, for* $w \in \left(-\frac{1}{32}, \frac{1}{32}\right)$ *the MBISS step-wise reproduces the space* $\mathbb{P}_3$ *of bivariate polynomials of degree* 3. *The latter means that in case the points at one level are sampled from a polynomial of degree* 3, *the points at the next level are samples of the same polynomial at refined grid values.*

We remark that, in view of Proposition 2, the MBISS has approximation order 4 for all $\omega \in \left(-\frac{1}{32}, \frac{1}{32}\right)$ and that for $\omega = 0$ the MBISS reduces to the better known Butterfly subdivision scheme presented in [12].

4.2 C^1 Quartic Interpolating Splines

Now we are able to construct the interpolating splines by approximating the values of f at the e-points of $Q^{\beta_1}_{4,2}f$, by one step of the MBISS. Indeed, according to (3), (4), and (5), considering the notations used for $Q^{\beta_1}_{4,2}f$, we define $f(e^{1,1}_{i,j})$ as (see Fig. 8)

$$\begin{aligned} f\left(e^{1,1}_{i,j}\right) &= \left(\tfrac{1}{2}-\omega\right)\left(f(v_{i,j})+f(v_{i+1,j+1})\right)+\omega\left(f(v_{i-1,j-1})+f(v_{i+2,j+2})\right) \\ &+\left(-\tfrac{1}{16}-\omega\right)\left(f(v_{i,j-1})+f(v_{i+2,j+1})+f(v_{i+1,j+2})+f(v_{i-1,j})\right) \\ &+\left(\tfrac{1}{8}+2\omega\right)\left(f(v_{i+1,j})+f(v_{i,j+1})\right) \end{aligned}$$

and similarly for the other e-points.

In this way, by combining the masks $\boldsymbol{\alpha}$, $\boldsymbol{\beta}$ and $\boldsymbol{\gamma}$ of $Q^{\beta_1}_{4,2}f$ with the masks of the MBISS, we obtain new masks $\boldsymbol{\alpha}'$, $\boldsymbol{\beta}'$ and $\boldsymbol{\gamma}'$ larger than the corresponding masks $\boldsymbol{\alpha}$, $\boldsymbol{\beta}$ and $\boldsymbol{\gamma}$ but still based on the same number of points (see Figs. 9, 10, 11). Such masks depend on the parameter β_1, coming from the quasi-interpolating

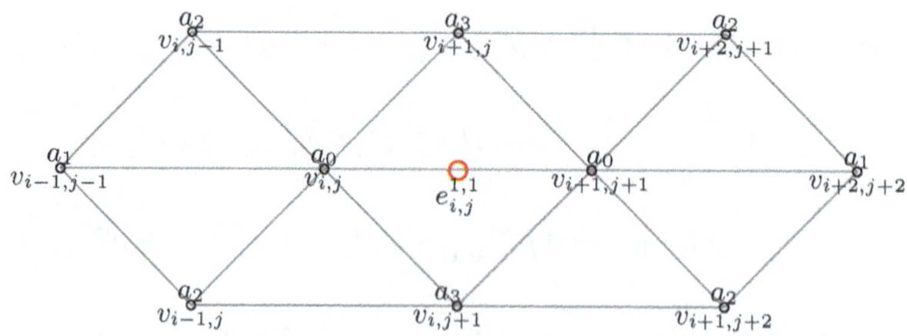

Fig. 8 Edge-point stencil of MBISS, with coefficients $a_0 = \frac{1}{2} - \omega$, $a_1 = \omega$, $a_2 = -\frac{1}{16} - \omega$, $a_3 = \frac{1}{8} + 2\omega$

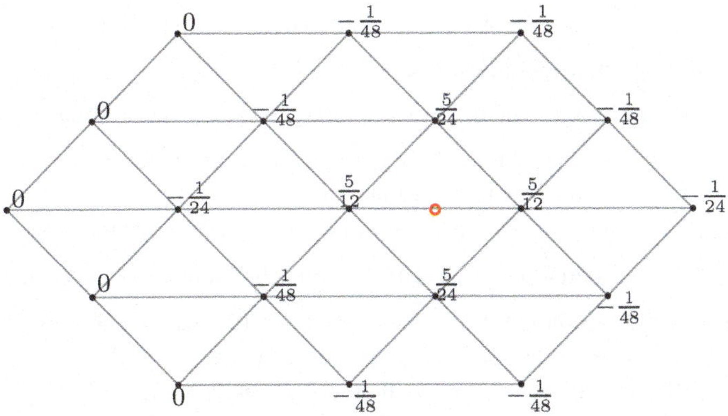

Fig. 9 The values in this figure provide the mask $\boldsymbol{\alpha}'$ used to evaluate $c\left(e_{i,j}^{1,1}\right)$

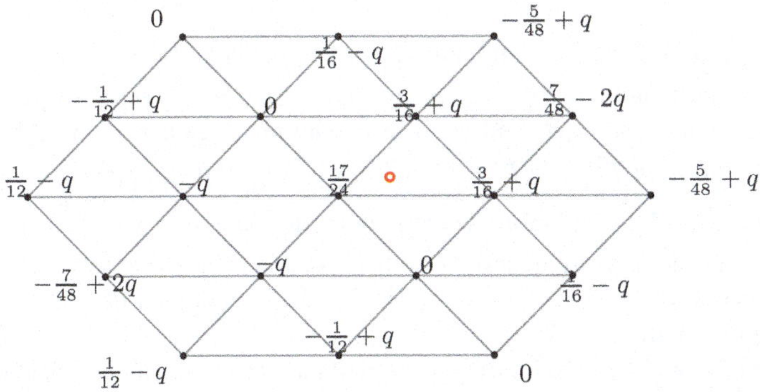

Fig. 10 The values in this figure provide the mask $\boldsymbol{\beta}'$ used to evaluate $c\left(z_{i,j}^{1,1}\right)$. The values depend on the parameter $q = \frac{3}{16}\beta_1 - \frac{5}{3}\omega + 4\beta_1\omega$

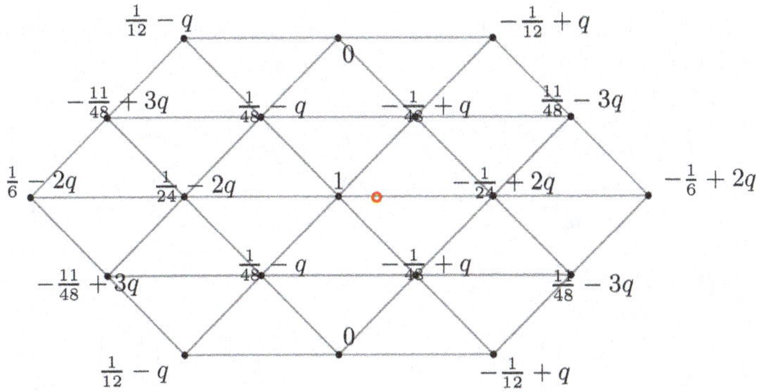

Fig. 11 The values in this figure provide the mask $\boldsymbol{\gamma}'$ used to evaluate $c\left(u_{i,j}^{1,1}\right)$. Also in this case its values depend on q

spline $Q_{4,2}^{\beta_1} f$ and ω coming from the subdivision scheme. Therefore, we denote the corresponding interpolating splines by $I_{4,1}^{\beta_1,\omega} f$.

We remark that while for the construction of $Q_{4,2}^{\beta_1} f$ we assume to know the values of f on $\mathcal{D}_2$, for the construction of $I_{4,1}^{\beta_1,\omega} f$ it is sufficient to know the function f on $\mathcal{D}_1$, the vertices of the triangulation.

Thanks to the approximation properties of $Q_{4,2}^{\beta_1}$ and of the MBISS here used, we have that $I_{4,1}^{\beta_1,\omega} f$ is a quartic spline with C^1 smoothness and the associated operator $I_{4,1}^{\beta_1,\omega}$ is exact on cubic polynomials and the error estimates of Theorem 1 hold.

Again, we continue by proposing some strategies in order to fix the free parameters. If we choose $\omega = 0$, that corresponds to the classical Butterfly subdivision scheme, we have only one free parameter. Since

$$\left\| I_{4,1}^{\beta_1,\omega} \right\|_\infty \leq \max\left\{ \left\| \boldsymbol{\alpha}' \right\|_1, \left\| \boldsymbol{\beta}' \right\|_1, \left\| \boldsymbol{\gamma}' \right\|_1 \right\} \leq \frac{145}{96},$$

we find that the value of β_1 that minimizes the upper bound is $\beta_1 = \frac{29}{72}$.

Another possibility is to consider the parameter $q := \frac{3}{16}\beta_1 - \frac{5}{3}\omega + 4\beta_1\omega$, appearing in the masks $\boldsymbol{\beta}'$ and $\boldsymbol{\gamma}'$. If we minimize again the upper bound for the infinity norm of $I_{4,1}^{\beta_1,\omega}$ $\left(\left\| I_{4,1}^{\beta_1,\omega} \right\|_\infty \leq \max\left\{ \left\| \boldsymbol{\alpha}' \right\|_1, \left\| \boldsymbol{\beta}' \right\|_1, \left\| \boldsymbol{\gamma}' \right\|_1 \right\} \right)$ we find the same value $\frac{145}{96}$ obtained before, corresponding to the choice $q = \frac{29}{384}$. Hence, we can choose ω and β_1 consequently. Another possible choice is to set $q = \frac{1}{12}$. In this case the masks $\boldsymbol{\beta}'$ and $\boldsymbol{\gamma}'$ have several zero coefficients in their definition, which is always convenient.

Obviously, other criteria for the selection of the free parameter can be considered.

5 Numerical Results

The performance of the operators defined in this paper are tested on two functions defined on the unit square. They are Franke's function

$$\begin{aligned} f_1(x) = {} & 0.75 e^{\left(-\frac{(9x_1-2)^2}{4} - \frac{(9x_2-2)^2}{4} \right)} + 0.75 e^{\left(-\frac{(9x_1+1)^2}{49} - \frac{9x_2+1}{10} \right)} \\ & + 0.5 e^{\left(-\frac{(9x_1-7)^2}{4} - \frac{(9x_2-3)^2}{4} \right)} - 0.2 e^{\left(-(9x_1-4)^2 - (9x_2-7)^2 \right)}, \end{aligned}$$

and the radial function

$$f_2(x) = 0.1 \left(1 + \cos\left(12\pi \cos\left(\pi \sqrt{x_1^2 + x_2^2} \right) \right) \right).$$

The latter is a highly oscillating function.

In order to estimate the maximal error (ME) as a function depending on a parameter h (ME_h), the error $|f - Qf|$ is evaluated at M points in a finite subset $G = \{(g_{1,i}, g_{2,j}) : (i, j) \in J\} \subset [0, 1]^2$. Moreover, the root mean square error ($RMSE$) is estimated as

$$RMSE_h := \sqrt{\frac{\sum_{(i,j)\in J} \left(f\left(g_{1,i}, g_{2,j}\right) - Qf\left(g_{1,i}, g_{2,j}\right)\right)^2}{M}}.$$

Regarding the value of M, the splines Qf have been evaluated by the de Casteljau's algorithm [16, p. 25] on 300 points in each of the triangles of the partition associated with the value h. Once computed ME_h and $RSME_h$, the numerical convergence orders are evaluated according to the expression $NCO := \log_2 \frac{ME_h}{ME_{h/2}}$.

As said before, f_2 is a highly oscillating function, therefore the initial value of h must be smaller than the one used for f_1.

Table 1 shows the values ME_h, $RMSH_h$ and NCO_h relative to $I_{4,1}^{\beta_1,\omega} f$ with $\beta_1 = \frac{29}{72}$ and $\omega = 0$. In Table 2 we report the results corresponding to the choice $q = \frac{1}{12}$.

Table 1 Numerical results relative to $I_{4,1}^{\beta_1,\omega} f$ with $\beta_1 = \frac{29}{72}$ and $\omega = 0$

	Test function f_1			Test function f_2		
h	ME_h	NCO	$RMSE_h$	ME_h	NCO	$RMSE_h$
1/4	3.44×10^{-1}	-	8.44×10^{-2}			
1/8	9.43×10^{-2}	1.87	2.48×10^{-2}			
1/16	2.89×10^{-2}	1.71	3.12×10^{-3}			
1/32	2.76×10^{-3}	3.39	2.50×10^{-4}			
1/64	1.81×10^{-4}	3.94	1.60×10^{-5}	1.05×10^{-1}	-	1.69×10^{-2}
1/128	1.16×10^{-5}	3.97	9.97×10^{-7}	1.11×10^{-2}	3.24	1.57×10^{-3}
1/256	7.22×10^{-7}	4.00	6.22×10^{-8}	6.88×10^{-4}	4.01	1.00×10^{-4}
1/512	4.51×10^{-8}	4.00	3.89×10^{-9}	4.27×10^{-5}	4.01	6.22×10^{-6}

Table 2 Numerical results relative to $I_{4,1}^{\beta_1,\omega} f$ with $q = \frac{1}{12}$

	Test function f_1			Test function f_2		
h	ME_h	NCO	$RMSE_h$	ME_h	NCO	$RMSE_h$
1/4	3.46×10^{-1}	-	8.54×10^{-2}			
1/8	8.97×10^{-2}	1.95	2.36×10^{-2}			
1/16	2.77×10^{-2}	1.70	2.96×10^{-3}			
1/32	2.63×10^{-3}	3.40	2.38×10^{-4}			
1/64	1.77×10^{-4}	3.89	1.57×10^{-5}	1.00×10^{-1}	-	1.63×10^{-2}
1/128	1.15×10^{-5}	3.95	9.91×10^{-7}	9.82×10^{-3}	3.35	1.48×10^{-3}
1/256	7.21×10^{-7}	4.00	6.22×10^{-8}	6.61×10^{-4}	3.89	9.78×10^{-5}
1/512	4.51×10^{-8}	4.00	3.89×10^{-9}	4.23×10^{-5}	3.97	6.18×10^{-6}

Table 3 Numerical results relative to $Q_{4,2}^{\beta_1} f$ with $\beta_1 = \frac{2}{5}$

	Test function f_1			Test function f_2		
h	ME_h	NCO	$RMSE_h$	ME_h	NCO	$RMSE_h$
1/4	1.69×10^{-1}	-	4.25×10^{-2}			
1/8	4.48×10^{-2}	1.92	8.06×10^{-3}			
1/16	7.90×10^{-3}	2.50	7.40×10^{-4}			
1/32	4.92×10^{-4}	4.01	4.52×10^{-5}			
1/64	2.96×10^{-5}	4.05	2.64×10^{-6}	4.04×10^{-2}	-	4.83×10^{-3}
1/128	1.82×10^{-6}	4.02	1.61×10^{-7}	2.39×10^{-3}	4.08	2.91×10^{-4}
1/256	1.13×10^{-7}	4.01	1.00×10^{-8}	1.25×10^{-4}	4.26	1.67×10^{-5}
1/512	7.05×10^{-9}	4.00	6.26×10^{-10}	7.31×10^{-6}	4.10	1.01×10^{-6}

Table 4 Numerical results for the quasi-interpolation operator given in [23]

	Test function f_1			Test function f_2		
h	ME_h	NCO	$RMSE_h$	ME_h	NCO	$RMSE_h$
1/4	7.27×10^{-2}	-	1.78×10^{-2}			
1/8	1.56×10^{-2}	2.22	1.76×10^{-3}			
1/16	1.28×10^{-3}	3.61	1.37×10^{-4}	4.02×10^{-1}	-	8.54×10^{-2}
1/32	1.02×10^{-4}	3.64	1.14×10^{-5}	8.86×10^{-2}	2.18	1.06×10^{-2}
1/64	1.06×10^{-5}	3.27	8.37×10^{-7}	7.66×10^{-3}	3.53	8.05×10^{-4}
1/128	7.70×10^{-7}	3.79	5.54×10^{-8}	4.51×10^{-4}	4.08	6.74×10^{-5}
1/256	4.97×10^{-8}	3.95	3.52×10^{-9}	3.73×10^{-5}	3.60	5.20×10^{-6}
1/512	3.13×10^{-9}	3.99	2.21×10^{-10}	2.79×10^{-6}	3.74	3.47×10^{-7}

Finally, for the sake of comparison, in Table 3 we report the results obtained by using the quasi-interpolating spline $Q_{4,2}^{\beta_1} f$ with $\beta_1 = \frac{2}{5} \in \left[\frac{13}{36}, \frac{41}{84}\right]$ and in Table 4 we report the results obtained by the quasi-interpolating spline proposed in [23] (see Table 5 and Table 3 in [2]).

The results are in accordance with the theoretical order of convergence. We remark that the approximating splines $Q_{4,2}^{\beta_1} f$ and the one proposed in [23] produce results similar to those obtained by $I_{4,1}^{\beta_1,\omega} f$ for the two different selections of the parameters. However, the efficiency of $I_{4,1}^{\beta_1,\omega}$ is higher than that of $Q_{4,2}^{\beta_1}$ and the operator proposed in [23] in terms of the number of evaluation points.

Moreover, Fig. 12 shows the interpolating splines $I_{4,1}^{\beta_1,\omega} f_1$, $I_{4,1}^{\beta_1,\omega} f_2$ and gives nice surfaces. They are comparable with those obtained in [2] and in [23] (see the figures there reported).

The approximation schemes here proposed have been developed to consider functions defined on the real plane, but the test functions f_1 and f_2 are defined on the unit square. To deal with triangles having a non interior vertex, the triangulation is extended as well as f_1 and f_2.

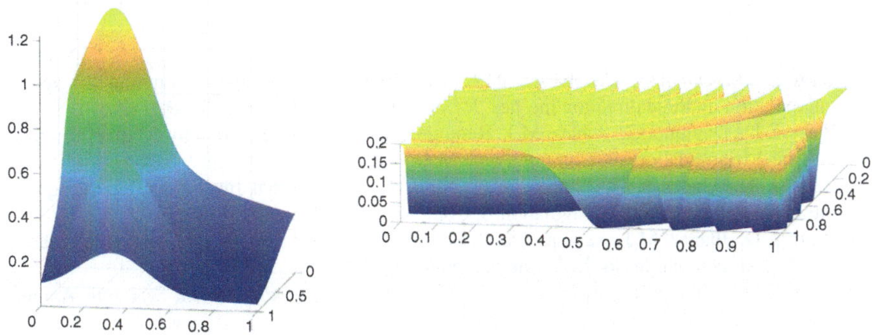

Fig. 12 The quartic C^1 splines $I_{4,1}^{\beta_1,\omega} f_1$ with $h = 1/64$ (left) and $I_{4,1}^{\beta_1,\omega} f_2$ (right) with $h = 1/256$. Their masks correspond to the values provided by $\beta_1 = \frac{29}{72}$, $\omega = 0$

6 Conclusions

In this paper, we have constructed and analysed C^1 quartic interpolating splines on type-1 triangulations, approximating regularly distributed data. A characteristic of the proposed methodology is that the Bernstein-Bézier coefficients in each triangle of the constructed quasi-interpolants are directly defined as appropriate linear combinations of point values at domain points that lie in a neighbourhood of the triangle to achieve C^1 regularity and approximation order four for enough regular functions. We have constructed such interpolating splines by combining a quasi-interpolating spline with one step of an interpolatory subdivision scheme. Numerical tests confirming the theoretical results have been provided for the proposed spline scheme.

We remark that, the approximation schemes constructed in this paper and based on regularly distributed point values can be used in two-stage methods, as in [9], by firstly computing a polynomial approximant on each triangle and then by sampling the necessary data values from the approximant on each triangle. Finally, in order to apply the approximation schemes here proposed to compact domains, it is possible to construct special rules near the boundary (see [21]) or extend the triangulation.

Acknowledgments The authors thank the referee for the helpful suggestions. The second, third and fifth authors are members of the INdAM Research group GNCS. The authors of the University of Granada are members of the research group FQM 191 *Matemática Aplicada* funded by the PAIDI program of the Junta de Andalucía. This work was partially supported by the program "Progetti di Ricerca 2019" of the Gruppo Nazionale per il Calcolo Scientifico (GNCS)—INdAM.

References

1. Barrera, D., Dagnino, C., Ibáñez, M.J., Remogna, S.: Point and differential C^1 quasi-interpolation on three direction meshes. J. Comput. Appl. Math. **354**, 373–389 (2019)
2. Barrera, D., Dagnino, C., Ibáñez, M.J., Remogna, S.: Quasi-interpolation by C^1 quartic splines on type-1 triangulations. J. Comput. Appl. Math. **349**, 225–238 (2019)
3. Barrera, D., Guessab, A., Ibáñez, M.J., Nouisser, O.: Optimal bivariate C^1 cubic quasi-interpolation on a type-2 triangulation. J. Comput. Appl. Math. **234**, 1188–1199 (2010)
4. Barrera, D., Ibáñez, M.J., Sablonnière, P., Sbibih, D.: On near-best discrete quasi-interpolation on a four-directional mesh. J. Comput. Appl. Math. **233**, 1470–1477 (2010)
5. Chui, C.K.: Multivariate Splines. In: CBMS-NSF Regional Conference Series in Applied Mathematics, Vol. 54. SIAM, Philadelphia (1988)
6. Charina, M., Conti, C., Jetter, K., Zimmermann, G.: Scalar multivariate subdivision schemes and box splines. Comput. Aided Geom. Des. **28**(5), 285–306 (2011)
7. Conti, C., Deng, C., Hormann, K.: Symmetric four-directional bivariate pseudo-spline symbols. Comput. Aided Geom. Des. **60**, 10–17 (2018)
8. Dagnino, C., Remogna, S., Sablonnière, P.: Error bounds on the approximation of functions and partial derivatives by quadratic spline quasi-interpolants on non-uniform criss-cross triangulations of a rectangular domain. BIT **53**, 87–109 (2013)
9. Davydov, O., Zeilfelder, F.: Scattered data fitting by direct extension of local polynomials to bivariate splines. Adv. Comput. Math. **21**, 223–271 (2004)
10. de Boor, C., Höllig, K., Riemenschneider, S.: Box Splines. Springer, New York (1993)
11. Dyn, N., Levin, D.: Subdivision schemes in geometric modelling. Acta Numer. **11**, 73–144 (2002)
12. Dyn, N., Levin, D., Gregory, J.A.: A butterfly subdivision scheme for surface interpolation with tension control. ACM Trans. Graph. **9**, 160–169 (1990)
13. Hed, S.: Analysis of subdivision schemes for surface generation. M.Sc. thesis, Tel Aviv University, 1992
14. Hering-Bertram, M., Reis, G., Zeilfelder, F.: Adaptive quasi-interpolating quartic splines. Computing **86**, 89–100 (2009)
15. Jia, R.Q.: Approximation order from certain spaces of smooth bivariate splines on a three-direction mesh. Trans. Am. Math. Soc. **295**, 199–212 (1986)
16. Lai, M.J., Schumaker, L.L.: Spline Functions on Triangulations. Cambridge University Press, Cambridge (2007)
17. Nürnberger, G., Zeilfelder, F.: Developments in bivariate spline interpolation. J. Comput. Appl. Math. **121**, 125–152 (2000)
18. Remogna, S.: Constructing good coefficient functionals for bivariate C^1 quadratic spline quasi-interpolants. In: Daehlen, M., et al. (eds.), Mathematical Methods for Curves and Surfaces. LNCS, vol. 5862, pp. 329–346. Springer, Berlin Heidelberg (2010)
19. Remogna, S.: Bivariate C^2 cubic spline quasi-interpolants on uniform Powel–Sabin triangulations of a rectangular domain. Adv. Comput. Math. **36**, 39–65 (2012)
20. Sablonnière, P.: Quadratic spline quasi-interpolants on bounded domains of $\mathbb{R}^d$, $d = 1, 2, 3$. Rend. Sem. Mat. Univ. Pol. Torino **61**, 229–246 (2003)
21. Sorokina, T., Zeilfelder, F.: Optimal quasi-interpolation by quadratic C^1 splines on four-directional meshes. In: Chui, C., et al. (eds.), Approximation Theory, Gatlinburg 2004, vol. XI, pp. 423–438. Nashboro Press, Brentwood (2005)
22. Sorokina, T., Zeilfelder, F.: Local quasi-interpolation by cubic C^1 splines on type-6 tetrahedral partitions. IMA J. Numer. Anal. **27**, 74–101 (2007)
23. Sorokina, T., Zeilfelder, F.: An explicit quasi-interpolation scheme based on C^1 quartic splines on type-1 triangulations. Comput. Aided Geom. Des. **25**, 1–13 (2008)
24. Wang, R.H.: Multivariate Spline Functions and Their Applications. Science Press, Beijing/New York; Kluwer Academic Publishers, Dordrecht/Boston/London (2001)

Approximation with Conditionally Positive Definite Kernels on Deficient Sets

Oleg Davydov

Abstract Interpolation and approximation of functionals with conditionally positive definite kernels is considered on sets of centers that are not determining for polynomials. It is shown that polynomial consistence is sufficient in order to define kernel-based numerical approximation of the functional with usual properties of optimal recovery. Application examples include generation of sparse kernel-based numerical differentiation formulas for the Laplacian on a grid and accurate approximation of a function on an ellipse.

Keywords Conditionally positive definite kernels · Numerical differentiation · Optimal recovery · Saddle point problem

1 Introduction

Let Ω be a set and P a finite dimensional space of functions on Ω. A function $K : \Omega \times \Omega \to \mathbb{R}$ is said to be a *conditionally positive definite kernel with respect to P* if for any finite set $X = \{x_1, \dots, x_n\} \subset \Omega$ the quadratic form $\sum_{i,j=1}^{n} c_i c_j K(x_i, x_j)$ is positive for all $c \in \mathbb{R}^n \setminus \{0\}$ such that $\sum_{i=1}^{n} c_i p(x_i) = 0$ for all $p \in P$ [9].

Given data (x_j, f_j), $j = 1, \dots, n$, with $x_j \in \Omega$, $f_j \in \mathbb{R}$, a sum of the form

$$\sigma(x) = \sum_{j=1}^{n} c_j K(x, x_j) + \tilde{p}, \qquad c_j \in \mathbb{R}, \quad \tilde{p} \in P \tag{1}$$

can be used to solve the interpolation problem

$$\sigma(x_i) = f_i, \qquad i = 1, \dots, n. \tag{2}$$

O. Davydov (✉)
University of Giessen, Giessen, Germany
e-mail: oleg.davydov@math.uni-giessen.de

G. E. Fasshauer et al. (eds.), *Approximation Theory XVI*, Springer Proceedings in Mathematics & Statistics 336, https://doi.org/10.1007/978-3-030-57464-2_3

Moreover, a solution of (2) satisfying the condition

$$\sum_{j=1}^{n} c_j p(x_j) = 0 \quad \text{for all } p \in P, \tag{3}$$

can always be found [9, p. 117]. This solution is unique if X is a *determining set* for P, that is $p \in P$ and $p|_X = 0$ implies $p = 0$.

In meshless finite difference methods, conditionally positive definite kernels with respect to spaces of polynomials are often used to produce numerical approximations of linear functionals

$$\lambda f \approx \sum_{i=1}^{n} w_i f(x_i), \qquad w_i \in \mathbb{R}, \tag{4}$$

such as the value $\lambda f = Df(x)$ of a differential operator D applied to a function f at a point $x \in \Omega$. If the interpolant $\sigma = \sigma_f$ satisfying (1)–(3) with $f_i = f(x_i)$ is uniquely defined, then the weights w_i of (4) can be obtained by the approximation $\lambda f \approx \lambda \sigma_f$, which leads to the conditions

$$\sum_{j=1}^{n} w_j K(x_i, x_j) + \tilde{p}(x_i) = \lambda' K(x_i), \quad i = 1, \dots, n, \quad \text{for some } \tilde{p} \in P, \tag{5}$$

$$\sum_{i=1}^{n} w_i p(x_i) = \lambda p \quad \text{for all } p \in P, \tag{6}$$

where $\lambda' K : \Omega \to \mathbb{R}$ is the function obtained by applying λ to the first argument of K. The weights w_i are in this case uniquely determined by the conditions (5)–(6). In particular, by introducing a basis for the space P, we can write both (1)–(3) and (5)–(6) as systems of linear equations with the same matrix which is non-singular as soon as X is a determining set for P. Solving this system is the standard way to obtain the weights w_i, see e.g. [7]. It is demonstrated in [1] that the weights satisfying (5)–(6) for a polyharmonic kernel K significantly improve the performance of meshless finite difference methods in comparison to the weights obtained by unconditionally positive definite kernels such as the Gaussian. In addition, these weights provide optimal recovery of λf from the data $f(x_i)$, $i = 1, \dots, n$, on spaces of functions of finite smoothness, see e.g. [9, Chapter 13].

An alternative interpretation of (5)–(6) is that the approximation (4) of λf is required to be exact for all $f = \sigma$ in the form (1) with coefficients c_j satisfying (3). Indeed, this can be easily shown with the help of the Fredholm alternative for matrices, see Theorem 1 below. In particular, (6) already expresses exactness of (4) for all elements of P. In the case when Ω is a domain in $\mathbb{R}^d$ and P is a space of d-variate polynomials, (6) can be used to obtain error bounds for the numerical differentiation with weights w_i, see e.g. [5, 6].

However, exactness (6) for $p \in P$ is sometimes achievable without X being a determining set for P. We then say that X is *P-consistent* for λ. The best known examples are the Gauss quadrature when $\lambda f = \int_a^b f(x)\,dx$ and the five point stencil for the two-dimensional Laplacian. Moreover, P-consistent sets with n significantly smaller than the dimension of P often can be used for the numerical discretization of the Laplace operator on gridded nodes in irregular domains, leading to sparser differentiation matrices [3].

In this paper we study numerical approximation formulas (4) obtained by requiring exactness conditions (5)–(6) on "deficient" sets X that are not determining for P.

Our main result (Theorem 2 and Corollary 1) shows that a unique formula satisfying these conditions exists as soon as X is P-consistent. Another consequence is that the coefficients c_j of the interpolant (1) are uniquely defined for any X (Corollary 2). Numerical differentiation formulas obtained in this way provide optimal recovery on native spaces of the kernels. We also discuss computational methods for the weights of the formula (4) and coefficients of the interpolant (1). In particular, a null space method can be used for the saddle point problems (5)–(6) or (1)–(3) even if in the case of deficient sets they do not satisfy restrictions usually required in the literature [2].

In the last section we describe two types of deficient sets that arise naturally in applications. First, deficient subsets of a grid may be used for numerical differentiation of the Laplacian (Sect. 3.1). Second, function values and differential operators on algebraic surfaces, in this case an ellipse, may be approximated using data located on the manifold, which are necessarily deficient sets for polynomials in the ambient space of degree at least the order of the surface (Sect. 3.2). In both cases, numerical results demonstrate a robust performance of the suggested numerical methods, and a reasonable approximation quality of the polyharmonic kernels we employ in the experiments.

2 Approximation on Deficient Sets

We assume that $K : \Omega \times \Omega \to \mathbb{R}$ is a conditionally positive definite kernel with respect to a linear space P of functions on Ω, with $\dim P = m$. Let $\{p_1, \ldots, p_m\}$ be a basis for P. By writing $\tilde{p} = \sum_{j=1}^m v_j p_j$, $v_j \in \mathbb{R}$, conditions (5)–(6) give rise to a linear system with respect to w_j and v_j, in block matrix form,

$$\begin{bmatrix} K_X & P_X \\ P_X^T & 0 \end{bmatrix} \cdot \begin{bmatrix} w \\ v \end{bmatrix} = \begin{bmatrix} a \\ b \end{bmatrix}, \tag{7}$$

where

$$K_X = [K(x_i, x_j)]_{i,j=1}^n, \quad P_X = [p_j(x_i)]_{i,j=1}^{n,m},$$

$$w = [w_j]_{j=1}^n, \quad v = [v_j]_{j=1}^m, \quad a = [\lambda' K(x_i)]_{i=1}^n, \quad b = [\lambda p_j]_{j=1}^m.$$

Condition (3) in matrix form is

$$P_X^T c = 0, \quad c = [c_j]_{j=1}^n,$$

that is c belongs to the null space $N(P_X^T)$ of P_X^T. Since

$$P_X v = [\tilde{p}(x_i)]_{i=1}^n, \quad \tilde{p} = \sum_{j=1}^m v_j p_j,$$

we see that the condition that X is a determining set for P is equivalent to $N(P_X) = 0$.

We show that the conditions (5)–(6) express the exactness of (4) for the sums σ conditional on (3), even when X is a *deficient set* for P, that is $N(P_X) \neq 0$. Recall that this condition is equivalent to $R(P_X^T) \neq \mathbb{R}^m$, where $R(A)$ denotes the range of a matrix A.

Theorem 1 *Let $X = \{x_1, \ldots, x_n\} \subset \Omega$. An approximation formula (4) satisfies the exactness condition $\lambda\sigma = \sum_{i=1}^n w_i \sigma(x_i)$ for all sums σ in the form (1) with coefficients c_j satisfying (3) if and only if (5)–(6) holds for the weights w_i, $i = 1, \ldots, n$.*

Proof The exactness condition is

$$\sum_{j=1}^n c_j \lambda' K(x_j) + \lambda p = \sum_{j=1}^n c_j \sum_{i=1}^n w_i K(x_i, x_j) + \sum_{i=1}^n w_i p(x_i)$$

for all $c = [c_j]_{j=1}^n$ satisfying (3) and all $p \in P$. In particular, for $c = 0$ we obtain (6), and rewrite the condition as

$$\sum_{j=1}^n c_j \Big(\lambda' K(x_j) - \sum_{i=1}^n w_i K(x_i, x_j)\Big) = 0 \quad \text{for all } c \in N(P_X^T).$$

By the Fredholm alternative for matrices this is equivalent to

$$\Big[\lambda' K(x_j) - \sum_{i=1}^n w_i K(x_i, x_j)\Big]_{j=1}^n \in R(P_X),$$

which is in turn equivalent to (5) in view of the symmetry of the kernel K. □

Linear systems of the type (7) have been extensively studied under the name of equilibrium equations [8, Section 4.4.6] or saddle point problems [2] because they arise in many application areas. Our approach below is a variation of the null space techniques described in [2, Section 6]. However, usual assumptions that $n \geq m$, K_X is positive semidefinite and P_X has full column rank are not satisfied in our case of interest.

As long as X is a deficient set, $R(P_X^T) \neq \mathbb{R}^m$ and hence the solvability of $P_X^T w = b$ cannot be guaranteed for all b. Nevertheless, should this last equation have a solution for w, there is a unique weight vector w satisfying (7).

Theorem 2 *There is a unique vector w satisfying (7) if and only if $b \in R(P_X^T)$.*

Proof The necessity of the condition $b \in R(P_X^T)$ is obvious. To show the sufficiency, assume that $P_X^T w_0 = b$ for some $w_0 \in \mathbb{R}^n$. Then the solution w must satisfy $P_X^T(w - w_0) = 0$ if it exists, so we look for w in the form

$$w = w_0 + \tilde{u}, \quad \tilde{u} \in N(P_X^T).$$

Let M be a matrix whose columns form a basis for $N(P_X^T)$. Then $\tilde{u} = Mu$ for some vector u, and we may write (7) equivalently as a linear system with respect to u and v,

$$K_X M u + P_X v = a - K_X w_0. \tag{8}$$

Since $M^T P_X = 0$, it follows that

$$M^T K_X M u = M^T(a - K_X w_0). \tag{9}$$

Since K is conditionally positive definite, the matrix $M^T K_X M$ is positive definite, and hence there is a unique u determined by the last equation. The existence of some $v \in \mathbb{R}^m$ such that (8) holds is equivalent to the claim that $K_X M u - a + K_X w_0 \in R(P_X)$. This claim follows from the Fredholm alternative since (9) implies that $K_X M u - a + K_X w_0 \perp N(P_X^T)$. Thus, u and v satisfying (8) exist, and u is uniquely determined. Then $w = w_0 + Mu$ is a unique vector satisfying (7). □

Remark 1 Theorem 2 is valid for any linear system (7) with arbitrary matrices A and B replacing K_X and P_X, respectively, and arbitrary a, b, as soon as A is definite on $N(B^T)$, that is $x^T A x \neq 0$ for all $x \in N(B^T) \setminus \{0\}$. Indeed, this condition implies that $M^T A M$ is non-singular and hence the argument in the proof goes through.

As long as the condition $b \in R(P_X^T)$ is satisfied, the weight vector w may be found by any solution method applicable to the system (7), for example via the pseudoinverse of its matrix when it is singular. Alternatively, we may use the null space matrix M of the above proof and find w from the linear system

$$\begin{bmatrix} M^T K_X \\ P_X^T \end{bmatrix} w = \begin{bmatrix} M^T a \\ b \end{bmatrix}, \tag{10}$$

which is in general overdetermined but has full rank because its solution w is unique. Indeed, any solution w of (10) satisfied (7) for some v since $M^T K_X w = M^T a$ implies $K_X w - a \perp N(P_X^T)$ and thus $K_X w - a \in R(P_X)$. We refer to [2, Section 6] for the computational methods for the null space matrix M. One obvious possibility is to employ the right singular vectors of P_X, see [8, Eq. (2.5.4)]. Should v be needed, it can be computed as a solution of the consistent linear system

$$P_X v = a - K_X w. \tag{11}$$

For example we can use

$$v = P_X^+(a - K_X w), \tag{12}$$

where P_X^+ denotes the Moore-Penrose pseudoinverse of P_X, is the unique v with the smallest 2-norm.

We formulate two immediate consequences of Theorem 2 for the numerical approximation of functionals and for the interpolation. Note that (1)–(3) can be written in the form (7) with w replaced by c, $a = [f_i]_{i=1}^n$, and $b = 0$. In particular, the condition $0 \in R(P_X^T)$ of Theorem 2 is trivially satisfied.

Corollary 1 *For any X and λ there is a unique numerical approximation formula (4) satisfying (5)–(6) as soon as (6) is solvable.*

Corollary 2 *For any data (x_j, f_j), $j = 1, \dots, n$, one or more interpolants σ satisfying (1)–(3) exist and their coefficients c_j, $j = 1, \dots, n$, are uniquely determined.*

Thanks to Theorem 1 we also obtain the property known for the case of a determining set X that the approximation $\lambda f \approx \sum_{i=1}^n w_i f(x_i)$ can be found by requiring $\lambda f \approx \lambda\sigma$ for any interpolant σ of Corollary 2 with $f_i = f(x_i)$.

Looking specifically at numerical differentiation, consider the case when $\Omega = \mathbb{R}^d$, $\lambda f = Df(x)$ for a linear differential operator D of order k and $x \in \mathbb{R}^d$, and $P = \mathbb{P}_q^d$, the space of d-variate polynomials of total order at most q (that is, total degree at most $q - 1$) for some $q \in \mathbb{N}$. Any kernel K that is conditionally positive definite with respect to P generates a native semi-Hilbert space $F(K, P)$ of functions on Ω with null space P, see e.g. [9]. By inspecting the arguments in Section 2 and Lemma 6 of [4], we see that thanks to Corollary 1, the optimal recovery property of the weights w_i defined by (5)–(6) remains valid for deficient sets $X = \{x_1, \dots, x_n\}$. More precisely, the worst case error of numerical differentiation formulas on the unit ball of $F_q(K) := F(K, \mathbb{P}_q^d)$,

$$E(u) := \sup_{\substack{f \in F_q(K) \\ \|f\|_{F_q(K)} \le 1}} \Big|Df(x) - \sum_{i=1}^n u_i f(x_i)\Big|,$$

can be computed as

$$E^2(u) = D'D''K(x,x) - \sum_{i=1}^{n} u_i\big(D'K(x,x_i) + D''K(x_i,x)\big) + \sum_{i,j=1}^{n} u_iu_jK(x_i,x_j). \tag{13}$$

The weight vector w has the *optimal recovery* property in the sense that it satisfies

$$E(w) = \min\big\{E(u) : u \in \mathbb{R}^n,\ Dp(x) = \sum_{i=1}^{n} u_i p(x_i) \text{ for all } p \in \mathbb{P}_q^d\big\} \tag{14}$$

as soon as the mixed partial derivatives of K exist at $(x,x) \in \mathbb{R}^d \times \mathbb{R}^d$ up to the order k in each of both d-dimensional variables, and X is such that there exists a vector $u \in \mathbb{R}^n$ with polynomial exactness

$$Dp(x) = \sum_{i=1}^{n} u_i p(x_i) \text{ for all } p \in \mathbb{P}_q^d.$$

We use here D' and D'' to indicate when $\lambda f = Df(x)$ acts on the first, respectively, the second argument of K.

Note that the equality-constrained quadratic minimization problem (14) provides an alternative way of computing the optimal weight vector on a deficient set. By Theorem 2 we know that its solution w is unique as soon as the feasible region is non-empty.

3 Examples

In this section we illustrate Corollaries 1 and 2 on particular examples where deficient sets X seem useful.

We consider the *polyharmonic kernels* $K_{s,d} : \mathbb{R}^d \times \mathbb{R}^d \to \mathbb{R}$, defined for all real $s > 0$ by $K_{s,d}(x,y) = \varphi_s(\|x-y\|_2)$, where

$$\varphi_s(r) := (-1)^{\lfloor s/2 \rfloor + 1} \begin{cases} r^s \log r, & \text{if } s \text{ is an even integer,} \\ r^s, & \text{otherwise.} \end{cases} \tag{15}$$

The kernel $K_{s,d}$ is conditionally positive definite with respect to $\mathbb{P}_q^d$ for all $q \geq \lfloor s/2 \rfloor + 1$. We cite [6, 9] and references therein for further information on these kernels. If $m = (s+d)/2$ is an integer and q is chosen equal to m, then the native space $F_m(K_{s,d})$ coincides with the Beppo-Levi space $BL_m(\mathbb{R}^d)$, see [9, Theorem 10.43]. For any $q \geq \lfloor s/2 \rfloor + 1$, the space $F_q(K_{s,d})$ can be described with the help of the generalized Fourier transforms as in [9, Theorem 10.21]. By the arguments in

Sect. 2, formulas (13) and (14) apply to $K_{s,d}$ as soon as $s > 2k$, where k is the order of the differential operator D.

3.1 Numerical Differentiation of Laplacian on a Grid

We are looking for numerical differentiation formulas of the type

$$\Delta f(0) \approx \sum_{\alpha \in Z_{d,r}} w_\alpha f(\alpha), \quad Z_{d,r} := \{\alpha \in \mathbb{Z}^d : \|\alpha\|_2 \le r\}, \quad r > 0, \tag{16}$$

where Δ is the Laplacian $\Delta f = \sum_{i=1}^d \partial^2 f/\partial x_i^2$. The set $Z_{d,r}$ for $0 \le r < 1$ consists of the origin only and hence is not useful for the approximation of the Laplacian. For $r = 1$ we have $Z_{d,1} = \{0, \pm e_1, \ldots \pm e_d\}$, where e_i is the i-th unit vector in $\mathbb{R}^d$, and

$$\Delta f(0) \approx -2df(0) + \sum_{i=1}^d f(e_i) + \sum_{i=1}^d f(-e_i) \tag{17}$$

is the classical numerical differentiation formula exact for all cubic polynomials $f = p \in \mathbb{P}_4^d$. Hence (6) is solvable for all $X = Z_{d,r}$, $r \ge 1$, if $P = \mathbb{P}_4^d$.

According to Corollary 1, we have computed the unique weights of the formula (16) satisfying (5)–(6) for the kernel $K_{7,d}$, $P = \mathbb{P}_4^d$ and $X = Z_{d,r}$ for all $d = 2, \ldots, 5$ and $r = 1, \sqrt{2}, \sqrt{3}, 2$. As a basis for $\mathbb{P}_4^d$ we choose ordinary monomials. However, the computation is performed using the rescaling of X as X/r according to the suggestion in [5, Section 6.1]

Table 1 presents information about the size $|X|$ of X, dimensions of the null spaces of P_X and P_X^T, the optimal recovery error (14) on $F_4(K_{7,d})$, the stability constant of the weight vector $\|w\|_1 = \sum_{i=1}^n |w_i|$, and the condition number *cond* of the system (10) we solved in order to compute the weights for $r \ne 1$. A smaller optimal recovery error indicates better approximation quality, whereas $\|w\|_1$ and *cond* measure the numerical stability of the formulas. Note that $\dim N(P_X^T) = 0$ for $r = 1$, which means that (17) is the only solution of (6) in this case, and hence it provides the optimal recovery on $F_4(K_{7,d})$. For $r = 2$ we have $\dim N(P_X) = 0$ and it follows that $Z_{d,2}$ is a determining set for $\mathbb{P}_4^d$. For $r = \sqrt{2}, \sqrt{3}$ we obtain examples of optimal recovery weights on deficient sets, with $\dim N(P_X)$ being the dimension of the affine space of weight vectors satisfying the polynomial exactness condition (6). These new weights seem to provide a meaningful choice for the two intermediate sets between the classical polynomial stencil on $Z_{d,1}$, and the standard polyharmonic weights on the determining set $Z_{d,2}$. Indeed, as expected, the optimal recovery error $E(w)$ reduces when $|X|$ increases, whereas the stability constant and condition numbers tend to increase.

Table 1 Numerical differentiation of Laplacian on a grid: $|X|$ is the cardinality of X, $dN = \dim N(P_X)$, $dNt = \dim N(P_X^T)$, $E(w)$ is given by (13), and *cond* is the condition number of the matrix of (10)

r	$\|X\|$	dN	dNt	$E(w)$	$\|w\|_1$	*cond*
$d = 2$, $X = Z_{2,r}$, $\dim \mathbb{P}_4^2 = 10$						
1	5	5	0	13.4	8.0	–
$\sqrt{2}$	9	2	1	10.6	13.5	2.0e+02
$\sqrt{3}$	9	2	1	10.6	13.5	2.0e+02
2	13	0	3	7.4	11.8	3.9e+02
$d = 3$, $X = Z_{3,r}$, $\dim \mathbb{P}_4^3 = 20$						
1	7	13	0	17.2	12.0	–
$\sqrt{2}$	19	4	3	12.3	22.7	3.8e+02
$\sqrt{3}$	27	3	10	12.4	24.8	2.5e+03
2	33	0	13	9.0	30.1	5.1e+03
$d = 4$, $X = Z_{4,r}$, $\dim \mathbb{P}_4^4 = 35$						
1	9	26	0	20.8	16.0	–
$\sqrt{2}$	33	8	6	14.0	31.8	5.7e+02
$\sqrt{3}$	65	4	34	13.9	39.7	6.9e+03
2	89	0	54	10.4	40.5	3.1e+04
$d = 5$, $X = Z_{5,r}$, $\dim \mathbb{P}_4^5 = 56$						
1	11	45	0	24.2	20.0	–
$\sqrt{2}$	51	15	10	15.6	40.9	7.7e+02
$\sqrt{3}$	131	5	80	15.4	56.4	1.3e+04
2	221	0	165	11.7	55.0	9.0e+04

3.2 Interpolation of Data on Ellipse

In this example we compute the kernel interpolant (1) satisfying (3) and $\sigma(x_i) = f(x_i)$, $i = 1, \ldots, n$, for the test function $f : \mathbb{R}^2 \to \mathbb{R}$ given by

$$f(x, y) = \sin(\pi x)\sin(\pi y).$$

We use the polyharmonic kernels $K_{s,2}$ and $P = \mathbb{P}_q^2$ for the pairs

$$(s, q) = (5, 3), (7, 4), (9, 5),$$

and choose sets X with $n = |X| = 5 \cdot 2^i$, $i = 0, 1, \ldots 6$, on the ellipse $\mathcal{E}$ with half-axes $a = 1$ and $b = 0.75$ centered at the origin. The sets are obtained by first choosing parameter values $t_i = ih$, $i = 0, \ldots, n-1$, where $h = 2\pi/n$, then adding to each t_i a random number ϵ_i with uniform distribution in the interval $[-0.3h, 0.3h]$, and selecting $x_i = \big(a\cos(t_i + \epsilon_i), b\sin(t_i + \epsilon_i)\big)$. The first two sets used in our experiments are shown in Fig. 1.

Since $X \subset \mathcal{E}$ and there exists a nontrivial quadratic polynomial $p \in \mathbb{P}_3^2$ that vanishes on $\mathcal{E}$,

$$p(x, y) = x^2/a^2 + y^2/b^2 - 1,$$

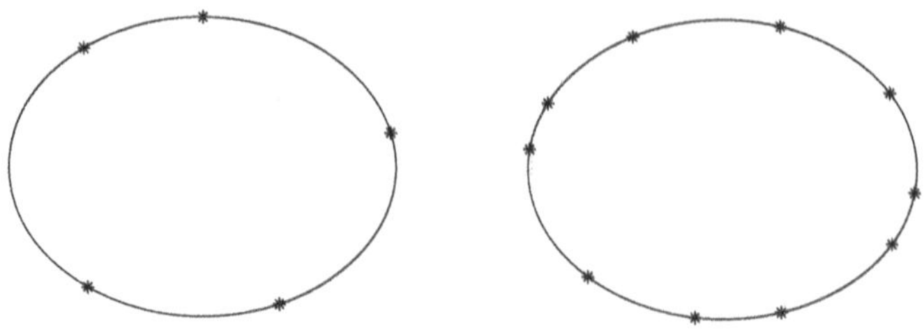

Fig. 1 Interpolation of data on ellipse: The sets X with $|X| = 5$ (left) and 10 (right)

all sets X are deficient for $\mathbb{P}_q^2$, $q \geq 3$. Nevertheless, according to Corollary 2, the coefficients c_j of the interpolant σ in (1) are uniquely determined and can be computed by solving the system (10). A polynomial $\tilde{p}$ of (1), $\tilde{p} = v_1 p_1 + \cdots + v_m p_m$, can be computed by solving (11). We will use the pseudoinverse as in (12), but in fact the polynomial $\tilde{p}$ is uniquely determined on the ellipse $\mathcal{E}$ as soon as $n \geq 2q-1$. Indeed, if both $\tilde{p}_1, \tilde{p}_2 \in \mathbb{P}_q^2$ satisfy (11), and $\tilde{p}_1 - \tilde{p}_2 = u_1 p_1 + \cdots + u_m p_m$, then $P_X u = 0$, which implies $(\tilde{p}_1 - \tilde{p}_2)|_X = 0$. Hence $x_1, \ldots, x_n$ are intersection points of the ellipse and the zero curve of $\tilde{p}_1 - \tilde{p}_2$, an algebraic curve of order $q-1$. By Bezout theorem, this curve must contain $\mathcal{E}$ as soon as $n > 2(q-1)$, which implies $\tilde{p}_1|_{\mathcal{E}} = \tilde{p}_2|_{\mathcal{E}}$.

Thus, $\sigma|_{\mathcal{E}}$ is well defined as soon as $|X| \geq 5$ for $q = 3$, $|X| \geq 7$ for $q = 4$ and $|X| \geq 9$ for $q = 5$. We are using $\sigma(x)$ as an approximation of $f(x)$ for $x \in \mathcal{E}$. Moreover, we also approximate the surface gradient

$$\nabla_{\mathcal{E}} f(x) := \nabla f(x) - \nabla f(x)^T \nu(x) \cdot \nu(x), \quad x \in \mathcal{E},$$

where $\nu(x)$ is the unit outer normal to $\mathcal{E}$ at x. The surface gradient $\nabla_{\mathcal{E}} f(x)$ can either be approximated by $\nabla_{\mathcal{E}} \sigma(x)$, or by using a numerical differentiation formula (4), with the same result. For each X, except of $|X| = 5$ for $q = 4, 5$, we evaluated the maximum error of the function and surface gradient,

$$max = \max_{x \in \mathcal{E}} |f(x) - \sigma(x)|,$$

$$maxg = \max_{x \in \mathcal{E}} \|\nabla_{\mathcal{E}} f(x) - \nabla_{\mathcal{E}} \sigma(x)\|_2,$$

by sampling the parameter t of the ellipse $\big(a \cos t, b \sin t\big)$, $t \in [0, 2\pi)$, equidistantly with the step $h/20$. The results are presented in Table 2, where we also included the condition number *cond* of the system (10). Note that we translate and scale X using its center of gravity z, and perform the computations with $K_{s,2}$ and ordinary monomials on the set $Y = (X - z)/\max\{\|x_i - z\|_2 : i = 1, \ldots, n\}$, in order to improve the condition numbers. The results in the table demonstrate a fast convergence of the interpolant σ and its surface gradient to f and $\nabla_{\mathcal{E}} f$. Note that although the condition numbers become high when the set X fills the ellipse

Table 2 Interpolation of a test function on an ellipse using the kernel $K_{s,2}$ and $P = \mathbb{P}_q^2$: $|X|$ is the cardinality of X, max and $maxg$ are the maximum errors of the function or the surface gradient, respectively, and $cond$ is the condition number of the matrix of (10)

	$s = 5, q = 3$			$s = 7, q = 4$			$s = 9, q = 5$		
$\|X\|$	max	$maxg$	$cond$	max	$maxg$	$cond$	max	$maxg$	$cond$
5	8.3e−01	2.5e+00	4.7e+00	–	–	–	–	–	–
10	2.5e−01	9.9e−01	3.0e+02	1.8e−01	7.9e−01	1.9e+02	1.6e−01	6.9e−01	2.3e+02
20	2.9e−03	2.0e−02	2.3e+04	1.6e−03	1.1e−02	9.8e+04	1.6e−03	1.5e−02	2.0e+05
40	4.6e−05	6.9e−04	2.5e+06	2.2e−06	3.6e−05	5.3e+07	6.1e−07	1.4e−05	6.3e+08
80	1.0e−06	3.1e−05	2.4e+08	1.4e−08	3.5e−07	2.3e+10	4.2e−10	1.5e−08	1.2e+12
160	2.2e−08	1.2e−06	1.2e+10	7.5e−11	4.2e−09	4.8e+12	8.1e−12	1.8e−09	4.6e+15
320	2.7e−10	3.1e−08	9.2e+11	1.0e−12	2.2e−10	2.9e+15	9.1e−12	4.9e−10	5.8e+17

more densely, they are moderate in comparison to significantly worse conditioned matrices arising if infinitely smooth kernels such as the Gaussian $K_{G,\varepsilon}(x-y) = \exp(-\varepsilon\|x-y\|_2^2)$, $\varepsilon > 0$, are employed, see also discussions in [7, Section 5.1.5].

References

1. Bayona, V., Flyer, N., Fornberg, B., Barnett, G.A.: On the role of polynomials in RBF-FD approximations: II. Numerical solution of elliptic PDEs. J. Comput. Phys. **332**, 257–273 (2017). https://doi.org/10.1016/j.jcp.2016.12.008
2. Benzi, M., Golub, G.H., Liesen, J.: Numerical solution of saddle point problems. Acta Numer. **14**, 1–137 (2005)
3. Davydov, O.: Selection of sparse sets of influence for meshless finite difference methods (2019). arxiv:1908.01567
4. Davydov, O., Schaback, R.: Error bounds for kernel-based numerical differentiation. Numer. Math. **132**(2), 243–269 (2016). DOI https://doi.org/10.1007/s00211-015-0722-9
5. Davydov, O., Schaback, R.: Minimal numerical differentiation formulas. Numer. Math. **140**(3), 555–592 (2018). https://doi.org/10.1007/s00211-018-0973-3
6. Davydov, O., Schaback, R.: Optimal stencils in Sobolev spaces. IMA J. Numer. Anal. **39**(1), 398–422 (2019). https://doi.org/10.1093/imanum/drx076
7. Fornberg, B., Flyer, N.: A Primer on Radial Basis Functions with Applications to the Geosciences. Society for Industrial and Applied Mathematics, Philadelphia (2015)
8. Golub, G.H., Van Loan, C.F.: Matrix Computations, 3rd edn. The Johns Hopkins University Press, Baltimore (1996)
9. Wendland, H.: Scattered Data Approximation. Cambridge University Press, Cambridge (2005)

Non-stationary Subdivision Schemes: State of the Art and Perspectives

Costanza Conti and Nira Dyn

Abstract This paper reviews the state of the art of non-stationary subdivision schemes, which are iterative procedures for generating smooth objects from discrete data, by repeated level dependent linear refinements. In particular the paper emphasises the potentiality of these schemes and the wide perspective they open, in comparison with stationary schemes based on level-independent linear refinements.

Keywords Subdivision schemes · Linear operators · Non-stationary schemes · Generation/reproduction of exponential polynomials · Analysis of convergence/smoothness

1 Introduction

Subdivision schemes were created originally to design geometrical models (see [4, 6, 30, 35],) but very soon they were recognised as methods for approximation (see [5, 36]). They are iterative methods for the generation of sets of points based on refinement rules that can be easily and efficiently implemented on a computer.

Since the 90s, subdivision schemes attracted many scientists for both the simplicity of their basic ideas and the mathematical elegance emerging in their analysis: they are defined by repeatedly applying simple and local refinement rules which have been extended to refine other objects such as vectors, matrices, manifold data, sets of points, curves, nets of functions. Therefore, the domain of application of subdivision is vast and they emerge in different contexts ranging from computer animation [31] to motion analysis [57].

C. Conti (✉)
Dipartimento di Ingegneria Industriale, University of Florence, Florence, Italy
e-mail: costanza.conti@unifi.it

N. Dyn
School of Mathematical Sciences, Tel Aviv University, Tel Aviv, Israel
e-mail: niradyn@tauex.tau.ac.il

G. E. Fasshauer et al. (eds.), *Approximation Theory XVI*, Springer Proceedings in Mathematics & Statistics 336, https://doi.org/10.1007/978-3-030-57464-2_4

The most studied subdivision schemes are *linear* and *stationary* (level independent). A nice aspect of linear subdivision schemes is that many of their properties can be translated into algebraic properties of Laurent polynomials. This makes their analysis easy and efficient. Moreover, since these schemes can be viewed as repeated multiplication by matrices, many analysis tools are based on linear algebra such as the "joint spectral radius" of two matrices (see [61]). Linear subdivision schemes are the subject of this survey paper. First we review the stationary schemes, and then in more details the non-stationary ones.

Stationary schemes are characterised by repeatedly applying the same simple and local refinement rule while the non-stationary (or level dependent) schemes apply a different rule in each level of refinement. Yet, changing rules with the levels is not a big difference from an implementation point of view, also in consideration that, realistically, only few subdivision iterations are executed. Contrary, from a theoretical point of view, non-stationary schemes are certainly more difficult to analyse. Level-dependent schemes were introduced to augment the class of limit functions defined through stationary schemes. For example, they allow the definition of C^∞ compactly supported functions like the Rvachev function (see, e.g. [39]) or exponential B-splines.

This type of limits shows that non-stationary schemes alleviate the limitations of stationary schemes that the smoothness of their limits of minimal compact support is bounded by the size of that support.

The non-stationary schemes are essentially different from the stationary ones: non-stationary schemes are able to generate conic sections, or to deal with level-dependent tension parameters for modifying the shape of a subdivision limit, while the stationary ones are not. An example of level-dependent subdivision schemes is given by Hermite schemes that allow to model curves and surfaces involving their gradient fields. They are interesting both in geometric modelling and biological imaging [1, 2, 14, 24, 65]. Additionally, non-stationary subdivision schemes play a role in the construction of non-stationary wavelet and framelets whose adaptivity makes them more flexible (see [13, 26, 42, 46, 67]). Last, but not least, level-dependent rules have the potential to overcome the standard limitations of subdivision surfaces such as artefacts and low regularity at extraordinary vertices/faces (see [64] for the limitations).

The paper is organised as follows: Sect. 2 provides a general description of the subdivision ideas together with classical examples of univariate and bivariate linear and stationary subdivision schemes. Also, the section presents a short description of the main subdivision applications and a review of the analysis tools of stationary linear schemes. Then, in Sect. 3 non-stationary subdivision schemes are discussed with emphasis on the motivation for their use. Section 4 is devoted to the analysis tools specific for non-stationary subdivision schemes, while the closing Sect. 5 presents open problems in the non-stationary setting.

2 Classical Subdivision Schemes

Subdivision schemes are efficient iterative methods for generating limit objects from discrete sets of data: Given $\mathcal{D}_0$—an initial set of data—the procedure iteratively defines a sequence of denser and denser sets of data $\{\mathcal{D}_k\}_{k\geq 0}$

$$\mathcal{D}_0 \underbrace{\longrightarrow}_{\text{ref. rule}} \mathcal{D}_1 \underbrace{\longrightarrow}_{\text{ref. rule}} \mathcal{D}_2 \cdots \underbrace{\longrightarrow}_{\text{ref. rule}} \mathcal{D}_k$$

by suitable *refinement rules* which can be linear or non-linear, level dependent or level independent, given by a formula or a geometric construction, just to mention some possibilities. Whenever $\lim_{k\to\infty} \mathcal{D}_k$ exists, in a sense to be explained later, it is the *subdivision limit* generated by the scheme.

At the early stage of the study of subdivision schemes, the initial set $\mathcal{D}_0$ consisted mainly of points, but in the last 30 years, subdivision was extended to more abstract settings, such as vector fields, manifold valued data, matrices, sets, curves or nets of functions. Examples of different possibilities are shown in the next figures after three refinement steps of a point subdivision scheme, a net subdivision scheme and a mesh subdivision scheme, respectively (Figs. 1, 2, 3).

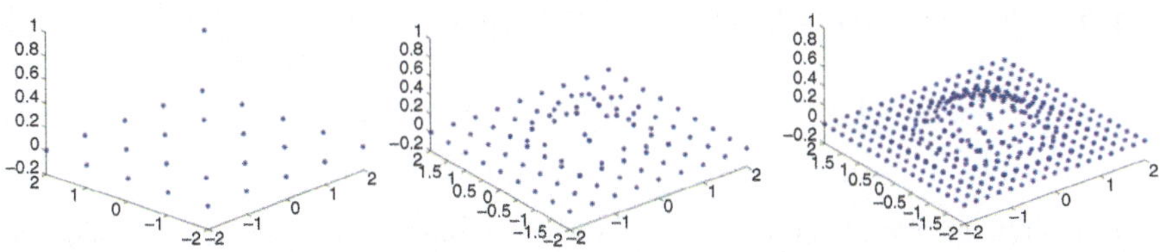

Fig. 1 Example of refinement of real values with limit a bivariate function

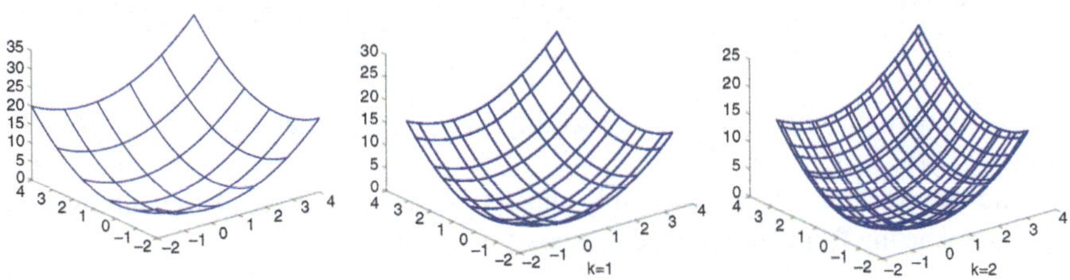

Fig. 2 Example of refinement of nets of curves with limit a surface

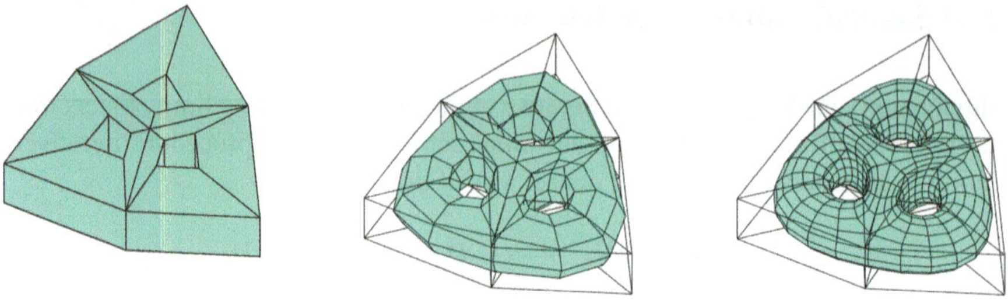

Fig. 3 Example of refinement of meshes with limit a surface

2.1 Binary, Linear, and Stationary Subdivision Schemes

The classical schemes are *binary, linear*, and *stationary*. We start with univariate schemes refining sequences of real values or of points in $\mathbb{R}^d$. The extension to the refinement of real values or of points given at the vertices of a regular mesh is the first step towards the bivariate case, which is of great importance for the generation of smooth surfaces.

Given a *mask* consisting of a finite set of real coefficients $\mathbf{a} = \{a_i, i \in I\}$, $I \subset \mathbb{Z}, |I| < \infty$, the associated linear *subdivision operator* transforming a sequence $\mathbf{p}$ of points in $\mathbb{R}$ into a refined sequence of points in $\mathbb{R}$ is

$$\mathcal{S}_{\mathbf{a}} : \ell(\mathbb{Z}) \to \ell(\mathbb{Z}) \qquad (\mathcal{S}_{\mathbf{a}}(\mathbf{p}))_i := \sum_{j \in \mathbb{Z}} a_{i-2j} p_j, \quad j \in \mathbb{Z}. \tag{1}$$

The refinement rule (1) encompasses two rules, one for the even indices, and one for the odd indices

$$(\mathcal{S}_{\mathbf{a}}(p))_{2i} := \sum_{j \in \mathbb{Z}} a_{2j} p_{i-j}, \quad (\mathcal{S}_{\mathbf{a}}(p))_{2i+1} := \sum_{j \in \mathbb{Z}} a_{2j+1} p_{i-j}, \quad j \in \mathbb{Z}.$$

In the following, without loss of generality, we assume that $I = \{0, \dots, N\}$, for some $N \in \mathbb{N}$.

The subdivision scheme is simply the repeated application of the subdivision operator starting from an initial sequence of points $\mathbf{p}^{[0]}$:

$$\begin{cases} \text{Input} \quad \mathbf{a}, \; p^{[0]} \\ \text{For} \quad k = 0, 1, \dots \\ \qquad\qquad p^{[k+1]} := S_{\mathbf{a}} p^{[k]} \end{cases} \tag{2}$$

The points in the sequence $\mathbf{p}^{[k]} = \{p_i^{[k]}\}_{i \in \mathbb{Z}}$ are attached to the *parametrization* $\{t_i^{[k]}\}_{i \in \mathbb{Z}}$ ($t_i^{[k]} < t_{i+1}^{[k]}$, $i \in \mathbb{Z}$), namely $p_i^{[k]}$ is attached to the *parameter value* $t_i^{[k]}$.

The scheme defined in (2), also denoted by $S_{\mathbf{a}}$, is called *convergent* if for any $\mathbf{p}^{[0]}$ there exists a continuous function $f_{\mathbf{p}^{[0]}}$, such that

$$\lim_{k\to\infty} \sup_{i\in\mathbb{Z}} |\, f_{\mathbf{p}^{[0]}}(t_i^{[k]}) - p_i^{[k]}\,| = 0, \tag{3}$$

with $f_{\mathbf{p}^{[0]}} \not\equiv 0$ for at leat one initial sequence $\mathbf{p}^{[0]} \not\equiv 0$. The limit is also denoted by $\mathcal{S}_{\mathbf{a}}^{\infty}(\mathbf{p}^{[0]})$. In case the limit function $f_{\mathbf{p}^{[0]}}$ is a $\mathcal{C}^{\ell}$ function for any $\mathbf{p}^{[0]}$ the scheme is said to be $\mathcal{C}^{\ell}$-regular.

We will restrict our attention to *non singular* subdivision schemes, i.e. convergent schemes such that

$$\mathcal{S}^{\infty}(\mathbf{p}^{[0]}) \equiv 0 \quad \Leftrightarrow \quad p_i^{[0]} = 0 \quad \text{for all} \quad i \in \mathbb{Z}.$$

The limit obtained starting with the *delta*-sequence $\boldsymbol{\delta} = \{\delta_{0,i}\}_{i\in\mathbb{Z}}$, $\phi_{\mathbf{a}} := \mathcal{S}_{\mathbf{a}}^{\infty}(\boldsymbol{\delta})$, usually called the *basic limit function* of the scheme, is of great importance. Indeed, by the linearity of the operator $S_{\mathbf{a}}$ we have that

$$f_{\mathbf{p}^{[0]}} = \sum_{j\in\mathbb{Z}} p_j^{[0]} \phi_{\mathbf{a}}(\cdot - j). \tag{4}$$

Thus, the smoothness of the scheme $S_{\mathbf{a}}$ is the smoothness of its basic limit function.

Most classical subdivision schemes are either *primal* or *dual*. In the primal case at each iteration the scheme retains or modifies the 'old' points and creates a 'new' point situated in the sequence in between two consecutive 'old' ones. In the dual case, S_a discards all given points after creating two new ones in between any pair of consecutive 'old' points. Algebraically, this is related to the choice of the *parameters* to which we attach the points generated by the scheme: the primal parametrization is such that $t_i^k = i\,2^{-k}$ for $k \geq 1$ and $t_i^{[0]} = i,\ i \in \mathbb{Z}$, while in the dual one $t_i^{[k]} = (i - \frac{1}{2})\,2^{-k}$ for $k \geq 1$ and $t_i^{[0]} = i,\ i \in \mathbb{Z}$. To unify the primal and the dual cases, we here consider the parameter values $t_i^{[k]} = (i + \tau)\,2^{-k}$ for $k \geq 1$ and $t_i^{[0]} = i,\ i \in \mathbb{Z}$ and call τ the *parametric shift* of the scheme. Note that in view of (1) and the parametrizations of the primal and dual cases, the support of ϕ_a is contained in $[0, N]$ (see e.g. [39]).

The parameterization is important for example when considering *reproduction* capabilities of subdivision schemes, discussed next.

A convergent subdivision scheme $S_{\mathbf{a}}$ with parameter shift τ *reproduces* a function space $\mathcal{V}$, if for any $g \in \mathcal{V}$, the initial sequence

$$\mathbf{p}^{[0]} := \{g(j + \tau) \in \mathbb{R}\}_{j\in\mathbb{Z}} \tag{5}$$

guarantees that $\mathcal{S}_{\mathbf{a}}^{\infty}(\mathbf{p}^{[0]}) \equiv g$. Moreover it *stepwise reproduces* $\mathcal{V}$ if at each step k, the refined sequence $\mathbf{p}^{[k]}$ is of the form

$$\mathbf{p}^{[k]} = \{g((j+\tau)\,2^{-k})\}_{j\in\mathbb{Z}}, \quad \text{for all} \quad k \ge 1. \tag{6}$$

From the above it obviously follows that stepwise-$\mathcal{V}$-reproduction implies $\mathcal{V}$-reproduction in case convergence is guaranteed.

Reproduction of polynomials of degree less or equal to n, namely corresponding to $\mathcal{V} \equiv \Pi_n$, is closely related to the *approximation order* of the subdivision scheme $\mathcal{S}_\mathbf{a}$. The approximation order measures the rate by which the limit functions generated by $\mathcal{S}_\mathbf{a}$ (from initial data sampled from a sufficiently smooth function f) get closer to f as the sampling density tends to zero. In other words, the approximation order of $\mathcal{S}_\mathbf{a}$ is the largest exponent r such that for all $f \in C^r$

$$\|f - \mathcal{S}_\mathbf{a}^\infty(\mathbf{f}^{[0]})(\frac{\cdot}{h})\|_\infty \le c\,h^r, \quad \text{for} \quad \mathbf{f}^{[0]} = \{f(ih)\}_{i\in\mathbb{Z}},$$

with c a constant independent of h.
It is easy to prove that subdivision schemes that reproduces Π_n have approximation order $r = n+1$ (see the proof in [37] for the 4-point scheme).

A weaker notion of reproduction is the notion of *generation* of a function space $\mathcal{V}$: It guarantees that for any $g \in \mathcal{V}$ and initial sequence (5)

$$\mathcal{S}_\mathbf{a}(\mathbf{p}^{[0]}) \in \mathcal{V}. \tag{7}$$

The generation of Π_n by $\mathcal{S}_a$ is a necessary condition for the scheme to be C^n-regular when ϕ_a is L_∞-stable (see [39, Theorem 4.16 and (4.20)]), namely when $C_1\|b\|_{L_\infty} \le \|\sum_{\alpha\in\mathbb{Z}} b_\alpha\phi_a(\cdot-\alpha)\|_{L_\infty} \le C_2\|b\|_{L_\infty}$ with $C_1,\ C_2$ positive constants independent of $b = \{b_\alpha\}_{\alpha\in\mathbb{Z}}$.

Extension of the univariate case to dimensions $s \ge 2$ is straightforward when the topology is that of the regular mesh $\mathbb{Z}^s$. Here we consider the case $d = 2$.

Bivariate linear, stationary and binary subdivision operators for regular meshes are defined similarly to (1) as

$$\mathcal{S}_\mathbf{a} : \ell(\mathbb{Z}^2) \to \ell(\mathbb{Z}^2) \qquad (\mathcal{S}_\mathbf{a}(\mathbf{p}))_\alpha = \sum_{\beta\in\mathbb{Z}^2} a_{\alpha-2\beta}\,p_\beta, \quad \alpha \in \mathbb{Z}^2. \tag{8}$$

In (8) there are four different refinement rules determined by the parity of the indices $\alpha = (\alpha_1, \alpha_2) \in \mathbb{Z}^2$. Hence, an equivalent form of (8) is

$$(\mathcal{S}_\mathbf{a}(\mathbf{p}))_{2\alpha+\epsilon} = \sum_{\beta\in\mathbb{Z}^2} a_{2\beta+\epsilon}\,p_{\alpha-\beta}, \quad \alpha \in \mathbb{Z}^2, \quad \epsilon \in \Xi_2,$$

where

$$\Xi_2 = \{0,1\}^2 = \{(0,0),\ (0,1),\ (1,0),\ (1,1)\}, \tag{9}$$

is the set of representative indices of a binary scheme. The subdivision limit is still a linear combination of shifts of its bivariate basic limit function

$$f_{\mathbf{p}^{[0]}} = \sum_{\beta \in \mathbb{Z}^2} p_\beta^{[0]} \phi_{\mathbf{a}}(\cdot - \beta), \quad \text{for} \quad \phi_{\mathbf{a}} := \mathcal{S}_{\mathbf{a}}^{\infty}(\boldsymbol{\delta}), \tag{10}$$

with $\boldsymbol{\delta} = \{\delta_{0,\alpha},\ \alpha \in \mathbb{Z}^2\}$ a bivariate sequence. The notions of convergence, regularity, generation, reproduction and approximation order are essentially the same as in the univariate case.

2.2 *Examples of Subdivision Schemes*

A famous example of univariate subdivision scheme is the Chaikin scheme [6] based on the simple rules

$$p_{2i}^{[k+1]} = \frac{1}{4} p_{i-1}^{[k]} + \frac{3}{4} p_i^{[k]} \quad p_{2i+1}^{[k+1]} = \frac{3}{4} p_i^{[k]} + \frac{1}{4} p_{i+1}^{[k]}, \quad i \in \mathbb{Z}, \tag{11}$$

corresponding to the mask

$$\mathbf{a} = \{\frac{1}{4}, \frac{3}{4}, \frac{3}{4}, \frac{1}{4}\}. \tag{12}$$

Figures 4 and 5 show the application of the rules in (11) to the initial $\boldsymbol{\delta}$-sequence and the component-wise application of the same rules to $2D$ initial points. A 'corner cutting' effect is evident.

Fig. 4 Three steps of the subdivision in (11) with initial points (in magenta)

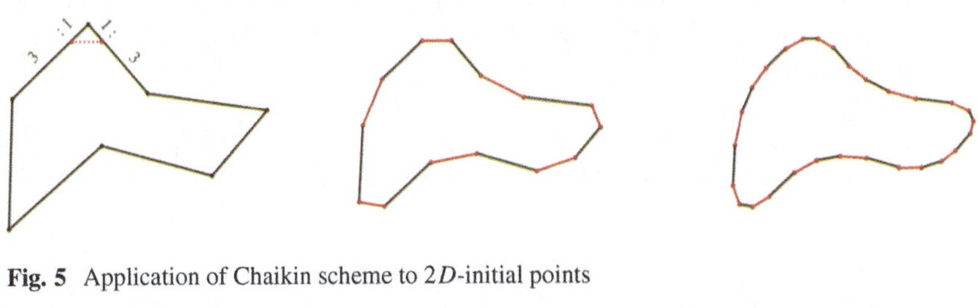

Fig. 5 Application of Chaikin scheme to $2D$-initial points

The Chaikin scheme is a quadratic spline subdivision scheme. Indeed, any degree-n spline with integer knots and smoothness C^{n-1} can be obtained as the limit of a subdivision scheme based on the rules

$$p_{2i}^{[k+1]} = \sum_{j\in\mathbb{Z}} \frac{1}{2^n}\binom{n+1}{2j} p_{i-j}^{[k]}, \qquad p_{2i+1}^{[k+1]} = \sum_{j\in\mathbb{Z}} \frac{1}{2^n}\binom{n+1}{2j+1} p_{i-j}^{[k]}, \quad i \in \mathbb{Z}. \tag{13}$$

The rules in (13) correspond to the masks

$$\mathbf{a}^n = \{\frac{1}{2^n}\binom{n+1}{i}, \ i = 0, \dots, n\}, \tag{14}$$

and reduce for $n = 2$ to (11) while (14) reduces to (12). For odd n the schemes are primal and for even n they are dual.
The regularity, polynomial reproduction and approximation order of spline subdivision schemes are known to be C^{n-1}, Π_0 and $r = 1$, respectively. Note that, placing the masks of the primal spline schemes symmetric relative to the origin, namely $a_{-i} = a_i$, $i = 0, \cdots, \frac{n+1}{2}$ the schemes produce Π_1, hence their approximation order is $r = 2$.

Important examples of subdivision schemes are *interpolatory* schemes where, for all k, $\mathbf{p}^{[k]}$ is contained in $\mathbf{p}^{[k+1]}$, so that the limit function is interpolating the input points. In contrast, the other types of schemes are called *approximating*.

A popular univariate example is the interpolatory 4-point scheme with rules

$$p_{2i}^{[k+1]} = p_i^{[k]}, \quad p_{2i+1}^{[k]} = -\frac{1}{16}p_{i-2}^{[k]} + \frac{9}{16}p_{i-1}^{[k]} + \frac{9}{16}p_i^{[k]} - \frac{1}{16}p_{i+1}^{[k]}, \ i \in \mathbb{Z}, \tag{15}$$

corresponding to the mask

$$\mathbf{a} = \{-\frac{1}{16}, \ 0, \ \frac{9}{16}, \ 1, \ \frac{9}{16}, \ 0, \ -\frac{1}{16}\}. \tag{16}$$

The four point scheme reproduces the polynomial space Π_3, is C^1 and has approximation order $r = 4$. It is a special instance of the family of 4-point schemes with tension parameter (see [37]) corresponding to $w = \frac{1}{16}$ and of the family of the interpolatory $2n + 2$-point schemes proposed by Dubuc-Deslauriers in [32] corresponding to $n = 1$. The schemes in the latter family (DD-family) have the refinement rules

$$p_{2i}^{[k+1]} = p_i^{[k]}, \ p_{2i+1}^{[k]} = \sum_{j=-n-1}^{n} \frac{(-1)^j(n+1)}{2^{4n+1}(2j+1)}\binom{2n+1}{n}\binom{2n+1}{n+j+1} p_{i-j}^{[k]}, \ i \in \mathbb{Z}, \tag{17}$$

with mask

Fig. 6 Three steps of the scheme with rules (15) with initial points $\boldsymbol{\delta}$ (in magenta)

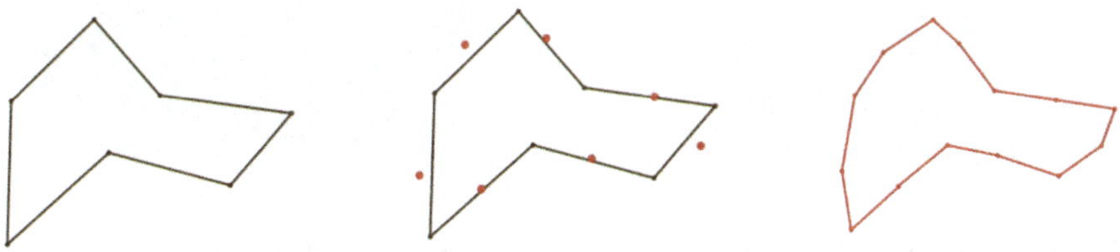

Fig. 7 One application of the 4-point scheme to $2D$-initial points

$$\begin{aligned}\mathbf{a}^n = \{&\tfrac{(-1)^n(n+1)}{2^{4n+1}(2n+1)}\binom{2n+1}{n}, \ \cdots, \ 0, \ \tfrac{n+1}{2^{4n+1}}\binom{2n+1}{n}\binom{2n+1}{n}, \ 1, \\ &\tfrac{n+1}{2^{4n+1}}\binom{2n+1}{n}\binom{2n}{n}, \ 0, \ \cdots, \ \tfrac{(-1)^n(n+1)}{2^{4n+1}(2n+1)}\binom{2n+1}{n}\}.\end{aligned} \tag{18}$$

It is easy to conclude from (17), that the scheme is based on $n+1$ points corresponding to the $n+1$ consecutive integer parameters on each side of $i+\frac{1}{2}$.

The DD $2(n+1)$-point scheme reproduces the polynomial space Π_{2n+1} and has approximation order $r = 2n+2$.

Figures 6 and 7 show the application of the rules in (15) to the $\boldsymbol{\delta}$ initial sequence and the component-wise application of the same rules to the same $2D$-initial points as in Fig. 5. The 'interpolation' effect is evident.

In the bivariate setting, two well known approximating subdivision schemes are the Doo-Sabin scheme and the Loop scheme. In the regular situation, namely when the meshes are $2^{-k}\mathbb{Z}^2, \ k \geq 0$, the first one is a tensor product of the Chaikin scheme while the second one is associated with the three direction box-splines defined by the directions $(1,0), \ (0,1), \ (1,1)$ repeated twice. The masks of these two schemes are respectively given in terms of the matrices as

$$\mathbf{a} = \begin{pmatrix} \frac{1}{16} & \frac{3}{16} & \frac{3}{16} & \frac{1}{16} \\ \frac{3}{16} & \frac{9}{16} & \frac{9}{16} & \frac{3}{16} \\ \frac{3}{16} & \frac{9}{16} & \frac{9}{16} & \frac{3}{16} \\ \frac{1}{16} & \frac{3}{16} & \frac{3}{16} & \frac{1}{16} \end{pmatrix} \quad \text{and} \quad \mathbf{a} = \begin{pmatrix} 0 & 0 & \frac{1}{16} & \frac{1}{8} & \frac{1}{16} \\ 0 & \frac{1}{8} & \frac{3}{8} & \frac{3}{8} & \frac{1}{8} \\ \frac{1}{16} & \frac{3}{8} & \frac{5}{8} & \frac{3}{8} & \frac{1}{16} \\ \frac{1}{8} & \frac{3}{8} & \frac{3}{8} & \frac{1}{8} & 0 \\ \frac{1}{16} & \frac{1}{8} & \frac{1}{16} & 0 & 0 \end{pmatrix}. \tag{19}$$

Figures 8 and 9 show the first and the second iteration of the rules based on the masks in (19) to the initial $\boldsymbol{\delta}$-sequence.

A bivariate interpolatory subdivision scheme related to the four point scheme is the butterfly scheme. The mask of the butterfly scheme is

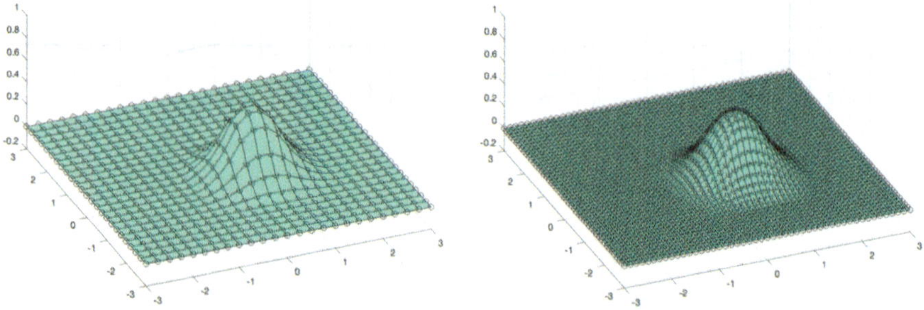

Fig. 8 Second and third iteration of Doo-Sabin scheme applied to the bivariate $\boldsymbol{\delta}$

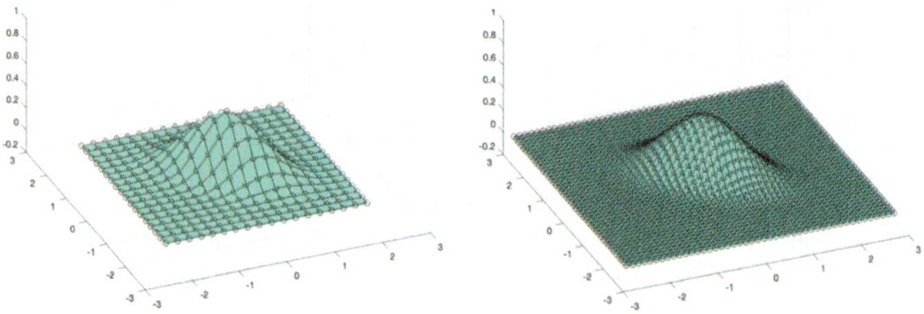

Fig. 9 Second and third iteration of Loop scheme applied to the bivariate $\boldsymbol{\delta}$

$$\mathbf{a} = \begin{pmatrix} 0 & 0 & 0 & 0 & -\frac{1}{16} & -\frac{1}{16} & 0 \\ 0 & 0 & -\frac{1}{16} & 0 & \frac{2}{16} & 0 & -\frac{1}{16} \\ 0 & -\frac{1}{16} & \frac{2}{16} & \frac{8}{16} & \frac{8}{16} & \frac{2}{16} & -\frac{1}{16} \\ 0 & 0 & \frac{8}{16} & 1 & \frac{8}{16} & 0 & 0 \\ -\frac{1}{16} & \frac{2}{16} & \frac{8}{16} & \frac{8}{16} & \frac{2}{16} & -\frac{1}{16} & 0 \\ -\frac{1}{16} & 0 & \frac{2}{16} & 0 & -\frac{1}{16} & 0 & 0 \\ 0 & -\frac{1}{16} & -\frac{1}{16} & 0 & 0 & 0 & 0 \end{pmatrix}. \tag{20}$$

Figure 10 shows the first and the second iteration of the Butterfly scheme applied to the bivariate $\boldsymbol{\delta}$. More complicated examples of interpolatory subdivision schemes can be found in [25], for example.

2.3 Main Applications

Subdivision schemes have a vast variety of applications. The most known is certainly in geometric modelling and computer aided geometric design (CAGD)

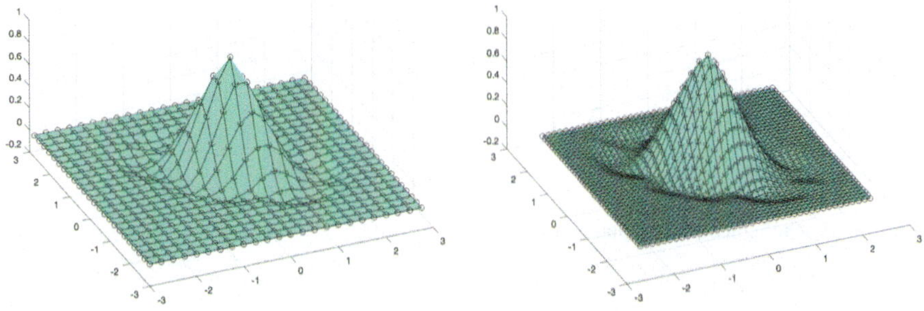

Fig. 10 Second and third iteration of the Butterfly scheme applied to the initial sequence δ

where they are used for the design of smooth curves and smooth surfaces of arbitrary topology. As already mentioned, other applications include construction of refinable functions, multiresolution and wavelets, image analysis through the generation of active contours and active surfaces, computer animation, isogeometric analysis and multigrid.

In the next two subsections we will briefly sketch the first two domains of application while application to image analysis is the subject of Sect. 3.3.

2.3.1 Geometric Modelling and CAGD

In the examples of Sect. 2.2 univariate subdivision schemes generate curves from an initial set of $2D$ points. Passing from curves to surfaces the setup becomes much more complicated since the topological relations between the data are richer than in the curve case (i.e., in the univariate case). In the surface case, a subdivision scheme deals with refinement of *meshes* consisting of vertices, faces and edges. The vertices are points in $3D$, the edges are pairs of vertices, and the faces are cyclic sets of edges (see Fig. 11).

Therefore, each subdivision scheme for surface generation in based on two refinement rules. A *topological* refinement rule describing the modification of the connectivity of the mesh with the added vertices and *geometric* refinement rules that describe where the new vertices, are located in $3D$. In a mesh faces and vertices are classified by the so-called vertex and face *valence*: The valence of a face counts the number of edges that delimit it whereas the valence of a vertex is the number of edges incident to it. Quadrilateral meshes consist of faces with valence 4 and regular vertices are of valence 4. In a triangular mesh all faces are triangles, and the regular vertices have valence 6. In a mesh with most faces and vertices of valence 4, the rest of the faces and vertices are the irregular ones. Similarly, in a mesh with most faces triangles and vertices of valence 6, the rest of the faces and vertices are the irregular ones. A mesh/region is called a regular mesh/region where all vertices and faces are regular. Non-regular vertices/faces are *extraordinary* and a mesh containing them is

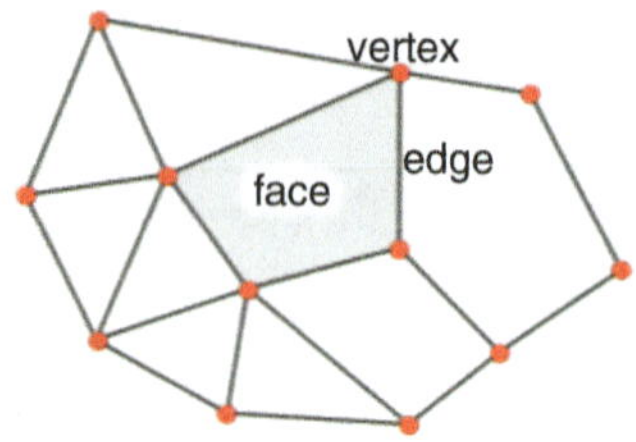

Fig. 11 A schematical representation of a mesh

said to be *irregular*. It is important to note that irregular meshes are necessary for the generation of surfaces of arbitrary topology.

The presence of an irregular element requires the definition of specific rules depending on the valence of the irregular element. The Doo-Sabin scheme and the Loop scheme provide rules for irregular vertices as well as the Catmull-Clark scheme (a tensor product cubic spline scheme in irregular regions). For details about subdivision schemes for surfaces we refer to the books [64, 68].

2.3.2 Generation of Refinable Functions and Wavelets

The link between subdivision schemes and wavelets is in the refinability property of basic limit functions. Indeed, any $\phi_{\mathbf{a}} = \mathcal{S}_{\mathbf{a}}^{\infty}(\boldsymbol{\delta})$ is *refinable* namely it satisfies the *refinement equation*

$$\phi_{\mathbf{a}} = \sum_{\alpha \in \mathbb{Z}^s} a_\alpha \phi_{\mathbf{a}}(2 \cdot -\alpha), \quad s \in \{1, 2\}, \tag{21}$$

with $\{a_\alpha\}_{\alpha \in \mathbb{Z}^s}$ the elements of the mask $\mathbf{a}$. Equation (21) follows from $(S_a \boldsymbol{\delta})_\alpha = a_\alpha$, $\alpha \in \mathbb{Z}^s$ and from (4) and (10) for $s = 1, 2$, respectively.

Equation (21) is the crucial ingredient to generate multiresolution analysis and wavelets even if, in most cases, the explicit expression of $\phi_{\mathbf{a}}$ is unknown. Nevertheless, several numerical procedures are possible for its computation. For example, in the univariate case ($s = 1$) using the refinement equation (21) k-times we easily see that

$$\phi_{\mathbf{a}} = \sum_{i \in \mathbb{Z}} a_i^{[k]} \phi_{\mathbf{a}}(2^k \cdot -i), \quad \text{where} \quad \mathbf{a}^{[0]} := \mathbf{a} \text{ and } \mathbf{a}^{[\ell]} := \mathcal{S}_{\mathbf{a}} \mathbf{a}^{[\ell-1]}, \ \ell = 1, \cdots, k.$$

Therefore, the computation of $\phi_{\mathbf{a}}$ at the dyadic points $j2^{-k}$, $j \in \mathbb{Z}$ is simply the convolution of the sequence $\mathbf{a}^{[k]}$ with values of $\phi_{\mathbf{a}}$. Note that $\phi_{\mathbf{a}}(i) \neq 0$ only for $i = 1, \ldots N-1$ since the support of $\phi_{\mathbf{a}}$ is contained in $[0, N]$ assuming that $\mathbf{a} = \{a_0, \ldots, a_N\}$)and $\phi_{\mathbf{a}}$ is continuous. Therefore, for $v = [\phi_{\mathbf{a}}(1), \ldots, \phi_{\mathbf{a}}(N-1)]$, we have

$$Av = v, \quad \text{with} \quad A_{i,j} = a_{2i-j}, \ i, j = 1, \ldots, N-1.$$

An alternative method for the computation of $\phi_{\mathbf{a}}$ is the so called *cascade algorithm*, involving the repeated application of the operator $\mathcal{T}_{\mathbf{a}}$,

$$\mathcal{T}_{\mathbf{a}} g = \sum_{\alpha \in \mathbb{Z}^s} a_\alpha g(2 \cdot - \alpha), \quad s \in \{1, 2\}.$$

Choosing as initial 'guess' any continuous compactly supported function ψ_0 satisfying $\sum_{\alpha \in \mathbb{Z}^s} \psi_0(x-1) \equiv 1$, the cascade algorithm generates the sequence $\{\psi_k\}_{k \geq 0}$ by repeated application of $\mathcal{T}_{\mathbf{a}}$, namely $\psi_{k+1} = \mathcal{T}_{\mathbf{a}} \psi_k$, $k \geq 0$, and it converges to $\phi_{\mathbf{a}}$.

We remark that the operator $\mathcal{T}_{\mathbf{a}}$ is adjoint of $\mathcal{S}_{\mathbf{a}}$ in the following sense:

$$\sum_{\alpha \in \mathbb{Z}^s} (\mathcal{S}_{\mathbf{a}}(\mathbf{p}))_\alpha f(2 \cdot - \alpha) = \sum_{\alpha \in \mathbb{Z}^s} p_\alpha (\mathcal{T}_{\mathbf{a}}(f)(\cdot - \alpha),$$

for any continuous and compactly supported function f and for any finitely supported sequence $\mathbf{p}$.

We can also calculate the Fourier transform $\hat{\phi}_{\mathbf{a}}$. Indeed, taking the Fourier transform of the refinement equation (21) we find

$$\hat{\phi}_{\mathbf{a}}(\xi) = \mathcal{H}_{\mathbf{a}}(\frac{\xi}{2}) \hat{\phi}_{\mathbf{a}}(\frac{\xi}{2}), \tag{22}$$

where $\mathcal{H}_{\mathbf{a}}(\xi) = \frac{1}{2^s} \sum_{\alpha \in \mathbb{Z}^s} a_\ell e^{2\pi i \, \ell \xi}$ is a trigonometric polynomial, due to the finite support of the mask $\mathbf{a}$. By repeated application of (22), we arrive at

$$\hat{\phi}_{\mathbf{a}}(\xi) = \prod_{k=1}^{\infty} \mathcal{H}_{\mathbf{a}}(\frac{\xi}{2^k}). \tag{23}$$

Orthonormal wavelets are derived from refinable functions whose integer shifts are orthonormal. Such refinable functions are defined by subdivision schemes with masks having special properties. These masks are closely related to masks of interpolating schemes. In particular the mask of the DD family are related to Daubechies orthonormal wavelets of compact support [27].

2.4 Analysis Tools

In this section we shortly review analysis tools for linear stationary subdivision schemes. As it can be observed in this section, in spite of the simplicity of the subdivision idea, analyzing convergence and regularity can be difficult. Indeed, even if the linearity of the operators allow for the use of linear algebra, e.g. *joint spectral radius* or *eigen-analysis*, these problems can be NP hard. On the contrary, the analysis of polynomial reproduction, approximation order and smoothing factors are based on elementary algebraic tools and are much simpler.

Certainly, an advantage of the *uniform framework* (i.e. dealing with uniformly distributed data) characterising 'classical' subdivision schemes, is that we can make use of standard mathematical tools of signal processing (e.g. discrete-time Fourier transform and z-transform) which simplify all formulations and derivations considerably. Indeed, a special role is played by the *subdivision symbol*, the Laurent polynomial with coefficients the elements of the mask **a**, i.e.

$$\mathcal{A}(\mathbf{z}) = \sum_{\alpha \in \mathbb{Z}^s} a_\alpha \mathbf{z}^\alpha, \quad \mathbf{z} \in \mathbb{C}^s \setminus \{0\}, \quad s = \{1, 2\}. \tag{24}$$

With the symbols the kth subdivision step reads as

$$\mathcal{P}^{[k+1]}(\mathbf{z}) = \mathcal{A}(\mathbf{z})\mathcal{P}^{[k]}(\mathbf{z}^2), \quad \text{where} \quad \mathcal{P}^{[k]}(\mathbf{z}) = \sum_{\alpha \in \mathbb{Z}^s} p_\alpha^{[k]} \mathbf{z}^\alpha, \quad k \geq 0.$$

Polynomial generation and reproduction translate into *algebraic* conditions on the subdivision symbol and its derivatives at the points of

$$\Xi_s' = \{e^{-i\pi\boldsymbol{\epsilon}}, \ \boldsymbol{\epsilon} \in \Xi_s\} \equiv \{-1, 1\}^s, \quad s \in \{1, 2\}. \tag{25}$$

With the help of the auxiliary polynomials

$$q_0(\mathbf{z}) := 1, \quad q_{\mathbf{j}}(\mathbf{z}) := \prod_{i=1}^{s} \prod_{\ell_i=0}^{j_i-1} (z_i - \ell_i), \quad \mathbf{j} \in \mathbb{N}_0^s, \quad s \in \{1, 2\}, \tag{26}$$

the polynomial generation/reproduction results are stated in the following proposition (see [8] for details). To state the proposition, we introduce the notion of a non-singular subdivision scheme, which is a scheme that generates zero limits if and only if the initial data is a zero sequence.

Proposition 1 ([8, Theorem 2.6]) *Let $S_{\mathbf{a}}$ be a convergent and non-singular subdivision scheme with mask* **a** *and symbol $\mathcal{A}(\mathbf{z})$. It* generates *polynomials of degree up to n, $n \in \mathbb{N}_0$, if and only if*

$$\mathcal{A}(\mathbf{1}_s) = 2^s, \quad (D^{\mathbf{j}}\mathcal{A})(\boldsymbol{\epsilon}) = 0 \quad \textit{for} \quad \boldsymbol{\epsilon} \in \Xi_s' \setminus \mathbf{1}_s, \quad |\mathbf{j}| \leq n, \tag{27}$$

where $D^{\mathbf{j}}$ is the **j***-th directional derivative ($\mathbf{j} \in \mathbb{Z}^s$) and $\mathbf{1}_s = (1, \cdots, 1) \in \mathbb{Z}^s$.*

Moreover, for a given parameter shift $\boldsymbol{\tau} \in \mathbb{R}^s$, it reproduces *polynomials of degree up to k if and only if*

$$(D^{\mathbf{j}}\mathcal{A})(\mathbf{1}_s) = 2^s q_{\mathbf{j}}(\boldsymbol{\tau}) \quad \textit{and} \quad (D^{\mathbf{j}}\mathcal{A})(\boldsymbol{\epsilon}) = 0 \quad \textit{for} \quad \boldsymbol{\epsilon} \in \Xi_s' \setminus \mathbf{1}_s, \quad |\mathbf{j}| \leq n.$$

Also, Π_n-reproduction implies approximation order $n+1$.

We remark that the algebraic conditions (27) are also called *sum rules of order n* or *zero-conditions* (see e.g. [47]) and [18], respectively).

Still of algebraic type is the investigation of existence of 'difference schemes' and 'smoothing factors' useful for the smoothness analysis of the basic limit functions. In the univariate setting ($s = 1$), a symbol contains k smoothing factors if there exists a Laurent polynomial $\mathcal{B}(z)$ such that

$$\mathcal{A}(z) = \left(\frac{1+z}{2}\right)^k \mathcal{B}(z).$$

The regularity of the scheme $S_{\mathbf{a}}$ is at least k, if the scheme associate with the symbol $\mathcal{B}(z)$ is convergent. A scheme $S_{\mathbf{a}}$ is convergent if and only if its symbol has the form $\mathcal{A}(z) = (1+z)\mathcal{B}(z)$ and the scheme $S_{\mathbf{b}}$ with symbol $\mathcal{B}(z)$ is contractive. A sufficient condition for that is (see e.g. [39])

$$\max\{\sum_{i\in\mathbb{Z}} |b_{2i}|, \sum_{i\in\mathbb{Z}} |b_{2i+1}|\} < 1.$$

In the bivariate situation, the construction of a difference scheme and the link between smoothing factors and smoothness of the limit is definitely more involved (see, [12], for example). To simplify, we can say that the existence of tensor-product type smoothing factors such as $(1+z_1)(1+z_2)$, $(1+z_1)(1+z_1z_2)$ or $(1+z_2)(1+z_1z_2)$ plus contractivity of the difference scheme implies C^1-regularity. For details we refer again to [39].

An apparently different approach to convergence and regularity analysis of subdivision schemes is given by the so called 'JSR approach'. Essentially, we associate to the binary scheme 2^s matrices constructed from the subdivision mask and the reproduced space of polynomials. Then, we compute their *joint spectral radius (JSR)* whose magnitude indicates the Hölder regularity of the scheme as explained. The JSR of a collection of matrices extends the classical notion of spectral radius of a matrix in the following sense.

Definition 1 Given a finite collection of square matrices $\mathcal{M}$, the JSR is

$$\rho(\mathcal{M}) := \lim_{m\to\infty} \max_{M_1,\dots,M_m\in\mathcal{M}} \left\| \prod_{j=1}^m M_j \right\|^{1/m}.$$

First introduced by Rota and Strang in 1960 [61], the JSR was almost forgotten, and then rediscovered in 1992 by Daubechies and Lagarias [28] in the context of the analysis of refinable functions. In general, unfortunately, even the numerical approximation of the JSR is a very challenging task making the JSR approach not always applicable. But, recently, an algorithm for the computation of the JSR has been proposed in [45] (see also [52], for a different approach) and a Matlab code is now available in [51]. We also observe that even if the difference schemes approach and the JSR approach appear to be intrinsically different, they characterize the subdivision regularity in terms of the same quantity. As demonstrated in [7] the two approaches differ only by the numerical schemes they provide for the estimation of the same quantity.

A completely different approach for estimating the regularity of S_a is by its Fourier transform. Indeed, the equality (23) can be used to determine the regularity of the basic limit function $\phi_{\mathbf{a}}$ (i.e. of the subdivision scheme $S_{\mathbf{a}}$), by estimating the decay of its Fourier transform. The latter approach is the one used by many authors (see [27, 34], for example).

Remark 1 The analysis tools presented in this section apply to regular regions or away from irregular elements. In case of meshes containing irregular vertices/faces a different approach to the analysis of subdivision scheme is needed. The appropriate tool to analyze the regularity of the generated limits in the vicinity of an irregular element involves the so called *characteristic map* and the spectral analysis of the local subdivision matrix. For all details we refer the interested reader to [58, 63, 66] and references therein.

3 Motivation for Non-stationary Subdivision Schemes

From the previous section we easily understand that the subdivision idea can also be implemented in a level dependent way, that is to say by using different masks at different iterations. Indeed, at level k, the operator $S_{\mathbf{a}}$ in (2) can be replaced by $S_{\mathbf{a}^{[k]}}$ leading to the *non-stationary* variant of subdivision

$$\begin{cases} \text{Input} \quad \{\mathbf{a}^{[k]}\}_{k\geq 0}, \ \mathbf{p}^{[0]} \\ \text{For} \quad k = 0, 1, \cdots \\ \qquad\quad \mathbf{p}^{[k+1]} := S_{\mathbf{a}^{[k]}}\mathbf{p}^{[k]} \end{cases} \tag{28}$$

Compared with the stationary ones, non-stationary subdivision schemes are not more complicated. Changing coefficients level by level is not a crucial implementation matter, considering that in practice, only few iterations are executed. Also, the definition of convergence and regularity as in (3) is not affected by the level dependence of the rules. Nevertheless, non-stationary subdivision schemes have different properties and enrich the class of subdivision limit functions. For example, applied to $2D$-points they can generate circles, ellipses, or other conics. Also, they allow the user to modify the shape of a subdivision limit by the help of level-dependent tension parameters. In the univariate case, they can generate exponential B-splines [38], C^∞ functions with compact support as the Rvachev-type function [39], or B-spline like functions with higher smoothness relative to the support size, [10, 15].

The algebraic formalism associated with non-stationary schemes is as in the stationary situation. The only difference is that now we deal with a sequence of symbols

$$\mathcal{A}^{[k]}(\mathbf{z}) = \sum_{\alpha \in \mathbb{Z}^s} a_\alpha^{[k]} \mathbf{z}^\alpha, \quad k \geq 0, \quad \mathbf{z} \in \mathbb{C}^s \setminus \{0\}, \quad s = \{1, 2\}. \tag{29}$$

Thus, the k-th subdivision step can be written as

$$\mathcal{P}^{[k+1]}(\mathbf{z}) = \mathcal{A}^{[k]}(\mathbf{z})\mathcal{P}^{[k]}(\mathbf{z}^2), \quad \text{with} \quad \mathcal{P}^{[k]}(\mathbf{z}) = \sum_{\alpha \in \mathbb{Z}^s} p_\alpha^{[k]} \mathbf{z}^\alpha, \quad k \geq 0.$$

The discussion on the use of the corresponding algebraic tools as well as of other associated tools like the JSR is postponed to Sect. 4. Here, we mention that in case the non-stationarity is characterized by the cyclic repetition of ℓ different masks the scheme is actually stationary with 2^ℓ-arity rather than 2. Indeed, for any $k = r \cdot \ell,\ r > 0$ we can consider ℓ steps simultaneously, and obtain

$$\mathcal{P}^{[k+\ell]}(\mathbf{z}) = \tilde{\mathcal{A}}(\mathbf{z})\mathcal{P}^{[k]}(\mathbf{z}^{2^\ell}), \text{ where } \tilde{\mathcal{A}}(\mathbf{z}) := \mathcal{A}^{[\ell-1]}(\mathbf{z})\mathcal{A}^{[\ell-2]}(\mathbf{z}^2)\cdots\mathcal{A}^{[0]}(\mathbf{z}^{2^{\ell-1}}),$$

implying that $\tilde{\mathcal{A}}(\mathbf{z})$ is the symbol of an arity-2^ℓ scheme that multiply by 2^ℓ the number of points at each step (see e.g.,[20]).

In the non-stationary case, when using the sequence of masks starting not with $\mathbf{a}^{[0]}$ but with any $\mathbf{a}^{[m]},\ m > 0$, we get different results according to the starting mask $a^{[m]}$, where m varies from 0 to ∞. The subdivision scheme in this case is

$$\begin{cases} \text{Input} \quad \{\mathbf{a}^{[k]}\}_{k\geq 0}, \ \mathbf{p}^{[0]} \\ \text{For} \quad k = 0, 1, 2, \cdots \\ \qquad\quad \mathbf{p}^{[k+1]} := S_{\mathbf{a}^{[m+k]}}\mathbf{p}^{[k]} \end{cases} \tag{30}$$

From the above we understand that in the level dependent case we have no longer a unique basic limit function but rather a *sequence of basic limit functions* $\{\phi_m,\ m \geq 0\}$ each defined as

$$\phi_m = \lim_{k\to\infty} S_{\mathbf{a}^{[k+m]}} \cdots S_{\mathbf{a}^{[m]}} \boldsymbol{\delta}, \tag{31}$$

where $\boldsymbol{\delta}$ is the sequence with value 1 at the origin, and zero on $\mathbb{Z}^s \setminus \{0\}$. Due to linearity and uniformity of the operators, the sequence of basic limit functions satisfies a system of 'generalized' refinement equations,

$$\phi_m = \sum_{\alpha \in \mathbb{Z}^s} a_\alpha^{[m]} \phi_{m+1}(2 \cdot -\alpha), \quad m \geq 0. \tag{32}$$

The system of generalized refinement equations (32) is the base to the generation of non-stationary multiresolution and non-stationary wavelets [3, 42].

The next subsections show the capabilities of level-dependent schemes in applications, e.g., in geometric design and in approximation [49, 50], in biological imaging [29, 65] and in the generation of non-stationary wavelets [13, 42, 67].

3.1 Reproduction of Conics and Quadrics and Use of Level Dependent Tension Parameters in CAGD

It is well known that B-spline curves and surfaces are central tools in computer-aided geometric design but also in computer graphics, due to the properties of B-splines, which guarantee, for example, that the B-spline curves/surfaces are in the convex hull of their control polygons/meshes. B-splines, unfortunately, are not capable to reproduce in an exact way conic sections which are needed very often. This is why different B-spline generalizations, like NURBS, have been proposed. The rational nature of NURBS is the reason why it is difficult to integrate or differentiate them. With NURBS it is possible to exactly represent conic sections but not all transcendental curves. Therefore, researchers have started to consider 'generalized B-splines' that is bell-shaped functions piecewise defined with segments in other spaces than rational polynomials. By selecting spaces of trigonometric or hyperbolic functions, for example, with generalized B-splines it is possible to represent polynomial curves, conic sections or transcendental curves. What is relevant to this paper is that several instances of generalized B-splines with integer knots can be see as limit functions of non-stationary subdivision schemes.

The computation of limit surfaces by a subdivision scheme is much simpler than the modelling of surfaces with NURBS (B-splines) since, in the latter case, the complete surface consists of NURB (B-splines) patches with geometric continuity between the patches. For details on connecting smoothly patches see [55, Chapter 13].

Note that meshes for modelling surfaces of arbitrary topology have irregular regions, and the refinement rules have to be adapted to the vicinity of irregular elements.

As an example we can consider the following non-stationary subdivision scheme generating exponential splines with segments in

$$\mathrm{span}\{e^{\theta t}, e^{-\theta t}, te^{\theta t}, te^{-\theta t}\}, \quad \theta \in \mathbb{R} \cup i\mathbb{R},$$

with θ a parameter to be chosen by the user (see [14] and [21]). These exponential splines are a special instance of L-splines (see [62]). The refinement rules are

$$
\begin{aligned}
p_{2i}^{[k+1]} &= \frac{1}{2(v^{[k]}+1)^2} p_{i-1}^{[k]} + \frac{4(v^{[k]})^2+2}{2(v^{[k]}+1)^2} p_i^{[k]} + \frac{1}{2(v^{[k]}+1)^2} p_{i+1}^{[k]}, \\
p_{2i+1}^{[k+1]} &= \frac{2v^{[k]}}{(v^{[k]}+1)^2} p_i^{[k]} + \frac{2v^{[k]}}{(v^{[k]}+1)^2} p_{i+1}^{[k]},
\end{aligned}
\tag{33}
$$

where the non-stationary parameter $v^{[k]}$ is defined as

$$
v^{[k]} = \frac{1}{2}\left(e^{\mathrm{i}\frac{\theta}{2^{k+1}}} + e^{-\mathrm{i}\frac{\theta}{2^{k+1}}}\right) = \sqrt{\frac{1+v^{[k-1]}}{2}}, \quad k \geq 0, \quad v^{[-1]} = \cos(\theta) > -1.
$$

The effect of the parameter θ on the exponential B-spline shape obtained when starting the subdivision process from the $\boldsymbol{\delta}$ sequence is illustrated in Fig. 12.

We remark that the above scheme is only generating exponential-polynomial spaces but is not reproducing them. Yet, in [19, 22, 40] and [54], exponential-polynomials reproducing schemes are provided. In the first two references, these schemes are shown to generate conics, cardiod, lemniscate, astroid or nephroid as shown in Fig. 13.

Similarly, bivariate non-stationary schemes reproducing quadrics are defined and investigated for example in [44, 48, 49, 53]. Since the corresponding refinement rules, in particular in case of extraordinary points, are non-trivial, we here simply present some of the pictures from [53] in Fig. 14.

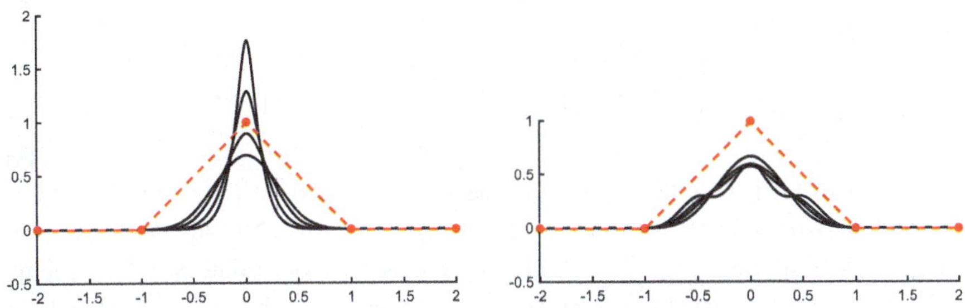

Fig. 12 Basic limit functions for the scheme in (33) with $\theta \in \{\mathrm{i}, 3\mathrm{i}, 5\mathrm{i}, 7\mathrm{i}\}$ (left) and $\theta \in \{3, 2.5, 2, 0\}$ (right) (from lower to taller functions). Initial control polygon represented by a dashed line

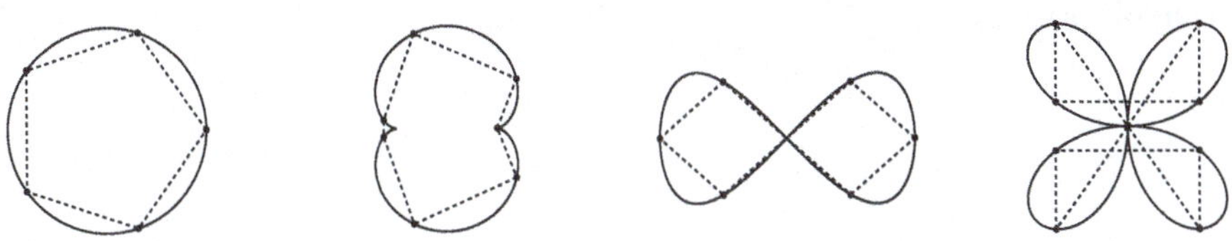

Fig. 13 Subdivision limit curves (full lines) and the initial control polygons (dashed line) connecting points from a circle, a nephroid a lemniscate and a quadrifolium

Fig. 14 First line: initial meshes. Second line: results obtained by applying 5 steps of the non-stationary scheme in [53]

To conclude this section, we shortly discuss how non-stationary tension parameters and level dependent rules can influence the shape of the subdivision limits. Let us consider the interpolatory non-stationary scheme with the first two odd rules

$$p_{2i+1}^{[k+1]} = \frac{1}{2}p_{i-1}^{[k]} + \frac{1}{2}p_i^{[k]}, \quad k = 0, 1, \ i \in \mathbb{Z}, \tag{34}$$

and then, for $k > 1$, for $\omega^{[k]}$ chosen at random from the interval $[\frac{3}{64}, \frac{1}{16}]$, the odd rules are given by

$$p_{2i+1}^{[k]} = -\omega^{[k]}p_{i-2}^{[k]} + (\frac{1}{2} + \omega^{[k]})p_{i-1}^{[k]} + (\frac{1}{2} + \omega^{[k]})p_i^{[k]} - \omega^{[k]}p_{i+1}^{[k]}, \quad k \geq 2, \ i \in \mathbb{Z}. \tag{35}$$

As shown in [10] by a JSR approach, the scheme based on (34) and (35) is C^1-convergent with Hölder exponent $\alpha \geq -\log_2 \frac{3}{8} \approx 1.4150$ and its basic limit function is supported in $[-\frac{3}{2}, \frac{3}{2}]$ while in the classical four point case the scheme is known to be C^1-convergent with Hölder exponent $2 - \epsilon$ for any $\epsilon > 0$ and the support is $[-3, 3]$ (see [32]). Figure 15 compares the two basic limit functions.

The last example shows that with a non-stationary interpolatory scheme it is possible to obtain a C^1 basic limit function of smaller support than in the stationary interpolatory case.

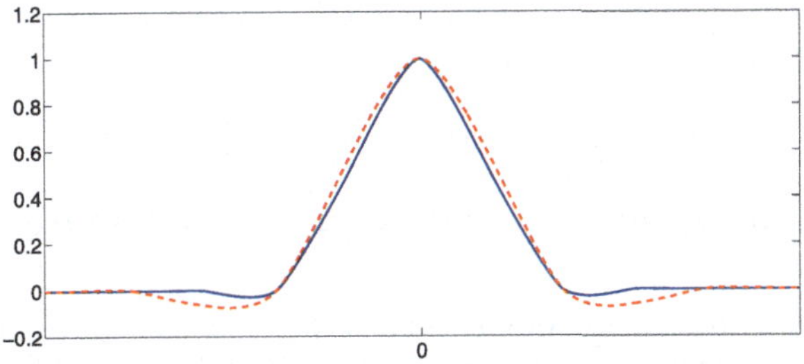

Fig. 15 Basic limit function of the 4-point scheme (red, dashed line), and of the scheme (34)–(35) (blue, solid line)

3.2 *Non-stationary Wavelets and Non-stationary Interpolatory Subdivision Schemes*

The construction of stationary orthonormal wavelets of compact support is closely related to the DD-family of subdivision schemes. Such a Daubechies wavelet is generated by the integer shifts of a refinable function, which is the basic limit function of a subdivision scheme. The mask of this scheme is derived from the mask of a corresponding DD-scheme, by taking an 'almost square root' of the symbol of the DD-scheme. This is possible since the symbols of the DD-schemes are non-negative on the unit circle (when z is replaced by $exp(i\theta), 0 \leq \theta < 2\pi$) [27]. This construction has two analogues in the non-stationary setting.

The first analogoue is derived from interpolatory schemes that reproduce spaces of exponential polynomials of finite dimension. In [40] non-stationary interpolatory schemes reproducing spaces of $2n$ exponentials are shown to converge and their smoothness is shown to be at least as that of the stationary DD-scheme reproducing all polynomials of degree less than $2n$. In [67] non-stationary wavelets are constructed from non-stationary interpolatory subdivision schemes by a similar procedure as in the stationary case, without a proof that this is indeed possible. These wavelets were already used in the analysis of signals that are better approximated by exponentials rather than by polynomials, such as signals that have their energy concentrated around specific frequencies. For example in neurophysiology, such wavelets are well-suited for the analysis of exponential pulses, corresponding to different neurons. Proofs that the above construction is possible are given in [42]. Also given, are proofs showing that the smoothness of the non-stationary wavelets related to spaces of real exponential polynomials is at least that of the corresponding stationary wavelets.

The second analogue is derived from non-stationary interpolatory subdivision schemes with masks of growing support. A simple example is the sequence of masks of the DD-schemes (17), with n the subdivision level (see Sect. 4.3). Following the construction in the stationary case, the basic limit function of the non-stationary scheme with masks 'almost square root' of the masks of the DD-schemes, is the 'father' wavelet. These wavelets, which are C^∞ compactly supported, are suitable for representing very smooth functions [13].

3.3 Image Segmentation: Active Contours and Active Surfaces

This section describes the use of non-stationary subdivision schemes in biological imaging and relies on the work done by the group of M. Unser at EPFL, Switzerland. *Active contours* or *snakes*, are tools for the segmentation of biomedical images. They consist of an initial curve that progresses towards the boundary of the object of interest guided by the minimization of an appropriate energy term. Relevant to our discussion is that subdivision schemes can also be used to describe a contour by the iterative application of refinement rules staring from an initial finite set of control points. The discrete nature of the initial representation is convenient in practice. It implicitly yields a continuously defined model whose properties depend on the used subdivision scheme: its approximation order, its capability of reproducing circular, elliptical, or polynomial shapes, its interpolating or approximating nature. In particular, the capability of modelling 'roundish' objects is facilitated by non-stationary schemes.

Therefore, as an alternative to the traditional approaches, in [2] subdivision schemes are used to model a curve driven by a small set of 'master' points, called control points, and a set of 'slave' points (generated by a specific subdivision scheme) that describe the curve. The advantages of the use of subdivision schemes are their simplicity of implementation and their multiresolution nature, so that the contour of a shape can be represented at varying resolutions and result into a snake be optimized in a coarse-to-fine fashion.

Based on similar ideas is the use of subdivision for the generation of active surfaces, also called 3D deformable models used for the extraction of volumetric structures. They consist of deformable surfaces that, starting from an initial user-provided configuration, evolve toward the boundary of the 3D object. The deformation can be manual or automatic. Certainly, a reasonable deformable model must depend on a small number of control points (to reduce the complexity of the deformation and to improves robustness), and must reproduce or approximate ellipsoids. In [1] the authors propose a 3D deformable model obtained by applying a tailored non-stationary subdivision scheme to a suitable coarse mesh with few control points. The approach presents several advantages: First, surfaces of arbitrary topological type can be handled; second, by simple modifications of the control points, easy and localized interactions can be achieved; third, the implementation is easy and cheap in virtue of the discrete nature of the scheme.

4 Analysis Tools for Non-stationary Subdivision Schemes

In this section we consider analysis tools of non-stationary schemes and highlight similarities and differences with the stationary case.

4.1 *Masks of Fixed Support*

First we consider analysis tools of non-stationary schemes for the case that all the masks $\{\mathbf{a}^{[k]}\}_{k\geq 0}$ have bounded support $\{0, ..., N\}$ for some positive integer N, which is the more common and studied situation. In this case the methods of analysis are related to the analysis of stationary cases via the notion of asymptotic similarity and asymptotic equivalence. We start by introducing the notion of asymptotic equivalence (see [38]).

Definition 2 Let $\ell \in \mathbb{N}$. The non-stationary schemes $\{S_{\mathbf{a}^{[k]}}\}_{k\geq 0}$ and $\{S_{\mathbf{b}^{[k]}}\}_{k\geq 0}$ are said *asymptotically equivalent of order* ℓ if they satisfy

$$\sum_{k=0}^{\infty} 2^{k\ell} \| S_{\mathbf{a}^{[k]}} - S_{\mathbf{b}^{[k]}} \|_{\infty} < \infty, \tag{36}$$

where $\| S_{\mathbf{a}^{[k]}} \|_{\infty} := \max_{\varepsilon \in \Xi_s} \left\{ \sum_{\alpha \in \mathbb{Z}^s} |a^{[k]}(2\alpha + \varepsilon)| \right\}$ and $\Xi_s := \{0, 1\}^s$.

Under an additional technical assumption on the schemes $\{S_{\mathbf{a}^{[k]}}\}_{k\geq 0}$ and $\{S_{\mathbf{b}^{[k]}}\}_{k\geq 0}$, the regularity of $\{S_{\mathbf{a}^{[k]}}\}_{k\geq 0}$ can be deduced from the known regularity of the asymptotically equivalent scheme $\{S_{\mathbf{b}^{[k]}}\}_{k\geq 0}$ with the method in [38]. Yet, in [38] only the convergence of non stationary schemes is derived by asymptotic equivalence of order $\ell = 0$ to a stationary scheme. The asymptotic equivalence of order $\ell \geq 1$ is too strong for analyzing smoothness. For that the notion of smoothing factors is introduced there.

Definition 3 Let the Laurent polynomials $\{\mathcal{A}^{[k]}(\mathbf{z})\}_{k\geq 0}$ be of the form

$$\mathcal{A}^{[k]}(\mathbf{z}) = \frac{1}{2}(1 + r_k \mathbf{z}^{\lambda}) \mathcal{B}^{[k]}(\mathbf{z}), \quad k \geq K \geq 0, \quad \lambda \in \mathbb{N}_0^s. \tag{37}$$

The factors $\{\frac{1}{2}(1 + r_k \mathbf{z}^{\lambda})\}_{k\geq K}$ are termed 'smoothing factors' if

$$r_k = 2^{\eta 2^{-k}}(1 + \epsilon_k) \quad \text{with} \quad \eta \in \mathbb{R} \text{ and } \sum_{k=K}^{\infty} |\epsilon_k| 2^k < \infty.$$

Theorem 1 ([38, Theorem 10]) *In the notation of Definition 3, if* $\{\mathcal{B}^{[k]}(\mathbf{z})\}_{k\geq 0}$ *corresponds to a* $C^m(\mathbb{R}^s)$ *non-stationary subdivision scheme then the basic limit* *functions of the non-stationary scheme with symbols* $\{\mathcal{A}^{[k]}(\mathbf{z})\}_{k\geq 0}$ *and their directional derivative in direction* λ *are also* C^m *smooth in* $\mathbb{R}^s$.

A direct consequence of Theorem 1 (see the remark below the statement of [38, Theorem 10]) is:

Corollary 1 *Let* $\{\mathcal{A}^{[k]}(\mathbf{z}) = \prod_{i=1}^{s} \frac{1}{2}(1 + r_{k,i}\mathbf{z}_i^{\lambda})\mathcal{B}(\mathbf{z})\}_{k\geq 0}$ *with s smoothing factors. If the stationary scheme corresponding to* $\mathcal{B}(\mathbf{z})$ *is* $C^m(\mathbb{R}^s)$ *and if* $\lambda_1, \cdots, \lambda_s$ *are linearly independent vectors, then the basic limit functions of the non-stationary scheme corresponding to* $\{\mathcal{A}^{[k]}(\mathbf{z})\}_{k\geq 0}$ *is* C^{m+1} *smooth in* $\mathbb{R}^s$.

In [41], the condition of asymptotical equivalence is weaken, in the univariate case, by requiring that the j-th derivatives of the symbols of the non-stationary scheme $\{S_{\mathbf{a}^{[k]}}\}_{k\geq 0}$ satisfy

$$|D^j \mathcal{A}^{[k]}(-1)| \leq C2^{-(\ell+1-j)k}, \quad j = 0, \ldots, \ell, \quad \ell \geq 0, \quad C \geq 0. \tag{38}$$

Moreover, they assume that the non-stationary scheme is asymptotically equivalent (of order 0) to some stationary scheme. The conditions in (38) are a generalization of the so-called sum rules in (27). In the stationary case, sum rules are known to be necessary for smoothness of subdivision (see e.g [5]), and also sufficient if the basic limit function of the scheme is L_∞-stable (see e.g. [39]).
In the spirit of (38) *approximate sum rules* are defined in [9]. They are a generalization of the notion of sum rules.

Definition 4 Let $\ell \geq 0$. The sequence of symbols $\{\mathcal{A}^{[k]}(\mathbf{z})\}_{k\geq 0}$ *satisfies approximate sum rules of order* $\ell + 1$, if

$$\mu_k := |\mathcal{A}^{[k]}(\mathbf{1}_s) - 2^s| \quad \text{and} \quad \delta_k := \max_{|\eta|\leq \ell} \max_{\epsilon\in\Xi'\setminus\{\mathbf{1}_s\}} |2^{-k|\eta|} D^\eta \mathcal{A}^{[k]}(\epsilon)| \tag{39}$$

satisfy

$$\sum_{k=0}^{\infty} \mu_k < \infty \quad \text{and} \quad \sum_{k=0}^{\infty} m^{k\ell}\, \delta_k < \infty\,. \tag{40}$$

Note that if the sequences $\{\mu_k\}_{k\geq 0}$ and $\{\delta_k\}_{k\geq 0}$ (called sum rule *defects*) are zero sequences, the corresponding non-stationary symbols satisfy sum rules of order $\ell + 1$.

We continue by introducing a weaker relation than asymptotical equivalence termed *asymptotic similarity* (generalization of the one given in [16]) relating the properties of non-stationary subdivision schemes to the corresponding properties of certain stationary schemes.

Definition 5 ([9]) For the mask sequence $\{\mathbf{a}^{[k]}\}_{k\geq 0}$ we denote by $\mathcal{L}$ the *set of masks which are accumulation points of this sequence*,

$$\mathbf{a} \in \mathcal{L}, \quad \text{if} \quad \exists \{k_n,\ n \in \mathbb{N}\} \text{ such that } \lim_{n\to\infty} \mathbf{a}^{[k_n]} = \mathbf{a}\,.$$

Definition 6 Two non-stationary schemes $\{S_{\mathbf{a}^{[k]}}\}_{k\geq 0}$ and $\{S_{\mathbf{b}^{[k]}}\}_{k\geq 0}$ are called *asymptotically similar*, if the corresponding sets of accumulation points coincide.

We already observed in Sect. 2.4 that in the stationary case, the rate of convergence of the corresponding subdivision scheme $S_{\mathbf{a}}$ and the Hölder regularity of the subdivision limits, can be given in terms of the joint spectral radius of the collection of certain matrices derived from the subdivision mask **a** and linked to the order of sum rules satisfied by the associated symbol $\mathcal{A}(\mathbf{z})$ (see also [52, 60]).

In the non-stationary setting the joint spectral radius has no straightforward generalization and is not directly applicable. Hence, in [9] a link between the stationary and non-stationary settings is established based on the sets of accumulation points $\mathcal{L}$ of $\{\mathbf{a}^{[k]}\}_{k\geq 0}$, and sufficient conditions for C^ℓ-convergence and Hölder regularity of non-stationary schemes are provided. As in the level independent case, each mask in the set $\mathcal{L}$ determines a set of transition matrices. The restrictions of all these transition matrices to a certain finite dimensional difference subspace (denoted by V_ℓ) is denoted by $\mathcal{T}_{\mathcal{L}}|_{V_\ell}$. Theorem 2 states that C^ℓ-convergence and Hölder regularity of non-stationary schemes is obtained via the joint spectral radius $\rho_{\mathcal{L}}$ of the collection of matrices $\mathcal{T}_{\mathcal{L}}|_{V_\ell}$.

Theorem 2 ([9, Theorem 2]) *Let $\ell \in \mathbb{N}$ and let $\{\delta_k\}_{k\geq 0}$ be defined in* (39). *Assume that the symbols of $\{S_{\mathbf{a}^{(k)}}\}_{k\geq 0}$ satisfy approximate sum rules of order $\ell + 1$ and that $\rho_{\mathcal{L}} := \rho\left(\mathcal{T}_{\mathcal{L}}|_{V_\ell}\right) < 2^{-\ell}$. Then the non-stationary scheme $\{S_{\mathbf{a}^{[k]}}\}_{k\geq 0}$ is C^ℓ-convergent and the Hölder exponent α of its limit functions satisfies*

$$\alpha \geq \min\left\{ -\log_2 \rho_{\mathcal{L}}\,,\ -\limsup_{k\to\infty} \frac{\log_2 \delta_k}{k} \right\}.$$

In the univariate case more results are available. In [23] and also in [14], the link between approximate sum rules, generation/reproduction of exponential polynomials and approximation order is investigated, in the univariate case. In fact, the authors show that the property of reproducing N exponential polynomials implies approximate sum rules of order N and even approximation order N if asymptotic similarity to a convergent stationary scheme is assumed. Moreover, under asymptotic similarity to a convergent stationary scheme and reproduction of one exponential polynomial, the property of generating N exponential polynomials implies approximate sum rules of order N. The property of generating exponential polynomials guarantees the existence of difference operators exactly as in the stationary case. Moreover, approximate sum rules of order N and asymptotic similarity to a stationary C^{N-1} subdivision scheme provide sufficient conditions for C^{N-1} regularity of non-stationary subdivision schemes.

These results are stated in the next theorems where for $\Lambda \subset \mathbb{C}$ and $\Gamma(\Lambda) = \{\nu(\lambda) : \lambda \in \Lambda\} \subset \mathbb{N}_0$, the space $EP_{\Gamma(\Lambda),\Lambda}$, is defined as

$$EP_{\Gamma(\Lambda),\Lambda} := \{x^j e^{\lambda x} \ : \ j = 0, \ldots, \nu(\lambda) - 1, \ \lambda \in \Lambda, \ \nu(\lambda) \in \Gamma(\Lambda)\}, \quad (41)$$

and denoted as $EP_{\Gamma,\Lambda}$, for short. Obviously, its dimension is

$$\dim\left(EP_{\Gamma,\Lambda}\right) = \sum_{\lambda \in \Lambda} \nu(\lambda).$$

Theorem 3 ([23, Theorem 10]) *Let $\{\mathcal{A}^{[k]}(z)\}_{k\geq 0}$ be the Laurent polynomials associated with a univariate non-stationary scheme which* reproduces *a space of univariate exponential polynomials $EP_{\Gamma,\Lambda}$. If $\dim\left(EP_{\Gamma,\Lambda}\right) = N$, then, for any $\ell = 0, \cdots, N-1$, we have*

$$|\mathcal{A}^{[k]}(1) - 2| = O(2^{-kN}), \qquad \left|\frac{d^\ell}{dz^\ell}\mathcal{A}^{[k]}(-1)\right| = O(2^{-k(N-\ell)}), \qquad k \to \infty. \quad (42)$$

Theorem 4 ([23, Theorem 13]) *Let $\{\mathcal{A}^{[k]}(z)\}_{k\geq 0}$ be the Laurent polynomials associated with a non-stationary subdivision scheme which* generates *the exponential polynomials space $EP_{\Gamma,\Lambda}$ of dimension N, and* reproduces *one $f \in EP_{\Gamma,\Lambda}$. Moreover, let $\lim_{k\to\infty} \mathbf{a}^{[k]} = \mathbf{a}$ with $S_{\mathbf{a}}$ a convergent stationary subdivision scheme. Then, for any $\ell = 0, \cdots, N-1$, we have*

$$|\mathcal{A}^{[k]}(1) - 2| = O(2^{-k}), \qquad \left|\frac{d^\ell}{dz^\ell}\mathcal{A}^{[k]}(-1)\right| = O(2^{-k(N-\ell)}), \qquad k \to \infty. \quad (43)$$

Theorem 5 ([14, Theorem 4.3]) *Assume that a convergent non-stationary scheme reproduces the exponential polynomials in the N-dimensional space $EP_{\Gamma,\Lambda}$. Assume further that $\lim_{k\to\infty} \mathbf{a}^{[k]} = \mathbf{a}$ with $S_{\mathbf{a}}$ a convergent stationary scheme. Then, for any initial data of the form $\mathbf{f}^{[0]} := \{f(2^{-m}i)\}_{i\in\mathbb{Z}}$ for an integer $m \geq 0$ with $f \in W^\gamma_\infty(\mathbb{R})$, $\gamma \in \mathbb{N}$, the approximation error is bounded by*

$$\|g_{\mathbf{f}^{[0]}} - f\|_\infty \leq c_f 2^{-\min(\gamma,N)m}, \quad (44)$$

with $c_f > 0$ denoting a constant depending on f but not on m.

Extension of Theorems 3, 4 to the multivariate setting is still to be done. Some extension of Theorem 5 is in [53].

To conclude we recall the conditions non-stationary schemes need to satisfy to generate and reproduce (in the sense of (7) and (6)) exponential-polynomial functions, that is functions in the space

$$EP_{\Gamma,\Lambda} := \{\mathbf{x}^{\boldsymbol{\gamma}} e^{\boldsymbol{\lambda}\cdot\mathbf{x}} : \boldsymbol{\gamma} \in \Gamma, \ \boldsymbol{\lambda} \in \Lambda\}, \quad \Gamma \subset \mathbb{N}_0^s, \quad \Lambda \subset \mathbb{C}^s.$$

In fact, both generation and reproduction of exponential-polynomials can still be characterised in terms of algebraic conditions involving the parameter values

$$\{\mathbf{t}_\alpha^{[k]} = 2^{-k}(\alpha + \boldsymbol{\tau}),\ \}_{\alpha\in\mathbb{Z}^s}, \quad \text{with} \quad \boldsymbol{\tau} = (\tau_1, \tau_2) \text{ in case } s = 2.$$

The conditions are in terms of the symbols $\{\mathcal{A}^{[k]}(\mathbf{z})\}_{k\geq 0}$ evaluated at

$$V_k=\{(v_1,\ldots,v_s)^T : v_j=\epsilon_j e^{-(2^{-(k+1)}\boldsymbol{\lambda}_j)},\ \boldsymbol{\lambda}=(\lambda_1,\ldots,\lambda_s)\in\Lambda,\ \boldsymbol{\epsilon}\in\{-1,1\}^s\ \},$$

and are collected in the following Theorem (taken from [11]) with the notation $\mathbf{v}^\tau = v_1^{\tau_1}\cdots v_s^{\tau_s}$ for $\gamma = (\gamma_1,\ldots,\gamma_s \in \mathbb{N}_0^s$. There a non-singular scheme is a scheme generating limits identically equal to zero, only from zero initial data.'

Theorem 6 ([11, Theorem 4.7]) *A non-singular subdivision scheme* $\{S_{\mathbf{a}^{[k]}}\}_{k\geq 0}$ *reproduces* $EP_{\Gamma,\Lambda}$ *if and only if there exists a parameter* $\boldsymbol{\tau}\in\mathbb{R}^s$ *such that for all* $\mathbf{v}\in V_k$, $k\geq 0$, $\boldsymbol{\gamma}\in\Gamma\subset\mathbb{N}_0$,

$$\mathbf{v}^{\boldsymbol{\gamma}} D^{\boldsymbol{\gamma}} \mathcal{A}^{[k]}(\mathbf{v}) = \begin{cases} 2\cdot\mathbf{v}^\tau q_{\boldsymbol{\gamma}}(\boldsymbol{\tau}), & \textit{for all } \mathbf{v} \textit{ corresponding to } \boldsymbol{\epsilon} = \mathbf{1}_s, \\ 0, & \textit{otherwise}, \end{cases} \tag{45}$$

where $q_{\boldsymbol{\gamma}}$ *is the polynomial of degree* $|\boldsymbol{\gamma}|$, $\boldsymbol{\gamma}\in\mathbb{N}_0^s$, *given by*

$$q_0(\mathbf{z}) := 1, \quad q_{\mathbf{j}}(z_1,\ldots,z_s) := \prod_{i=1}^{s}\prod_{\ell_i=0}^{\gamma_i-1}(z_i-\ell_i), \quad \boldsymbol{\gamma} = (\gamma_1,\ldots,\gamma_s). \tag{46}$$

4.2 Non-stationary Schemes with Extraordinary Vertices/Faces

The analysis of a level-dependent subdivision scheme in the neighborhood of an irregular vertex/face is very challenging. The main difficulties are due to the fact that any approach based on the spectral analysis of the subdivision matrix and on the study of the characteristic map is not applicable. Indeed, no general tools to analyze non-stationary subdivision schemes at irregular vertices/faces were available till very recently. The only contributions to this analysis are the very recent paper [17] and the work of Jena et al. in [48], where a specific scheme is considered. In [17] a general procedure to check if a non-stationary subdivision scheme is convergent in the neighborhood of an extraordinary vertex/face is given. Moreover, sufficient conditions for the limit surface to be tangent plane continuous at the limit point of an extraordinary vertex/face are also given in that paper. Below we report both results.

We recall that the problem of extraordinary points occurs in the generation of surfaces that is in the case $s = 2$ and that we can restrict our analysis to meshes with a single extraordinary element surrounded by ordinary vertices (see [63]).

At each step, in the neighborhood of an irregular vertex/face, a subdivision algorithm relating the vertices of the k-th level mesh with those of the next level $k+1$, can be conveniently encoded in the rows of a local subdivision matrix M_k

whose dimension depends on the valency of the vertex. If the scheme is level-independent each step of the subdivision algorithm can be conveniently encoded in the rows of one local subdivision matrix M. The dimension of this matrix depends on the valency of the extraordinary vertex, too.

Theorem 7 ([17, Theorem 4.1]) *Let $\mathcal{S}$ be a non-singular, non-stationary subdivision scheme whose action in an irregular region is described by a matrix sequence $\{M_k\}_{k\geq 0}$. Let $\mathcal{S}$ be also rotationally symmetric. Moreover, let $\bar{\mathcal{S}}$ be a rotationally symmetric, stationary subdivision scheme associated with the matrix M in the same irregular region. If,*

(i) *$\bar{\mathcal{S}}$ is convergent both in regular and irregular regions,*
(ii) *$\mathcal{S}$ is asymptotically equivalent to $\bar{\mathcal{S}}$ in regular region,*
(iii) *in the irregular regions, for all $k \geq 0$, the matrices M_k and M satisfy $\|M_k - M\|_\infty \leq \frac{C}{\sigma^k}$ with C a constant $(0 < C < \infty)$ and $\sigma > 1$,*

then, for all initial data the non-stationary subdivision scheme $\mathcal{S}$ is convergent, both in regular regions and in the irregular one.

To understand the next result we recall from [17] that the iterated refinement of a surface subdivision scheme in the neighborhood of an irregular element generates a sequence of surface rings $\{\mathbf{R}_k\}_{k\geq 1}$ corresponding to regular points which covers all of the surface except for the 'central' point which is the limit of the extraordinary vertex or face.

Theorem 8 ([17, Theorem 4.2]) *Let $\mathcal{S}$ be as in Theorem 7. Assume in the regular patch ring $\mathbf{R}_{k+1}$ the action of $\mathcal{S}$ is described by a vector $\boldsymbol{\Phi}_{k+1}(u, v)$ consisting of all the basic limit functions of $\mathcal{S}$ whose support intersect $\mathbf{R}_{k+1}$. Moreover, let $\bar{\mathcal{S}}$ be as in Theorem 7 associated with a matrix M in the same irregular region. Under the conditions:*

(i) *$\bar{\mathcal{S}}$ is C^1-convergent in regular regions and its symbol $\mathcal{A}(\mathbf{z})$ contains the factor $(1 + z_1)(1 + z_2)$;*
(ii) *in regular regions $\mathcal{S}$ is defined by the symbols $\{\mathcal{A}^{(k)}(\mathbf{z})\}_{k\geq 0}$ where each $\mathcal{A}^{(k)}(\mathbf{z})$ contains the factor $(1 + z_1)(1 + z_2)$;*
(iii) *in regular regions $\mathcal{S}$ is asymptotically equivalent of order 1 to $\bar{\mathcal{S}}$;*
(iv) *the eigenvalues of M are $\lambda_0 = 1$, $0 < \lambda_1 < 1$, and the rest have absolute value less than λ_1;*
(v) *in the irregular regions, for all $k \geq 0$, the matrices M_k and M satisfy, $\|M_k - M\|_\infty \leq \frac{C}{\sigma^k}$ with C some constant $(0 < C < \infty)$ and $\sigma > \frac{1}{\lambda_1} > 1$;*
(vi) *the entries of $\boldsymbol{\Phi}_{k+1}(u, v)$ sum up to 1;*

the surface generated by $\mathcal{S}$ is normal continuous.

4.3 *Masks of Growing Support*

This section is devoted to a short description of non-stationary univariate subdivision schemes based on masks with growing supports. This is an important example of the potential strength of non-stationary schemes, since it allows for the generation of basic limit functions with high regularity and small support. For details concerning the analysis of these types of schemes and some of their applications we refer the reader to [13] and [33]. The analysis of smoothness of the schemes in these papers is based on the growing number of smoothing factors in their symbols and on Fourier analysis. The application is the generation of C^∞ multiresolution analysis with high approximation order and the generation of C^∞ compactly supported wavelets [13, 43].

The most famous example of a subdivision scheme of this type is given by the one based on the masks in (14) with n the subdivision level. As shown in [33], the basic limit function ϕ_0 is the Rvachev's up-function which is C^∞ and supported on [0, 2], [56]. The first three steps of this scheme are depicted in the next Fig. 16.

A similar example of C^∞ compactly-supported basic limit functions can be obtained if each $\mathcal{A}^{[k]}(\mathbf{z})$ is a product of k smoothing factors (see Definition 3). In this example the support is also [0, 2].

Another nice example is given by the interpolatory non-stationary scheme based on the masks (18) again with n the subdivision level (see [13, 33, 43]). The basic limit function ϕ_0 is a function which is C^∞ and supported in $[-3, 3]$. The first three steps of this scheme are shown in Fig. 17.

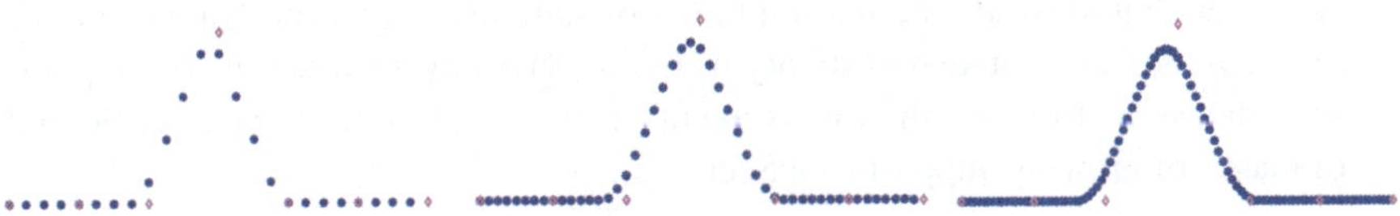

Fig. 16 Three steps of the scheme with masks (14) with n the subdivision level (initial points in magenta)

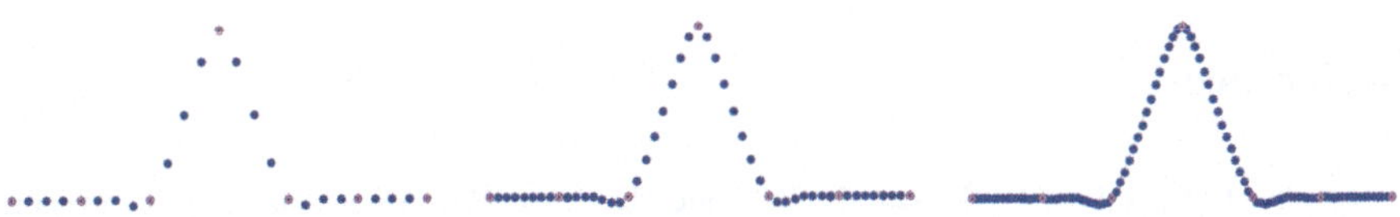

Fig. 17 Three steps of the scheme with masks (18) with n the subdivision level (initial points in magenta)

5 Open Problems in Non-stationary Subdivision

This closing section provides a short overview of open problems—specifically for non-stationary subdivision schemes—that are important to consider in the near future. Yet, due to space reasons, it will not be a detailed description as the one in the recent paper [59] related to the stationary case. Topics are listed in order of difficulty, with respect to the authors' point of view.

- Bivariate results: from Sect. 4.1 it is evident that many results on convergence/regularity and approximation order are available in the univariate case only. Their extension to the bivariate setting is important. Also, construction of bivariate non-stationary interpolatory subdivision schemes and wavelets based on them is a topic that deserves further study;
- Applications: exponential reproducing non-stationary schemes could be used more extensively in image processing and highly smooth wavelets, as in [13], could be applied to real-world problems where the analysed functions are of high smoothness;
- Artefacts and unexpected behaviour of subdivision curves/surfaces: it would be important to better investigate the use of non-stationary tension parameters to tune and control subdivision surfaces;
- New tools for analysis of non-stationary schemes: we believe that to escape from the notions of asymptotic similarity or asymptotic equivalence would give a great impulse to non-stationary schemes;
- Increase the smoothness at extraordinary vertices of subdivision surfaces: we suppose that the possibility of changing the rule coefficients with the iterations can be crucial to overcome the limitation of stationary schemes that are limited to C^1-smoothness at extraordinary vertices. The key idea for increasing the smoothness, is to allow the involvement of more and more points, i.e. the use of masks of growing support (see Sect. 4.3).

Acknowledgments The first author thanks the Research Italian network on Approximation (RITA) and Indam-GNCS for supporting this research activity.

References

1. Badoual, A., Novara, P., Romani, L., Schmitter, D., Unser, M.: A non-stationary subdivision scheme for the construction of deformable models with sphere-like topology. Graphical Models **94**, 38–51 (2017)
2. Badoual, A., Schmitter, D., Uhlmann, V., Unser, U.: Multiresolution subdivision snakes. IEEE Trans. Image Process. **26**(3), 1188–1201 (2017)
3. Bruni, V., Cotronei, M., Pitolli, F.: A family of level-dependent biorthogonal wavelet filters for image compression. J. Comput. Appl. Math. **367**, 112467 (2020)
4. Catmull, E., Clark, J.: Recursively generated B-splines surfaces on arbitrary topological meshes. Comput. Aided Des. **10**(6), 350–355 (1978)

5. Cavaretta, A.S., Dahmen, W., Micchelli, C.A.: Stationary subdivision. Mem. Am. Math. Soc. **93**(453), 1Ű-185 (1991)
6. Chaikin, G.: An algorithm for high speed curve generation. Comput. Graph. Image Process. **3**, 346–349 (1974)
7. Charina, M.: Vector multivariate subdivision schemes: comparison of spectral methods for their regularity analysis. Appl. Comput. Harmon. Anal. **32**, 86–108 (2012)
8. Charina, M., Conti, C.: Polynomial reproduction of multivariate scalar subdivision schemes. J. Comput. Appl. Math. **240**, 51–61 (2013)
9. Charina, M., Conti, C., Guglielmi, N., Protasov, V.: Regularity of non-stationary subdivision: a matrix approach. Numer. Math. **135**, 639–678 (2017)
10. Charina, M., Conti, C., Guglielmi, N., Protasov, V.: Limits of level and parameter dependent subdivision schemes: A matrix approach. Appl. Math. Comput. **272**, 20–27 (2016)
11. Charina, M., Conti, C., Romani, L.: Reproduction of exponential-polynomials by multivariate non-stationary subdivision schemes with a general dilation matrix. Numer. Math. **39**, 395–424 (2013)
12. Charina, M., Conti, C., Sauer, T.: Regularity of multivariate vector subdivision scheme. Numer. Algorithms **39**, 97–113 (2005)
13. Cohen, A., Dyn, N.: Non-stationary subdivision schemes and multiresolution analysis. SIAM J. Math. Anal. **26**, 1745–1769 (1996)
14. Conti, C., Cotronei, M., Romani, L.: Beyond B-splines: exponential pseudo-splines and subdivision schemes reproducing exponential polynomials. Dolomites Res. Notes Approximation **10**, 31–42 (2017)
15. Conti, C., Gori, L., Pitolli, F.: Totally positive functions through non-stationary subdivision schemes. J. Comput. Appl. Math. **200**, 255–265 (2007)
16. Conti, C., Dyn, N., Manni, C., Mazure, M.L.: Convergence of univariate non-stationary subdivision schemes via asymptotic similarity. Comput. Aided Geom. Des. **37**, 1–8 (2015)
17. Conti, C., Donatelli, M., Romani, L., Novara, P.: Convergence and normal continuity analysis of non-stationary subdivision schemes near extraordinary vertices and faces. Constructive Approximation **50**, 457–496 (2019)
18. Charina, M., Conti, C., Jetter, K., Zimmermann, G.: Scalar multivariate subdivision schemes and box splines. Comput. Aided Geom. Des. **28**(5), 285–306 (2011)
19. Conti, C., Romani, L.: Affine combination of B-splines subdivision masks and its non-stationary counterparts. BIT Numer. Math. **50**(2), 269–299 (2010)
20. Conti, C., Romani, L.: Dual univariate m-ary subdivision schemes of de Rham-type. J. Math. Anal. Appl. **407**, 443–456 (2013)
21. Conti, C., Romani, L.: A new family of interpolatory non-stationary subdivision schemes for curve design in geometric modeling, CP1281. In: Simos, T.E., Psihoyios, G., Tsitouras, Ch. (eds.) ICNAAM, Numerical Analysis and Applied Mathematics, International Conference 2010 (2010)
22. Conti, C., Romani, L.: Algebraic conditions on non-stationary subdivision symbols for exponential-polynomial reproduction. J. Comput. Appl. Math. **236**, 543–556 (2011)
23. Conti, C., Romani, L., Yoon, J.: Approximation order and approximate sum rules in subdivision. J. Approximation Theory **207**, 380–401 (2016)
24. Conti, C., Romani, L., Unser, M.: Ellipse-preserving Hermite interpolation and subdivision. J. Math. Anal. Appl. **426**, 211–227 (2015)
25. Conti, C., Zimmermann, G.: Interpolatory rank-1 vector subdivision schemes. CAGD **21**, 341–351 (2004)
26. Cotronei, M., Sissouno, N.: A note on Hermite multiwavelets with polynomial and exponential vanishing moments. Appl. Numer. Math. **120**, 21–34 (2017)
27. Daubechies, I.: Ten lectures on wavelets. In: CBMS Conf. Series in Appl. Math., vol. 61. SIAM, Philadelphia (1992)
28. Daubechies, I., Lagarias, J.C.: Sets of matrices all infinite products of which converge. Linear Algebra Appl. **161**, 227–263 (1992)

29. Delgado-Gonzalo, R., Thevenaz, P., Seelamantula, C.S., Unser, M.: Snakes with an ellipse-reproducing property. IEEE Trans. Image Process. **21**, 1258–1271 (2012)
30. De Rham, G.: Sur une courbe plane. J. Math. Pures Appl. **35**(9), 25–42 (1956)
31. DeRose, T., Kass, M., Truong, T.: Subdivision surfaces in character animation. In: Proceeding SIGGRAPH '98, Proceedings of the 25th Annual Conference on Computer Graphics and Interactive Techniques, pp. 85–94 (1998)
32. Deslauriers, G., Dubuc, S.: Symmetric iterative interpolation processes. Constr. Approx. **5**(1), 4–68 (1989)
33. Derfel, G., Dyn, N., Levin, D.: Generalized refinement equations and subdivision processes. J. Approx. Theory **80**, 272–297 (1995)
34. Dong, B., Shen, Z.: Pseudo-splines, wavelets and framelets. Appl. Comput. Harmon. Anal. **22**, 78–104 (2007)
35. Doo, D., Sabin, M.: Behavior of recursive division surfaces near extraordinary points. Comput. Aided Design **10**(6), 356–360 (1978)
36. Dyn, N.: Subdivision schemes in computer-aided geometric design. In: Light, W. (ed.) Advances in Numerical Analysis II Wavelets, Subdivision Algorithms and Radial Basis Functions, pp. 36–104 Clarendon Press, Oxford (1992)
37. Dyn, N., Levin, D., Gregory, J.A.: A 4-point interpolatory subdivision scheme for curve design. Comput. Aided Geom. Des. **4**(4), (1987)
38. Dyn, N., Levin, D.: Analysis of asymptotically equivalent binary subdivision schemes. J. Math. Anal. Appl. **193**, 594–621 (1995)
39. Dyn, N., Levin, D.: Subdivision schemes in geometric modeling. Acta Numer. **11**, 73–144 (2002)
40. Dyn, N., Levin, D., Luzzatto, A.: Exponentials reproducing subdivision schemes. Found. Comput. Math. **3**, 187–206 (2003)
41. Dyn, N., Levin, D., Yoon, J: Analysis of univariate non-stationary subdivision schemes with application to gaussian-based interpolatory schemes. SIAM J. Math. Anal. **39**, 470–488 (2007)
42. Dyn, N., Kounchev, O., Levin, D., Render, H.: Regularity of generalized Daubechies wavelets reproducing exponential-polynomials with real-valued parameters. Appl. Comput. Harmon. Anal. **37**, 288–306 (2014)
43. Dyn, N., Ron, A.: Multiresolution analysis by infinitely differentiable compactly supported functions. Appl. Comput. Harmon. Anal. **2**, 15–20 (1995)
44. Fang, M., Ma, W., Wang, G.: A generalized surface subdivision scheme of arbitrary order with a tension parameter. Comput. Aided Des. **49**, 8–17 (2014)
45. Guglielmi, N., Protasov, V.: Invariant polytopes of sets of matrices with application to regularity of wavelets and subdivisions. SIAM J. Matrix Anal. Appl. **37**(1), 18–52 (2016)
46. Han, B.: Framelets and wavelets. Algorithms, analysis, and applications. Applied and Numerical Harmonic Analysis. Birkhäuser/Springer, Cham (2017)
47. Han, B., Jia, R.Q.: Characterization of Riesz bases of wavelets generated from multiresolution analysis. Appl. Comput. Harmon. Anal. **23**(3), 321–345 (2007)
48. Jena, M.K., Shunmugaraj, P., Das, P.C.: A non-stationary subdivision scheme for generalizing trigonometric spline surfaces to arbitrary meshes. Comput. Aided Geom. Des. **20**, 61–77 (2003)
49. Lee, Y.-J., Yoon, J.: Non-stationary subdivision schemes for surface interpolation based on exponential-polynomials. Appl. Numer. Math. **60**, 130–141 (2010)
50. Lu, Y., Wang, G., Yang, X.: Uniform hyperbolic polynomial B-spline curves. Comput. Aided Geom. Des. **19**(6), 335–343 (2002)
51. Mejstrik, T.: Improved invariant polytope algorithm and applications. II revision in ACM Trans. Math. Softw. (2019)
52. Möller, C., Reif, U.: A tree-based approach to joint spectral radius determination. Linear Algebra Appl. **463**, 154–170 (2014)
53. Novara, P., Romani, L., Yoon, J.: Improving smoothness and accuracy of modified butterfly subdivision scheme. Appl. Math. Comput. **272**, 64–79 (2016)

54. Novara, P., Romani, L.: Building blocks for designing arbitrarily smooth subdivision schemes with conic precision. J. Comput. Appl. Math. **279**, 67–79 (2015)
55. Prautzsch, H., Boehm, W., Paluszny, M.: Bézier and B-Spline Techniques. Springer, Berlin, Heidelberg (2002)
56. Rvachev, V.A.: Compactly supported solutions of functional-differential equations and their applications. Russ. Math. Surv. **45**(1), 87–120 (1990)
57. Rahman, I.U., Drori, I., Stodden, V.C., Donoho, D.L., Schröder, P.: Multiscale representations for manifold-valued data. Multiscale Model. Simul. **4**(4), 1202–1232 (2005)
58. Reif, U.: A unified approach to subdivision algorithms near extraordinary vertices. Comput. Aided Geom. Des. **12**, 153–174 (1995)
59. Reif, U., Sabin M.A.: Old problems and new challenges in subdivision. J. Comput. Appl. Math. **349**, 523–531 (2019)
60. Rioul, O.: Simple regularity criteria for subdivision schemes. SIAM J. Math. Anal. **23**(6), 1544–1576 (1992)
61. Rota, G.C., Strang, W.G. : A note on the joint spectral radius. Indag. Math. **22**(4), 379–381 (1960)
62. Schumaker, L.: Spline Functions: Basic Theory, 3rd edn. Cambridge Mathematical Library, Cambridge University Press (2007)
63. Peters, J., Reif, U.: Analysis of algorithms generalizing B-spline subdivision. SIAM J. Numer. Anal. **35**(2), 728–748 (1998)
64. Peters, J., Reif, U.: Subdivision Surfaces. Springer (2008)
65. Uhlmann, V., Delgado-Gonzalo, R., Conti, C., Romani, L., Unser, M.: Exponential Hermite splines for the analysis of biomedical images. In: ICASSP, IEEE International Conference on Acoustics, Speech and Signal Processing - Proceedings 685–3874, pp. 1631–1634 (2014)
66. Umlauf, G.: Analyzing the characteristic map of triangular subdivision schemes. Constr. Approx. **16**, 145–155 (2000)
67. Vonesch, C., Blu, T., Unser, M.: Generalized Daubechies wavelet families. IEEE Trans. Signal Process. **55**, 4415–4429 (2007)
68. Warren, J., Weimer, H.: Subdivision Methods for Geometric Design: A constructive Approach. Morgan Kaufman, San Mateo, CA (2002)

Cubature Rules Based on Bivariate Spline Quasi-Interpolation for Weakly Singular Integrals

Antonella Falini, Tadej Kanduč, Maria Lucia Sampoli, and Alessandra Sestini

Abstract In this paper we present a new class of cubature rules with the aim of accurately integrating weakly singular double integrals. In particular we focus on those integrals coming from the discretization of Boundary Integral Equations for 3D Laplace boundary value problems, using a collocation method within the Isogeometric Analysis paradigm. In such setting the regular part of the integrand can be defined as the product of a tensor product B-spline and a general function. The rules are derived by using first the spline quasi-interpolation approach to approximate such function and then the extension of a well known algorithm for spline product to the bivariate setting. In this way efficiency is ensured, since the locality of any spline quasi-interpolation scheme is combined with the capability of an ad–hoc treatment of the B-spline factor. The numerical integration is performed on the whole support of the B-spline factor by exploiting inter-element continuity of the integrands.

Keywords Cubature rules · Singular and nearly singular integrals · Boundary Element Methods · Tensor product B-splines · Spline quasi-interpolation · Spline product · Isogeometric Analysis

A. Falini
Department of Computer Science, University of Bari, Bari, Italy
e-mail: antonella.falini@uniba.it

T. Kanduč
Faculty of Mathematics and Physics, University of Ljubljana, Ljubljana, Slovenia
e-mail: tadej.kanduc@fmf.uni-lj.si

M. L. Sampoli
Department of Information Engineering and Mathematics, University of Siena, Siena, Italy
e-mail: marialucia.sampoli@unisi.it

A. Sestini (✉)
Department of Mathematics and Computer Science, University of Florence, Florence, Italy
e-mail: alessandra.sestini@unifi.it

G. E. Fasshauer et al. (eds.), *Approximation Theory XVI*, Springer Proceedings in Mathematics & Statistics 336, https://doi.org/10.1007/978-3-030-57464-2_5

1 Introduction

The accurate and efficient numerical evaluation of singular integrals is one of the crucial steps in the numerical simulation of differential problems that can be modeled by Boundary Integral Equations (BIEs) [12]. This is the case when relying on Boundary Element Methods (BEMs), which were introduced in the 1980s for the numerical solution of several differential problems, either stationary and evolutive, see for example [7, 24] and references therein. The main features of BEMs are the reduction of the problem dimension and the easiness of application to problems on unbounded domains. On the other hand it is well known that one of the major efforts with any BEM formulation consists in having to deal with singular and nearly singular integrals, which require special numerical treatment in order to preserve the theoretical convergence order of the numerical solution produced by the adopted discretization.

In this paper we focus on cubature rules for weakly singular integrals. Since the interest in integrals of this kind comes from the isogeometric formulation of BEMs, let us briefly recall their main ideas. The first formulation of BEMs considered a piecewise linear approximation of the boundary of the domain, but more accurate curvilinear BEMs already appeared in the 1990s. In the latter methods the boundary of a 2D domain is described through a planar parametric curve. In the parameter domain of the curve a set of Lagrangian functions is defined for the discretization of the considered BIE. The basis of the discretization space where the missing Cauchy data are approximated is just obtained by lifting such functions to the physical boundary of the domain using its parametric representation. Such methodology is common to collocation and Galerkin approaches and can be extended also to the isogeometric formulation of a BEM. This is characterized by the significant assumption that the boundary is parametrically represented in B-spline or NURBS form and the discretization space V is defined through B-splines instead of Lagrangian functions. This makes possible to increase the smoothness of functions belonging to V at desired joints between adjacent elements, often guaranteeing a remarkable reduction of the number of degrees of freedom necessary to attain a certain level of accuracy [3]. Note that additional flexibility can be achieved by relying on generalized B-splines, see for example [14] and references therein, that can be used for the description of the geometry and/or the definition of the discretization space V [4]. Furthermore, it has been already shown in the literature that for a 2D IgA–BEM the element–by–element assembly strategy is not anymore strictly necessary [1]. This computational advantage is obtained since the required integrals, even when singular, can be approximated by rules formulated directly on the support of the B-spline explicitly appearing in the integrand as one of the basis functions generating V [6].

The literature on numerical approximation of singular integrals is quite vast and it is difficult to cover all the results on this issue, see for instance the book [19] or the more recent paper [2, 9] and references therein. As our interest for singular integrals directly descends from their occurrence within the Isogeometric formulation of

BEMs (IgA–BEMs), we limit our attention to the integrals of this kind arising in 3D problems. Singularity removal is often proposed for the numerical treatment of the occurring multivariate weakly singular integrals. For example in [13] where the 3D Stokes problem is considered, the singularity is removed by exploiting carefully chosen known solutions of the analyzed partial differential equation. In other papers these integrals are reformulated by using a suitable coordinate transformation, see for example [22] for Duffy and [23] for polar transformations. In these cases the additional emerging transformation term approximately cancels out the singularity of the kernel and the resulting integrals become regular. In [10] an adaptive Gaussian quadrature rule is presented and it is shown that it is able to tackle singular and also near singular integrals. However all these approaches do not exploit the smoothness of B-splines, taking only into account their piecewise polynomial nature. For this reason, the related cubature rules are always applied after splitting the integration domain into elements with a consequent increase of the computational cost. Instead, in this paper, the B-spline factor is explicitly treated and the cubature rule is applied on the whole B-spline support, not suffering from inter-element smoothness decrease of B-splines. The rules here proposed are an extension to the bivariate setting of the quadrature formulas for singular integrals introduced in [6]. Their key ingredients are a spline quasi-interpolation approach and the spline product formula [16], both considered in their tensor–product formulation. By exploiting the integration on the whole B-spline support, they are attractive for IgA-BEM also in the 3D case, where a replacement of element-by-element assembly with a function-by-function strategy is even more advantageous.

The paper is organized as follows. First we introduce cubature rules for weakly singular integrals, showing their effectiveness when the considered kernel is multiplied by a general function and a B-spline. Then the combination with suitable multiplicative or subtractive techniques specific of the 3D setting is analyzed, in order to show that they become applicable to deal with specific singular integrals of interest in the IgA-BEM setting.

2 The Problem

In this paper we focus on cubature rules for singular integrals of the following type,

$$\int_{R_{\mathbf{I}}} \mathcal{K}(\mathbf{s},\, \mathbf{t})\, B_{\mathbf{I},\mathbf{d}}(\mathbf{t})\, f_{\mathbf{s}}(\mathbf{t})\, d\mathbf{t}, \qquad \mathbf{s} \in R_{\mathbf{I}}^{E}, \tag{1}$$

where $B_{\mathbf{I},\mathbf{d}}$ is an assigned bivariate B-spline of bi-degree $\mathbf{d} := (d_1, d_2)$ with support in the rectangle $R_{\mathbf{I}}$, $R_{\mathbf{I}}^{E} \supset R_{\mathbf{I}}$, and

$$\mathcal{K}(\mathbf{s},\, \mathbf{t}) := \frac{1}{\sqrt{(\mathbf{t}-\mathbf{s})^T A(\mathbf{s})(\mathbf{t}-\mathbf{s})}}, \qquad \mathbf{t} = (t_1, t_2), \quad \mathbf{s} = (s_1, s_2), \tag{2}$$

with $A(\mathbf{s})$ denoting a symmetric and positive definite matrix (which ensures that the singularity appears just at $\mathbf{t} = \mathbf{s}$). Concerning the smoothness requirements for f_s, since our rules are based on the tensor product formulation of (a variant of) an Hermite quasi-interpolation scheme, it is reasonable to assume f_s belonging to $C^{1,1}(R_{\mathbf{I}})$, that is to the space of bivariate functions g such that $\frac{\partial^{i+j} g}{\partial t_1^i \partial t_2^j}$ is continuous in $R_{\mathbf{I}}$ for $i, j \leq 1$. We refer to [20] for an introduction on basic properties and definitions of B-splines and in particular on their tensor product bivariate extension. We observe that for $\mathbf{s} \in R_{\mathbf{I}}$ the integral in (1) is weakly singular and it becomes nearly singular when $\mathbf{s} \in R_{\mathbf{I}}^E \setminus R_{\mathbf{I}}$, with the maximal distance from $R_{\mathbf{I}}$ of $\mathbf{s} \in R_{\mathbf{I}}^E \setminus R_{\mathbf{I}}$ sufficiently small to exclude regular integrals. This is in contrast to other approaches proposed in the literature (see for instance [21]), where typically different integration methods are used for singular and nearly singular integrals. We also note that our rules numerically compute the integral in (1) by approximating only the factor $f_{\mathbf{s}}$. This is particularly useful when the function $f_{\mathbf{s}}$ is more regular in R_I than $B_{\mathbf{I},\mathbf{d}}$, since usually it can be better approximated than the whole product $B_{\mathbf{I},\mathbf{d}} f_{\mathbf{s}}$ [6].

We outline that the kernel $\mathcal{K}$ is of interest for BEMs when $A(\mathbf{s})$ is the matrix containing the coefficients at $\mathbf{t} = \mathbf{s}$ of the first fundamental form associated to a differentiable parametric surface $\mathbf{X} = \mathbf{X}(\mathbf{t}), \mathbf{t} \in \mathcal{D} \subset \mathbb{R}^2$,

$$A(\mathbf{t}) = \begin{bmatrix} (\mathbf{X}_{t_1} \cdot \mathbf{X}_{t_1})(\mathbf{t}) & (\mathbf{X}_{t_1} \cdot \mathbf{X}_{t_2})(\mathbf{t}) \\ (\mathbf{X}_{t_1} \cdot \mathbf{X}_{t_2})(\mathbf{t}) & (\mathbf{X}_{t_2} \cdot \mathbf{X}_{t_2})(\mathbf{t}) \end{bmatrix}. \tag{3}$$

Indeed in this case the quadratic homogeneous polynomial

$$P_{\mathbf{s}}(\mathbf{t}) := (\mathbf{t} - \mathbf{s})^T A(\mathbf{s})(\mathbf{t} - \mathbf{s}) \tag{4}$$

collects the lowest order non-zero terms of the Taylor expansion at $\mathbf{t} = \mathbf{s}$ of $\|\mathbf{X}(\mathbf{t}) - \mathbf{X}(\mathbf{s})\|_2^2$. So $\mathcal{K}(\mathbf{s}, \mathbf{t})$ is a local approximation of

$$\mathcal{G}(\mathbf{s}, \mathbf{t}) := \frac{1}{\|\mathbf{X}(\mathbf{t}) - \mathbf{X}(\mathbf{s})\|_2}, \tag{5}$$

which is, up to a multiplicative constant, the kernel appearing in the single layer potential,

$$\int_{R_{\mathbf{I}}} \mathcal{G}(\mathbf{s}, \mathbf{t})\, B_{\mathbf{I},\mathbf{d}}(\mathbf{t})\, g_{\mathbf{s}}(\mathbf{t})\, d\mathbf{t}, \tag{6}$$

for 3D Laplace problems, written in intrinsic coordinates. The B-spline factor in (6) corresponds to a basis function of the tensor product spline space V used for the discretization, while $g_{\mathbf{s}}$ appears in the formulation as the Jacobian of the domain transformation to the parametric domain. Note that $\mathcal{G}$ is substantially the kernel associated also with the Helmholtz problem, missing only an additional regular trigonometric factor appearing in the fundamental solution of such equation.

In this work we consider the so-called singularity extraction procedure, based on either a subtractive or a multiplicative technique, to derive a more convenient formulation of the singular integral. Following this procedure, the integral in (6) is transformed into an integral with the same kind of singularity but with a more standard kernel, possibly added to a regular integral.

Denoting with $\mathcal{G}_a$ the approximating kernel having the same kind of singularity of $\mathcal{G}$ at $\mathbf{t} = \mathbf{s}$, with the subtractive technique the integral in (6) is decomposed in the following sum,

$$\int_{R_{\mathbf{I}}} \mathcal{G}_a(\mathbf{s},\, \mathbf{t})\, B_{\mathbf{I},\mathbf{d}}(\mathbf{t})\, g_{\mathbf{s}}(\mathbf{t})\, d\mathbf{t} + \int_{R_{\mathbf{I}}} (\mathcal{G}(\mathbf{s},\, \mathbf{t}) - \mathcal{G}_a(\mathbf{s},\, \mathbf{t}))\; B_{\mathbf{I},\mathbf{d}}(\mathbf{t})\, g_{\mathbf{s}}(\mathbf{t})\, d\mathbf{t} \tag{7}$$

where the second integral is regular if $\mathcal{G}_a$ is suitably defined. The first integral in (7) is still weakly singular and it becomes equal to the integral in (1) if $\mathcal{G}_a = \mathcal{K}$ is chosen and $f_{\mathbf{s}} = g_{\mathbf{s}}$ is set. In this case the regularity of $f_{\mathbf{s}}$ is that of the Jacobian of $\mathbf{X}$. Then, considering the IgA paradigm, we can observe that it can be low (anyway at least $C^{1,1}$ if $\mathbf{X}$ is a regular $C^{2,2}$ NURBS parameterization) only at the original knots involved in the CAGD representation of $\mathbf{X}$, and not at the other knots used to define the discretization space V. Furthermore, without loss of generality, we can assume that the original knots have maximal multiplicity, so that the possible reduction of regularity of $f_{\mathbf{s}}$ can appear only at the boundary of R_I. With the multiplicative technique, setting $\rho_{\mathbf{s}}(\mathbf{t}) := \mathcal{G}(\mathbf{s},\, \mathbf{t})/\mathcal{G}_a(\mathbf{s},\, \mathbf{t})$, and $f_{\mathbf{s}}(\mathbf{t}) := \rho_{\mathbf{s}}(\mathbf{t})\, g_{\mathbf{s}}(\mathbf{t})$, we obtain

$$\int_{R_{\mathbf{I}}} \mathcal{G}(\mathbf{s},\, \mathbf{t})\, B_{\mathbf{I},\mathbf{d}}(\mathbf{t})\, g_{\mathbf{s}}(\mathbf{t})\, d\mathbf{t} = \int_{R_{\mathbf{I}}} \mathcal{G}_a(\mathbf{s},\, \mathbf{t})\, B_{\mathbf{I},\mathbf{d}}(\mathbf{t})\, f_{\mathbf{s}}(\mathbf{t})\, d\mathbf{t}, \tag{8}$$

where the function $f_{\mathbf{s}}$ is regular, again if $\mathcal{G}_a$ is suitably defined. If in particular $\mathcal{G}_a = \mathcal{K}$, we get

$$\rho_{\mathbf{s}}(\mathbf{t}) = \frac{\sqrt{(\mathbf{t}-\mathbf{s})^T A(\mathbf{s})(\mathbf{t}-\mathbf{s})}}{\|\mathbf{X}(\mathbf{t}) - \mathbf{X}(\mathbf{s})\|_2}, \tag{9}$$

with A defined as in (3). Note that this reformulation of the singular integral in (6) can be considered as a bivariate generalization of the standard one proposed in the literature for dealing with univariate singular kernels, where $\mathcal{G}_a$ is just defined as $\mathcal{G}_a(s,t) = 1/|s-t|$. In the bivariate setting the function $\rho_{\mathbf{s}}$ defined in (9) is continuous at $\mathbf{t} = \mathbf{s}$, since it can be verified that $\lim_{\mathbf{t}\to\mathbf{s}} \rho_{\mathbf{s}}(\mathbf{t})$ exists and is equal to 1. Unfortunately $\rho_{\mathbf{s}}$ is not smoother than C^0 at such point for a general surface $\mathbf{X}$. Thus, when the integral of interest is that defined in (6) and $\mathbf{X}$ is a general surface, we would need to consider higher order approximations of $\mathcal{G}$ instead of $\mathcal{K}$, in order to deal with functions $f_{\mathbf{s}}$ more regular at $\mathbf{t} = \mathbf{s}$ when they are obtained by using the multiplicative technique. Note that also adopting the subtractive technique this can be useful to increase the regularity of the integrand of the regular integral in (7). To keep the presentation of our rules concise, this technical but important aspect is not addressed in this paper.

3 Cubature Rules Based on Tensor-Product Spline Quasi-Interpolation

Quasi-Interpolation (QI) is a general approach for approximating a function or a given set of discrete data with low computational cost, see for instance [18] and references therein. For a chosen finite dimensional approximating space and a suitable local basis generating it, the coefficients of the approximation are locally computed with explicit formulas by using linear functionals depending on the function and possibly also on its derivatives and/or integrals. Since there is already an explicit B-spline factor in the considered integral in (1), it is particularly beneficial for us to approximate the function f_s using a spline quasi-interpolation operator. That way the B-spline factor is preserved in the expression for the numerical integration and the spline product algorithm can be readily applied [16]. The easiest extension of a univariate QI scheme to the bivariate setting relies on its tensor-product formulation which anyway performs function approximation on a rectangular domain, requiring information at the vertices of a quadrilateral grid of the domain. We add that in the bivariate spline setting there has recently been a lot of interest for QI schemes on special type triangulations or even on general ones adopting macroelements, see for example [5, 11] and references therein. However, since for application to cubature the analytic expression of the function to be approximated is available and our integration domain is rectangular, for our purposes the tensor-product extension is more suitable. In particular we adopt a tensor-product derivative free QI scheme which is a natural choice for numerical integration.

Denoting with $S_{p,T}$ the space of univariate splines with degree p and with T the associated extended knot vector defined in the reference domain $[-1, 1]$, $-T = \{\xi_0 \leq \cdots, \xi_{p-1} \leq \xi_p \leq \cdots \leq \xi_{m+1} \leq \cdots \leq \xi_{m+p+1}\}$, with $\xi_j < \xi_{j+p+1}$ and $\xi_p = -1, \xi_{m+1} = 1$—a spline $\sigma \in S_{p,T}$ can be represented by using the standard B-spline basis, $\mathcal{B}_{j,p}$, $j = 0, \ldots, m$,

$$\sigma(\cdot) = \sum_{j=0}^{m} \lambda_j \, \mathcal{B}_{j,p}(\cdot).$$

Thus a univariate derivative free QI scheme to approximate a univariate function f can be compactly written as follows,

$$\boldsymbol{\lambda} = C\mathbf{f}, \tag{10}$$

where $\boldsymbol{\lambda} := (\lambda_0, \ldots, \lambda_m)^T$ is the vector of the spline coefficients; C is a $(m + 1) \times (K + 1)$ banded matrix characterizing the scheme; $\mathbf{f} := (f(\tau_0,) \ldots, f(\tau_K))^T$, with $-1 \leq \tau_0 < \cdots < \tau_K \leq 1$ completing the characterization of the scheme. On this concern observe that, if $C_{i,j}$ $j = i - L, \ldots, i + U$ are the non vanishing elements in C, it must be required that $\tau_{i-L}, \ldots, \tau_{i+U}$ belong to the support of $\mathcal{B}_{i,p}$. Furthermore a certain polynomial reproduction capability of the scheme must be required to ensure a suitable convergence order.

Within this kind of QI schemes, we refer to the derivative free variant of the Hermite QI method introduced in [15]. Such variant requires in input only the values of f at the spline breakpoints, since the derivative values required in the original scheme are approximated with suitable finite differences [15].

In the tensor product formulation of the scheme we have to define a spline σ in the space $S_{p_1,\mathcal{T}_1} \times S_{p_2,\mathcal{T}_2}$,

$$\sigma(t_1, t_2) = \sum_{i=0}^{m_1} \sum_{j=0}^{m_2} \lambda_{i,j}\, \mathcal{B}_{i,p_1}(t_1)\mathcal{B}_{j,p_2}(t_2).$$

Setting $\mathbf{t} := (t_1, t_2)$ and $\mathbf{I} := \{(i, j), i = 0, \ldots, m_1, j = 0, \ldots, m_2\}$ we can compactly write

$$\sigma(\mathbf{t}) = \sum_{\mathbf{i} \in \mathbf{I}} \lambda_{\mathbf{i}}\, \mathcal{B}_{\mathbf{I},\mathbf{p}}(\mathbf{t}),$$

where $\mathcal{B}_{\mathbf{I},\mathbf{p}}(\mathbf{t}) := \mathcal{B}_{i,p_1}(t_1)\mathcal{B}_{j,p_2}(t_2)$. Using for example the lexicographical ordering for the elements of $\mathbf{I}$ and the Kronecker product between matrices, the tensor product extension of the scheme can be expressed as follows,

$$\boldsymbol{\lambda} = (A_1 \otimes A_2)\mathbf{f}, \tag{11}$$

where now $\mathbf{f} = \left(f(\tau_0^{(1)}, \tau_0^{(2)}),\ f(\tau_0^{(1)}, \tau_1^{(2)}), \cdots,\ f(\tau_{K_1}^{(1)}, \tau_{K_2}^{(2)})\right)^T$ with f denoting a bivariate function and $\boldsymbol{\lambda}$ is the vector $\boldsymbol{\lambda} := \left(\lambda_{(0,0)}, \lambda_{(0,1)}, \ldots, \lambda_{(m_1,m_2)}\right)^T$.

In order to extend to the bivariate setting the quadrature rule for singular integrals containing a B-spline weight developed in [6], we need two additional ingredients: a bivariate generalization of the spline product formula and explicit analytical formulas to compute specific singular integrals. In more detail, we first consider the tensor product generalization of the algorithm in [16] to express the product $\sigma\, B_{\mathbf{I},\mathbf{d}}$ in the bivariate B-spline basis of the product space. Such space has bi-degree (p_1+d_1, p_2+d_2) and the related extended knot vectors in each coordinate direction are obtained by merging T_k and $\mathcal{T}_k$, for $k = 1, 2$, knot vectors in each direction k for $B_{\mathbf{I}}$ and σ, respectively. The other necessary step for approximating the integral in (1) consists in the computation of the so-called *modified moments*,

$$\mu_{\mathbf{i}}(\mathbf{s}) := \int_{R_{\mathbf{I}}} \mathcal{K}(\mathbf{s}, \mathbf{t})\, B_{\mathbf{i}}^{(\Pi)}(\mathbf{t})\, d\mathbf{t}, \qquad \mathbf{i} \in \mathbf{I}^{(\Pi)},$$

where $B_{\mathbf{i}}^{(\Pi)}, \mathbf{i} \in \mathbf{I}^{(\Pi)}$, denotes the B-spline basis of the product space. For this aim we need again to generalize to the bivariate setting the univariate recursion for B-splines whose usage in this context was introduced in [1], see also [8].

The final approximation of the integral in (1) is then simply given by the product $\boldsymbol{\mu}(\mathbf{s})^T \boldsymbol{\lambda}^{(\Pi)}$, where $\boldsymbol{\mu}(\mathbf{s})$ is the vector containing the above modified moments ordered in lexicographical way and $\boldsymbol{\lambda}^{(\Pi)}$ is a vector of the same length whose entries are the coefficients expressing $\sigma B_{I,\mathbf{d}}$ in the B-spline basis of the product space.

4 Numerical Results

This section is devoted to check the performance of our cubature rules.

In the experiments we always assume that the bi-degree $\mathbf{d} = (d, d)$ of the B-spline factor in the integrand of (1) is equal to $(2, 2)$ or $(3, 3)$ and that $R_I = [-1, 1]^2$. For simplicity, we consider a uniform distribution of the $d+1$ breakpoints of the B-spline in each coordinate direction. In order to deal either with nearly singular and singular integrals, we consider the source points $\mathbf{s} = (s_1, s_2) \in \mathcal{S}^2$ with $\mathcal{S} := \{-1.1, -1, -0.5, 0, 0.5, 1, 1.1\}$.

The tests are performed on a uniform $N \times N$ grid for the breakpoints of the quasi-interpolating spline σ, with N ranging from 6 to 14 with step 2. The bi-degree $\mathbf{p} = (p, p)$ of the quasi-interpolant is set to $(2, 2)$ or $(3, 3)$.

Example 1 In the first example we consider the quadratic bivariate polynomial function $f_{\mathbf{s}}(\mathbf{t}) = f(\mathbf{t}) = t_1^2 + t_2^2$. The aim of the test is to check the exactness of the proposed cubature rule, since the integration rule is based on the chosen tensor product QI scheme, which is exact on polynomials of bi-degree (ℓ_1, ℓ_2) with $\ell_k \le p$. For this example the matrix A defining the kernel $\mathcal{K}$ in (2) is just a constant matrix with all unit entries. We verified that already with $N = 6$ we get a maximum relative error of $1.54\text{e} - 13$ for $\mathbf{s} \in \mathcal{S}^2$ restricted to the interior of R_I. It becomes $7.56\text{e} - 12$, and $9.60\text{e} - 12$ when $\mathbf{s} \in \mathcal{S}^2$ is restricted to the boundary of R_I and to values external to R_I, respectively.

Example 2 In order to check the convergence order, in this example we consider A equal to the identity and the analytic function $f_{\mathbf{s}}(\mathbf{t}) = f(\mathbf{t}) = \exp(t_1 t_2)$. The results are collected in Table 1, where in particular the maximal absolute errors `errmax1`, `errmax2` and `errmax3` are reported, varying the number $N \times N$ of cubature nodes uniformly distributed in R_I. The results show a very good behavior of the rules for the considered test function and matrix.

Table 1 Example 2. Maximal absolute cubature error and convergence order for $\mathbf{s} \in \mathcal{S}^2$ outside (errmax1, o_1), on the boundary (errmax2, o_2) and inside (errmax3, o_3) the integration domain R_I, for $p = 2, 3$ and $d = 2, 3$.

N	errmax1	o_1	errmax2	o_2	errmax3	o_3	errmax1	o_1	errmax2	o_2	errmax3	o_3
$d = 2$	p = 2						p = 3					
6	2.5704e − 05	–	4.3428e − 05	–	8.3210e − 05	–	1.0520e − 06	–	2.1322e − 06	–	2.1322e − 06	–
8	8.4609e − 06	3.9	1.6115e − 05	3.5	1.6697e − 05	5.6	2.7380e − 07	4.7	5.4119e − 07	4.8	5.4278e − 07	4.8
10	3.6045e − 06	3.8	6.9256e − 06	3.8	6.9256e − 06	3.9	9.9469e − 08	4.5	1.9417e − 07	4.6	1.9417e − 07	4.6
12	1.7283e − 06	4.0	3.3031e − 06	4.1	3.3031e − 06	4.1	4.4251e − 08	4.4	8.5289e − 08	4.5	8.5289e − 08	4.5
14	9.1746e − 07	4.1	1.7456e − 06	4.1	1.7456e − 06	4.1	2.2321e − 08	4.4	4.2435e − 08	4.5	4.2435e − 08	4.5
$d = 3$	p = 2						p = 3					
6	5.0578e − 06	–	1.5198e − 05	–	2.5845e − 05	–	3.3475e − 07	–	8.3595e − 07	–	8.3595e − 07	–
8	2.6660e − 06	2.2	5.9122e − 06	3.3	5.9122e − 06	5.1	8.7285e − 08	4.7	2.1109e − 07	4.8	2.1156e − 07	4.8
10	1.1965e − 06	3.6	2.6836e − 06	3.5	2.6836e − 06	3.5	3.1949e − 08	4.5	7.6082e − 08	4.6	7.6082e − 08	4.6
12	5.7522e − 07	4.0	1.2883e − 06	4.0	1.2883e − 06	4.0	1.4385e − 08	4.4	3.3872e − 08	4.4	3.3873e − 08	4.4
14	3.0410e − 07	4.1	6.8169e − 07	4.1	6.8170e − 07	4.1	1.0270e − 08	2.2	1.7292e − 08	4.4	1.7292e − 08	4.4

Example 3 This example considers the case of the matrix A defined as in (3), with $\mathbf{X}$ being the standard parameterization for the lateral surface of a cylinder of radius $r = 2$

$$\mathbf{X}(\mathbf{t}) = (r\cos(\pi t_1/4),\ r\sin(\pi t_1/4),\ t_2)\,,$$

which implies that R_I is mapped to a quarter of the lateral cylindrical surface with height 2. The factor $f_{\mathbf{s}}$ in (1) is assigned as the product between $\rho_{\mathbf{s}}$ which is defined in (9) and the Jacobian $J(\mathbf{t})$, with

$$J(\mathbf{t}) := \|\mathbf{X}_{t_1}(\mathbf{t}) \times \mathbf{X}_{t_2}(\mathbf{t})\|_2. \tag{12}$$

This means that the integral with the form in (1) considered for this experiment has been obtained from (6) by using the multiplicative strategy introduced in (8) with $\mathcal{G}_a = \mathcal{K}$, obtaining in this case a $C^{1,1}$ smooth function $\rho_{\mathbf{s}}$ also when $\mathbf{s} \in R_I$.

Figure 1 shows the convergence behavior of the absolute cubature errors `errmax1`, `errmax2` and `errmax3` for the four considered choices of the pair

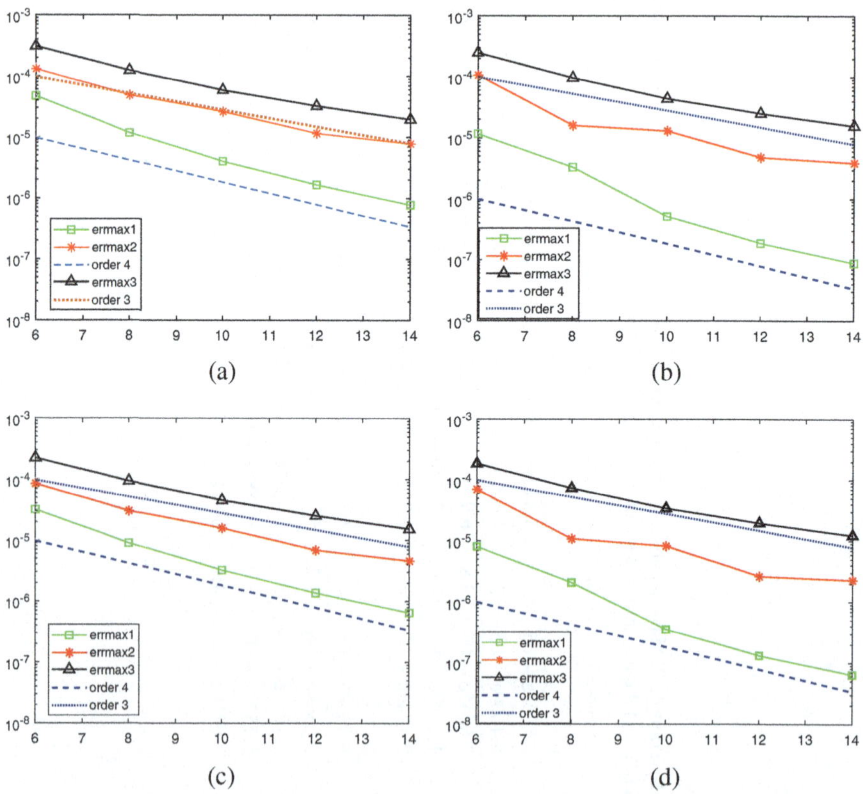

Fig. 1 Example 3. The convergence behavior of the absolute cubature errors `errmax1`, `errmax2` and `errmax3` for $d = 2$, $p = 2$ (**a**), $d = 2$, $p = 3$ (**b**), $d = 3$, $p = 2$ (**c**) and $d = 3$, $p = 3$ (**d**)

(d, p). Comparing left and right images of the figure and first referring to `ermax2` and `errmax3` (i.e. when the rules are applied to singular integrals), we can observe that there is not significant advantage in using $p = 3$ instead of $p = 2$, either from the point of view of the convergence order or from that of the initial ($N = 6$) and final ($N = 14$) accuracy. This is a different behavior with respect to Example 2 where the function f was highly smooth everywhere. Referring to `errmax1` (i.e. for nearly–singular integrals) however, this comment does not hold anymore.

We observe that for the maximum considered value of N, $N = 14$, we achieve a value for `ermax3` of the order of 10^{-5} which corresponds to a relative error of the same order; at a first sight this could seem not satisfactory but we remark that the portion of the cylindrical surface taken into account for the integration is quite large. Indeed, repeating the experiment mapping R_I to a smaller portion of the surface, the relative error decreases. Finally, comparing top and bottom images we can also conclude that different regularity of the B-spline factor in (1) associated with different choices of d does not significantly influence the accuracy of our rules.

Example 4 In the last example, we consider an integral of interest for the BIE formulation of the 3D Helmholtz problem $\Delta u + k^2 u = 0$, where k is the wave number defined as $k = 2\pi/\lambda$, with λ denoting the wavelength of the electromagnetic radiation. The boundary of the domain of the differential problem is assumed equal to a section of a one sheet hyperboloid which can be parametrically represented as follows,

$$\mathbf{X}(\mathbf{t}) = (\cos(\pi t_1/4)\sqrt{1+t_2^2},\ \sin(\pi t_1/4)\sqrt{1+t_2^2},\ t_2).$$

As in the previous example, the integration domain R_I is mapped to a quarter of the boundary of the considered section of hyperboloid whose height is 2. The matrix A is again defined by the formula in (3) but now the function f_s is assigned as follows,

$$f_{\mathbf{s}}(\mathbf{t}) = J(\mathbf{t})\cos(k\|\mathbf{X}(\mathbf{t}) - \mathbf{X}(\mathbf{s})\|_2),$$

with $k = \pi/2$ and J defined as in (12). Note that such function is $C^{1,1}$ also at $\mathbf{t} = \mathbf{s}$. The so defined expression of (1) is the real part of the weakly singular integral to be computed when the subtractive decomposition in (7) is applied for the Helmholtz kernel on the considered domain and the IgA–BEM collocation approach is adopted for the numerical solution. The results for this example are shown in Fig. 2. From the figure we note that in this case increasing p from 2 to 3 produced a better accuracy. The errors for the same value of N are a bit worse than those obtained in Example 3. This is due to the more oscillating nature of the function $f_{\mathbf{s}}$. For a different approach to be applied in the nearly singular case with highly oscillating functions see for instance [17].

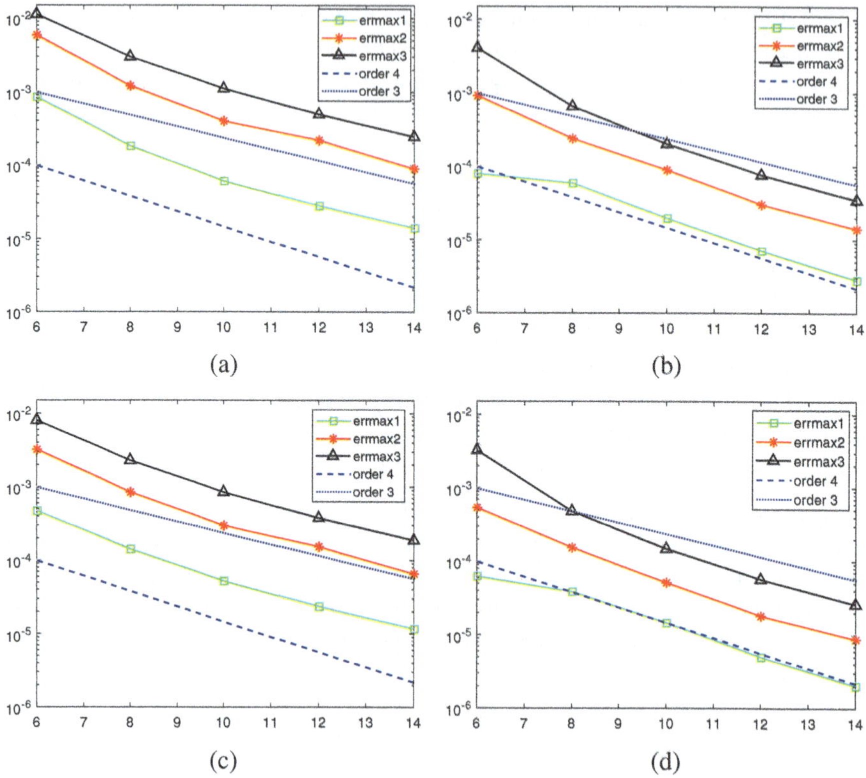

Fig. 2 Example 4. The convergence behavior of the absolute cubature errors `errmax1`, `errmax2` and `errmax3` for $d = 2,\ p = 2$, (**a**), $d = 2,\ p = 3$ (**b**), $d = 3,\ p = 2$ (**c**) and $d = 3,\ p = 3$ (**d**)

5 Conclusions

In this paper cubature rules for weakly singular double integrals containing an explicit B-spline factor are presented. The key ideas for these formulas are the extension of a derivative free spline quasi-interpolation scheme and of an algorithm for spline product to the bivariate setting. Numerical results, also of interest in the IgA-BEM setting, confirm good performances of the proposed rules.

Acknowledgments The authors are all members of Gruppo Nazionale per il Calcolo Scientifico (GNCS) of the Istituto Nazionale di Alta Matematica (INdAM). The support of GNCS through "Progetti di ricerca 2019" program is gratefully acknowledged. The first author is also thankful to the INdAM-GNCS funding "Finanziamento Giovani Ricercatori 2020".

References

1. Aimi, A., Calabrò, F., Diligenti, M., Sampoli, M.L., Sangalli, G., Sestini, A.: New efficient assembly in isogeometric analysis for symmetric Galerkin boundary element method. CMAME **331**, 327–342 (2018)
2. Aimi, A., Calabrò, F., Falini, A., Sampoli, M.L.: Sestini, A.: Quadrature formulas based on spline quasi-interpolation for hypersingular integrals arising in IgA-CMAME **372**, 113441 (2020)
3. Aimi, A., Diligenti, M., Sampoli, M.L., Sestini, A.: Isogeometric analysis and symmetric Galerkin BEM: a 2D numerical study. Appl. Math. Comput. **272**, 173–186 (2016)
4. Aimi, A., Diligenti, M., Sampoli, M.L., Sestini, A.: Non-polynomial spline alternatives in isogeometric symmetric Galerkin BEM. Appl. Numer. Math. **116**, 10–23 (2017)
5. Barrera, D., Dagnino, C., Ibáñez, M.J., Remogna, S.: Point and differential C^1 quasi-interpolation on three directional meshes. JCAM **354**, 373–389 (2019)
6. Calabrò, F., Falini, A., Sampoli, M.L., Sestini, A.: Efficient quadrature rules based on spline quasi-interpolation for application to IGA-BEMs. JCAM **338**, 153–167 (2018)
7. Costabel, M.: Developments in boundary element methods for time-dependent problems. In: Jentsch, L., Troltzsch, F. (eds.) Problems and Methods in Mathematical Physics, vol. 134, pp. 17–32. Springer, Leipzig (1994)
8. Falini, A., Giannelli, C., Kanduč, T., Sampoli, M.L., Sestini, A.: A collocation IGA-BEM for 3D potential problems on unbounded domains, submitted (2020).
9. Gao, X.-W.: An effective method for numerical evaluation of general 2D and 3D high order singular boundary integrals. Comput. Methods Appl. Mech. Eng. **199** 2856–2864 (2010)
10. Gong, Y.P., Dong, C.Y.: An isogeometric boundary element method using adaptive integral method for 3D potential problems. JCAM **319**, 141–158 (2017)
11. Grošelj, J., Speleers, H.: Three recipes for quasi-interpolation with cubic Powell–Sabin splines. CAGD **67**, 47–70 (2018)
12. Hsiao, G.C., Wendland, W.L.: Boundary Integral Equations. Springer, Berlin (2008)
13. Klaseboer, E. Fernandez, C., Khoo, B.: A note on true desingularisation of boundary integral methods for three-dimensional potential problems. Eng. Anal. Bound. Elem. **33**(6), 796–801 (2009)
14. Manni, C., Pelosi, F., Sampoli, M.L: Generalized B-splines as a tool in isogeometric analysis. CMAME **200**, 867–881 (2011)
15. Mazzia, F., Sestini, A.: The BS class of Hermite spline quasi–interpolants on nonuniform knot distributions. BIT **49**, 611–629 (2009)
16. Mørken, K.: Some identities for products and degree raising of splines. Constr. Approx. **7**, 195–208 (1991)
17. Occorsio, D., Serafini, G.: Cubature formulae for nearly singular and highly oscillating integrals. Calcolo **55**(1), 4 (2018)
18. Sablonnière, P.: Recent progress in univariate and multivariate polynomial or spline quasi–interpolants. In: de Brujn, M.G., Mache, D.H., Szabados, J. (eds.) Trends and Applications in Constructive Approximation, pp. 229–245 Birkhäuser, Basel (2005).
19. Sladek, V., Sladek, J.: Singular Integrals in Boundary Element Methods. WIT Press, Southampton (1998)
20. Schumaker, L.: Spline Functions: Basic Theory, 3rd ed. Cambridge University Press, Cambridge (2007)
21. Scuderi, L.: A new smoothing strategy for computing nearly singular integrals in 3D Galerkin BEM. JCAM **225**, 406–427 (2009)
22. Tan, F., Lv, J., Jiao, Y., Liang, J., Zhou, S.: Efficient evaluation of weakly singular integrals with Duffy-distance transformation in 3D BEM. Eng. Anal. Bound. Elem. **104**, 63–70 (2019)

23. Taus, M., Rodin, G.J., Hughes, T.J.R.: Isogeometric analysis of boundary integral equations: high-order collocation methods for the singular and hyper-singular equations. Math. Models Methods Appl. Sci. **26**, 1447–1480 (2016)
24. Wendland, W.I.: On some mathematical aspects of boundary element methods for elliptic problems. In: The Mathematics of Finite Elements and Applications, vol. V. Academic Press, London (1985)

On DC Based Methods for Phase Retrieval

Meng Huang, Ming-Jun Lai, Abraham Varghese, and Zhiqiang Xu

Abstract In this paper, we develop a new computational approach which is based on minimizing the difference of two convex functions (DC) to solve a broader class of phase retrieval problems. The approach splits a standard nonlinear least squares minimizing function associated with the phase retrieval problem into the difference of two convex functions and then solves a sequence of convex minimization subproblems. For each subproblem, the Nesterov accelerated gradient descent algorithm or the Barzilai-Borwein (BB) algorithm is adopted. In addition, we apply the alternating projection method to improve the initial guess in [20] and make it much more closer to the true solution. In the setting of sparse phase retrieval, a standard ℓ_1 norm term is added to guarantee the sparsity, and the subproblem is solved approximately by a proximal gradient method with the shrinkage-threshold technique directly. Furthermore, a modified Attouch-Peypouquet technique is used to accelerate the iterative computation, which leads to more effective algorithms than the Wirtinger flow (WF) algorithm and the Gauss-Newton (GN) algorithm and etc. Indeed, DC based algorithms are able to recover the solution with high probability when the measurement number $m \approx 2n$ in the real case and $m \approx 3n$ in the complex case, where n is the dimension of the true solution. When $m \approx n$, the ℓ_1-DC based algorithm is able to recover the sparse signals with high probability. Our main results show that the DC based methods converge to a critical point linearly. Our study is a deterministic analysis while the study for the Wirtinger

M. Huang
Department of Mathematics, Hong Kong University of Science and Technology, Hong Kong, China
e-mail: menghuang@ust.hk

M.-J. Lai (✉) · A. Varghese
Department of Mathematics, The University of Georgia, Athens, GA, USA
e-mail: mjlai@uga.edu; mjlai@math.uga.edu

Z. Xu
Institute of Computational Mathematics, Academy of Mathematics and Systems of Science, Chinese Academic Sciences, Beijing, China
e-mail: xuzq@lsec.cc.ac.cn

G. E. Fasshauer et al. (eds.), *Approximation Theory XVI*, Springer Proceedings in Mathematics & Statistics 336, https://doi.org/10.1007/978-3-030-57464-2_6

flow (WF) algorithm and its variants, the Gauss-Newton (GN) algorithm, the trust region algorithm is based on the probability analysis. Finally, the paper discusses the existence and the number of distinct solutions for phase retrieval problem.

Keywords Phase retrieval · Sparse signal recovery · DC methods · Nonlinear least squares · Non-convex analysis

1 Introduction

1.1 Phase Retrieval

The phase retrieval problem has been extensively studied in the last 40 years due to its numerous applications, such as X-ray diffraction, crystallography, electron microscopy, optical imaging and etc., see, e.g. [11, 16, 18, 28, 29, 31, 35]. In particular, an explanation of the image recovery from the phaseless measurements and a survey of recent research results can be found in [25]. Mathematically, the phaseless retrieval problem or simply called phase retrieval problem can be stated as follows. Given measurement vectors $\mathbf{a}_i \in \mathbb{R}^n$ (or $\in \mathbb{C}^n$), $i = 1, \ldots, m$ and the measurement values $b_i \geq 0, i = 1, \cdots, m$, we would like to recover an unknown signal $\mathbf{x} \in \mathbb{R}^n$ (or $\in \mathbb{C}^n$) through a set of quadratic equations:

$$b_1 = |\langle \mathbf{a}_1, \mathbf{x} \rangle|^2, \ldots, b_m = |\langle \mathbf{a}_m, \mathbf{x} \rangle|^2. \tag{1}$$

Noting that for any constant $c \in \mathbb{R}^n$ (or $\in \mathbb{C}^n$) with $|c| = 1$, it holds $|\langle \mathbf{a}_i, c\mathbf{x} \rangle|^2 = |\langle \mathbf{a}_i, \mathbf{x} \rangle|^2$ for all i. Thus we can only hope to recover $\mathbf{x}$ up to a unimodular constant. We say the measurements $\mathbf{a}_1, \cdots, \mathbf{a}_m$ are generic if $A = (\mathbf{a}_1, \ldots, \mathbf{a}_m)$ corresponds to a point in a non-empty Zariski open subset of $\mathbb{R}^{n\times m}$ (or $\mathbb{C}^{n\times m}$). Also, $b_1, \cdots, b_m$ are essential if there exist n values $b_{j_1}, \cdots, b_{j_n}$ are all positive. One fundamental problem in phase retrieval is to give the minimal m for which there exists $A = (\mathbf{a}_1, \ldots, \mathbf{a}_m)$ can recover $\mathbf{x}$ up to a unimodular constant. For the real case, it is well known that the minimal measurement number m is $2n - 1$ (cf. [4]). For the complex case $\mathbb{C}^n$, this question remains open. Conca, Edidin, Hering and Vinzant [14] proved $m \geq 4n - 4$ generic measurements $\mathbf{a}_1, \ldots, \mathbf{a}_m$ have phase retrieval property for $\mathbb{C}^n$ and they furthermore show that $4n-4$ is sharp if n is in the form of 2^k+1, $k \in \mathbb{Z}_+$. In [38], for the case $n = 4$, Vinzant present $11 = 4n - 5 < 4n - 4$ measurement vectors which have phase retrieval property for $\mathbb{C}^4$. It implies that $4n-4$ is not sharp for some dimension n. Similar results about the minimal measurement number for sparse phase retrieval can be found in [39].

There are many computational algorithms available to find a true signal $\mathbf{x}$ up to a phase factor. It is common folklore that for given $\mathbf{a}_i, i = 1, \ldots, m$, we may not be able to find a solution $\mathbf{x}$ from any given vector $\mathbf{b} = (b_1, \cdots, b_m)^\top$, e.g. a perturbation of the exact observations $\mathbf{b}^*$. We shall give this fact a mathematical explanation (see Theorem 1 in the next section). Thus, the phase retrieval problem is usually formulated as follows:

$$\min_{\mathbf{x}\in\mathbb{R}^n \text{ or } \mathbb{C}^n} \sum_{i=1}^{m}(|\langle \mathbf{a}_i, \mathbf{x}\rangle|^2 - b_i)^2. \tag{2}$$

Although it is not a convex minimization problem, the objective function is differentiable. Hence, many computational algorithms can be developed and they are very successful actually. A gradient descent method (called Wirtinger flow in the complex case) is developed by Candès et al. in [12]. They show that the Wirtinger flow algorithm converges to the true signal up to a global phase factor with high probability provided $m \geq O(n \log n)$ Gaussian measurements. Lately, many variants of Wirtinger flow algorithms were developed, such as Thresholded WF[9], Truncated WF [13], Reshaped WF [45], and Accelerated WF [8] etc. In [20], Gao and Xu propose a Gauss-Newton (GN) algorithm to find a minimizer of (2). They proved that, for the real signal, the GN algorithm can converge to the global optimal solution quadratically with $O(n \log n)$ measurements starting from a good initial guess. Indeed, Gao and Xu also provide a initialization procedure which is much better than the initialization algorithm given in [12] numerically. Another approach to minimize (2) is called the trust region method which was studied in [36], and the geometric analysis of the landscape function $f(\mathbf{x}) = \sum_{i=1}^{m}(|\langle \mathbf{a}_i, \mathbf{x}\rangle|^2 - b_i)^2$ is also given. To recover sparse signals from the measurements (1), a standard approach is adding the ℓ_1 term $\lambda\|\mathbf{x}\|_1$ to (2) or using the proximal gradient method as discussed in [34].

1.2 Our Contribution

In this paper, we consider a broader class of phase retrieval problem which includes standard phase retrieval as a special case. We aim to recover $\mathbf{x} \in \mathbb{R}^n$ (or $\in \mathbb{C}^n$) from nonlinear measurements

$$b_i = f(\langle \mathbf{a}_i, \mathbf{x}\rangle), \quad i = 1, \ldots, m, \tag{3}$$

where $f : \mathbb{C} \to \mathbb{R}_+$ is a twice differentiable convex function and satisfies the following coercive condition:

$$f(x) \to \infty \text{ when } |x| \to \infty.$$

If we take $f(x) = |x|^2$, then it reduces to the standard phase retrieval. For another example, we can take $f(x) = |x|^4$ and etc. To guarantee the unique recovery of $\mathbf{x}$, it has been proved that the number of measurements satisfies $m \geq n + 1$ for the real case ($2n + 1$ for the complex case, respectively) (see Theorem 2.1 in [24]).

Recovering $\mathbf{x}$ from the nonlinear observation is also raised in many areas, such as neural networks (cf. [6, 32]).

To reconstruct $\mathbf{x}$ from (3), it is standard to formulate it as

$$\min_{\mathbf{x}\in\mathbb{R}^n \text{ or } \mathbb{C}^n} \sum_{i=1}^{m} (f(\langle \mathbf{a}_i, \mathbf{x}\rangle) - b_i)^2. \tag{4}$$

We approach it by using the standard technique for a difference of convex minimizing functions. Indeed, for the case $\mathbf{x} \in \mathbb{R}^n$ and $\mathbf{a}_i \in \mathbb{R}^n$, let $F(\mathbf{x}) = \sum_{i=1}^{m}(f(\langle \mathbf{a}_i, \mathbf{x}\rangle) - b_i)^2$ be the minimizing function. As it is not convex, we then write it as

$$F(\mathbf{x}) = \sum_{i=1}^{m} (f(\langle \mathbf{a}_i, \mathbf{x}\rangle) - b_i)^2 := F_1(\mathbf{x}) - F_2(\mathbf{x}), \tag{5}$$

where $F_1(\mathbf{x}) = \sum_{i=1}^{m} f^2(\langle \mathbf{a}_i, \mathbf{x}\rangle) + b_i^2$ and $F_2(\mathbf{x}) = \sum_{i=1}^{m} (2b_i f(\langle \mathbf{a}_i, \mathbf{x}\rangle))$. Note that f is a convex function with $f(x) \geq 0$ for all $x \in \mathbb{R}$. Then F_1 and F_2 are convex functions. The minimization (4) will be approximated by

$$\mathbf{x}^{(k+1)} := \arg\min_{\mathbf{x}} F_1(\mathbf{x}) - \nabla F_2(\mathbf{x}^{(k)})^\top (\mathbf{x} - \mathbf{x}^{(k)}) \tag{6}$$

for any given $\mathbf{x}^{(k)}$. We call this algorithm as DC based algorithm following from the ideas in [21], where the sparse solutions of under-determined linear system were studied. Although DC based algorithms have been studied for a long time (see e.g. [37, 41, 42] and the references therein), this is the first time to use a DC based algorithm to solve phase retrieval problem and achieve the best numerical performance compared to others methods from the knowledge of the authors.

The above minimization (6) is a convex problem with differentiable function for each k. We solve it by using the standard gradient descent method with Nestrov's acceleration (cf. [30]) or the Barzilai-Borwein (BB) method (cf. [5]). There are several nice properties of this DC based approach. We can show that

$$F(\mathbf{x}^{(k+1)}) \leq F(\mathbf{x}^{(k)}) - \ell \|\mathbf{x}^{(k+1)} - \mathbf{x}^{(k)}\|^2$$

for some constant $\ell > 0$. That is, $F(\mathbf{x}^{(k)}), k \geq 1$ is strictly decreasing sequence. Furthermore, we can prove the sequence $\{\mathbf{x}^{(k)}\}_{k=1}^{\infty}$ converges to a critical point $\mathbf{x}^*$. Using the Kurdyka-Łojasiewicz inequality, we can show $\|\mathbf{x}^{(k)} - \mathbf{x}^*\| \leq C\tau^k$ for $\tau \in (0, 1)$. If the function $F(\mathbf{x})$ has the property that any local minimizer $\mathbf{x}^\star$ is a global minimizer over a neighborhood $N(\mathbf{x}^\star)$ and the initial point $\mathbf{x}^{(1)}$ is within $N(\mathbf{x}^\star)$, then the DC based algorithm will converge to the global minimizer linearly. Actually, the function $F(\mathbf{x})$ indeed has such a property for real phase retrieval problem and such initial point can be obtained based on the initialization scheme discussed in [20]. Our numerical experiments show that the DC based algorithm can recover the true solutions when $m \approx 2n$ in the real case and $m \approx 3n$ in the complex case. See Sect. 6 for our numerical simulations.

Furthermore, we develop an ℓ_1-DC based algorithm to recover sparse signals. That is, starting from $\mathbf{x}^{(k)}$, we solve

$$\mathbf{x}^{(k+1)} := \arg\min \lambda \|\mathbf{x}\|_1 + F_1(\mathbf{x}) - \nabla F_2(\mathbf{x}^{(k)})^\top (\mathbf{x} - \mathbf{x}^{(k)}) \tag{7}$$

using a proximal gradient method, where $\lambda > 0$ is a parameter. The convergence of the ℓ_1-DC based algorithm can be established similar to the DC based algorithm. To accelerate the convergence of the ℓ_1-DC based algorithm, we use Attouch-Peypouquet's acceleration method (cf. [2]). To have a better initialization, we use the projection technique (cf. [17]). In addition, the hard thresholding operator is used to project each iteration onto the set of sparse vectors. With these updates, the algorithm works very well. The numerical experiments show that the modified ℓ_1-DC based algorithm can recover sparse signals as long as $m \approx n$ provided $s \ll n$, where s and n are the sparsity and dimension of signals.

In summary, to establish the convergence of the DC based algorithms, we follow the well-known approach based on the Kurdyka-Łojasiewicz inequality (cf. [1, 3, 41, 43]). Due to the nice properties of F_1 and F_2 in the setting of phase retrieval, we are able to specify the exponent θ in the Kurdyka-Łojasiewicz function and hence, the rate of convergence is precisely given, see Theorem 4. More precisely, for phase retrieval in the real case, we will show that F is strongly convex at the minimizer due to the positive definiteness of the Hessian in Appendix. In the complex case, the Hessian is no long positive definite but nonnegative definite near the minimizers. For sparse phase retrieval, we are no longer able to determine θ. Thus in order to establish the convergence rate in this settings, we break the neighborhood of the minimizers into two parts: within the ball or outside the ball of the given tolerance. Note that it is easy to check if an iterative point $\mathbf{x}^{(k+1)}$ is within the ball or not by checking the minimal value $F(\mathbf{x}^{(k+1)}) \le \epsilon$ or not. For the iterative points outside the ϵ-ball, we establish the convergence rate; for the iterative points within the ball, we no longer need to consider it, see Theorem 7.

1.3 Organization

The paper is organized as follows. Firstly, using tools of algebraic geometry, we give the existence of solutions in phase retrieval problem and give an estimate of how many distinct solutions in Sect. 2. In Sect. 3, we give the analysis of convergence for our DC based algorithms. Accelerated gradient descent methods including Nesterov's and Attouch-Peypouquet's accelerated techniques as well as the BB technique for inner iterations will be discussed in Sect. 4. Furthermore, we will study the ℓ_1-DC based algorithm for recover sparse signals and discuss the convergence in Sect. 5. Our numerical experiments are collected in Sect. 6, where we give the performance of our DC based algorithms and compare it with the Gauss-

Newton algorithm for general signals and sparse signals. Particularly, we show that the DC based algorithm is able to recover signals when $m \approx 2n$. In addition, our ℓ_1-DC based algorithm with the update techniques is able to recover sparse signals when $m \approx n$.

2 On Existence and Number of Phase Retrieval Solutions

In this section, we shall discuss the existence of solution for phase retrieval and give an estimate for the number of distinct solutions. To beginning, we first recall PhaseLift (cf. [10]) which shows the connection between phase retrieval and low-rank matrix recovery.

Let $X = \mathbf{x}\mathbf{x}^\top$ and $A_j = \mathbf{a}_j\mathbf{a}_j^\top$, $j = 1, \ldots, m$. Then the constrains in (1) can be rewritten as

$$b_j = \mathrm{tr}(A_j X), \quad j = 1, \ldots, m, \tag{8}$$

where $\mathrm{tr}(\cdot)$ is the trace operator.

Note that the scaling of $\mathbf{x}$ by a unimodular constant c would not change X. Indeed, $(c\mathbf{x})(c\mathbf{x})^\top = |c|^2\mathbf{x}\mathbf{x}^\top = \mathbf{x}\mathbf{x}^\top = X$. Conversely, given a positive semi-definite matrix X with rank 1, there exists a vector $\mathbf{x}$ such that $X = \mathbf{x}\mathbf{x}^\top$. So the phase retrieval problem can be recast as a matrix recovery problem (cf. [10]): Find $X \in \mathcal{M}_1$ satisfying linear measurements: $\mathrm{tr}(A_j X) = b_j$, $j = 1, \ldots, m$, where $\mathcal{M}_r = \{X \in \mathbb{R}^{n\times n} : \mathrm{rank}(X) = r\}$. In mathematical formulation, it aims to solve the following low rank matrix recovery problem:

$$\min \ \mathrm{rank}(X) \quad \text{s.t.} \quad \mathrm{tr}(A_j X) = b_j, \ j = 1, \ldots, m \quad \text{and} \quad X \succeq 0. \tag{9}$$

As we will show in Theorem 1, for given $b_j \geq 0$, $j = 1, \ldots, m$ there may not exist a matrix $X \in \mathcal{M}_r$ with $r < n$ satisfying the constraint conditions exactly unless b_j are exactly the measurement values from a matrix X. Thus to find the solution X, we reformulate the above problem as follows:

$$\min \sum_{i=1}^{m} |\mathrm{tr}(A_j X) - b_j|^2 \quad \text{s.t.} \quad X \in \mathcal{M}_r \quad \text{and} \quad X \succeq 0. \tag{10}$$

Since $\mathcal{M}_r$ is a closed set, the above least squares problem will have a bounded solution if the following coercive condition holds:

$$\sum_{i=1}^{m} |\mathrm{tr}(A_j X) - b_j|^2 \to \infty \text{ when } \|X\|_F \to \infty. \tag{11}$$

In the case that the above coercive condition does not hold, one has to use other conditions to ensure that the minimizer of (10) is bounded. For example, if there is a matrix X_0 which is orthogonal to A_j in the sense that $\mathrm{tr}(A_j X_0) = 0$ for all $j = 1, \ldots, m$, then the coercive condition will not hold as one can let $X = \ell X_0$ with $\ell \to \infty$.

We are now ready to discuss the existence of solution for phase retrieval problem. Let $\mathcal{M}_r$ be the set of $n \times n$ matrices with rank r and $\overline{\mathcal{M}_r}$ be the set of all matrices with rank no more than r. It is known that dimension of $\mathcal{M}_r$ is $2nr - r^2$ (cf. Proposition 12.2 in [22]). Since $\overline{\mathcal{M}_r}$ is the closure of $\mathcal{M}_r$ in the Zariski topology (cf. [44]) and hence the dimension of $\overline{\mathcal{M}_r}$ is also $2nr - r^2$. Furthermore, it is clear that $\overline{\mathcal{M}_r}$ is an algebraic variety. In fact, $\overline{\mathcal{M}_r}$ is an irreducible variety which is a standard result in algebraic geometry. To make the paper self-contain, we present a short proof.

Lemma 1 *$\overline{\mathcal{M}_r}$ is an irreducible variety.*

Proof Denote by $GL(n)$ the set of invertible $n \times n$ matrices. Consider the action of $GL(n) \times GL(n)$ on $M_n(R)$ given by: $(G_1, G_2) \cdot M \mapsto G_1 M G_2^{-1}$, for all $G_1, G_2 \in GL(n)$. Fix a rank r matrix M. Then the variety $\mathcal{M}_r$ is the orbit of M. Hence, we have a surjective morphism, a regular algebraic map described by polynomials, from $GL(n) \times GL(n)$ onto $\mathcal{M}_r$. Since $GL(n) \times GL(n)$ is an irreducible variety, so is $\mathcal{M}_r$. Hence, the closure $\overline{\mathcal{M}_{r_g}}$ of the irreducible set $\mathcal{M}_{r_g}$ is also irreducible (cf. Example I.1.4 in [23]). □

Define a map

$$\mathcal{A} : \mathcal{M}_1 \to \mathbb{R}^m$$

by projecting any matrix $X \in \mathcal{M}_1$ to $(b_1, \cdots, b_m)^\top \in \mathbb{R}^m$ in the sense that

$$\mathcal{A}(X) = (\mathrm{tr}(A_1 X), \cdots, \mathrm{tr}(A_m X))^\top.$$

Given the map $\mathcal{A}$, we define the range $\mathcal{R}_+ = \{\mathcal{A}(X) : X \in \overline{\mathcal{M}_1}, X \succeq 0\}$ and the range $\mathcal{R} = \{\mathcal{A}(X) : X \in \overline{\mathcal{M}_1}\}$. It is clear that the dimension of $\mathcal{R}_+$ is less than or equal to the dimension of $\mathcal{R}$. Since each entry $\mathrm{tr}(A_j X)$ of the map $\mathcal{A}$ is a linear polynomial about the entries of X, then the map $\mathcal{A}$ is a regular. We expect that $\dim(\mathcal{R})$ is less than or equal to the dimension of the $\mathcal{M}_1$ which is equal to $2n - 1$. If $m > 2n - 1$, then $\mathcal{R}$ is not able to occupy the whole space $\mathbb{R}^m$. The Lebesgue measure of the range $\mathcal{R}$ is zero and hence, a randomly choosing vector $\mathbf{b} = (b_1, \cdots, b_m)^\top \in \mathbb{R}^m$, e.g. $\mathbf{b} \in \mathbb{R}_+^m$ will not be in $\mathcal{R}$ with probability one and hence, not in $\mathcal{R}_+$. Thus, there will not be a solution $X \in \mathcal{M}_1$ such that $\mathcal{A}(X) = \mathbf{b}$.

Certainly, these intuitions should be made more precise. To this end, we first recall the following result from Theorem 1.25 in Sec 6.3 of [33].

Lemma 2 *Let $f : X \to Y$ be a regular map and X, Y are irreducible varieties with $\dim(X) = n$ and $\dim(Y) = m$. If f is surjective, then $m \le n$. Furthermore, it holds:*

(a) *for any* $y \in Y$ *and for any component* F *of the fiber* $f^{-1}(y)$, $\dim(F) \geq n - m$;
(b) *there exists a nonempty open subset* $U \subset Y$ *such that* $\dim(f^{-1}(y)) = n - m$ *for* $y \in U$.

We are now ready to prove

Theorem 1 *If one randomly chooses a vector* $\mathbf{b} = (b_1, \cdots, b_m)^\top \in \mathbb{R}_+^m$ *with* $m > 2n - 1$, *the probability of finding a solution* X *satisfying the minimization (9) is zero. In other words, for almost all vectors* $\mathbf{b} = (b_1, \cdots, b_m)^\top \in \mathbb{R}_+^m$ *the solution of (9) is a matrix with rank more than or equal to* 2.

Proof Let $X = \overline{\mathcal{M}_1}$ and $Y = \{\mathcal{A}(M), M \in \overline{\mathcal{M}_1}\}$. From Lemma 1, we know X is an irreducible variety. Since Y is the continuous image of the irreducible variety $\overline{\mathcal{M}_1}$, it is also an irreducible variety. Note that $\mathcal{A}$ is a regular map. By Lemma 2, we have $\dim(Y) \leq \dim(\overline{\mathcal{M}_1}) = 2n - 1 < m$. Thus, Y is a proper lower dimensional closed subset in $\mathbb{R}^m$. For almost all points in $\mathbb{R}^m$, they do not belong to Y. In other words, for almost all points $\mathbf{b} = (b_1, \cdots, b_n) \in \mathbb{R}^m$, there is no matrix $M \in \overline{\mathcal{M}_1}$ such that $\mathcal{A}(M) = \mathbf{b}$ and hence, no matrix $M \in \overline{\mathcal{M}_1}$ with $M \succeq 0$ such that $\mathcal{A}(M) = \mathbf{b}$. □

Note that the above discussion is still valid after replacing $\mathcal{M}_1$ by $\mathcal{M}_r$ with $r < n$. Under the assumption that $m > 2nr - r^2$, we can show that the generalized phase retrieval problem [40] does not have a solution for randomly chosen $\mathbf{b} = (b_1, \cdots, b_m) \in \mathbb{R}^m$ with probability one.

Next we define the subset $\chi_\mathbf{b} \subset \overline{\mathcal{M}_1}$ by

$$\chi_\mathbf{b} = \left\{ M \in \overline{\mathcal{M}_1} : \mathcal{A}(M) = \mathbf{b} \text{ and } \mathcal{A}^{-1}(\mathcal{A}(M)) \text{ is zero dimensional} \right\}.$$

Here, a set S is said to be zero dimensional if the dimension of real points is zero for $S \subset \mathbb{R}^n$ or the dimension of complex points is zero for $S \subset \mathbb{C}^n$. As we are working over the fields like $\mathbb{R}$ or $\mathbb{C}$, it means that if the fiber is zero-dimensional, then it has only finite number of real or complex points. For those $\mathbf{b} \in \mathbb{R}_+^m$ with $\chi_\mathbf{b} \neq \emptyset$, we are interested in the upper bound of the number of solutions which satisfy the minimization (9). To do so, we need more results from algebraic geometry.

Lemma 3 ([22] Proposition 11.12.) *Let* X *be a quasi-projective variety and* $\pi : X \to \mathbb{R}^m$ *be a regular map; let* Y *be closure of the image. For any* $p \in X$, *let* $X_p = \pi^{-1}\pi(p)) \subseteq X$ *be the fiber of* π *through* p, *and let* $\mu(p) = \dim_p(X_p)$ *be the local dimension of* X_p *at* p. *Then* $\mu(p)$ *is an upper-semi-continuous function of* p *in the Zariski topology on* X, *i.e. for any* m *the locus of points* $p \in X$ *such that* $\dim_p(X_p) > m$ *is closed in* X. *Moreover, if* $X_0 \subseteq X$ *is any irreducible component,* $Y_0 \subseteq Y$ *the closure of its image and* μ *the minimum value of* $\mu(p)$ *on* X_0, *then*

$$\dim(X_0) = \dim(Y_0) + \mu. \tag{12}$$

As we have shown in the proof of Theorem 1, we have $\dim(\mathcal{R}) \leq \dim(\overline{\mathcal{M}_r})$. Next, we give a more precise characterization about these dimensions.

Lemma 4 *Assume* $m > \dim(\overline{\mathcal{M}_r})$. *Then* $\dim(\overline{\mathcal{M}_r}) = \dim(\mathcal{R})$ *if and only if* $\chi_{\mathbf{b}} \neq \emptyset$ *for some* $\mathbf{b} \in \mathcal{R}$.

Proof Assume $\dim(\overline{\mathcal{M}_r}) = \dim(\mathcal{R})$. From Lemma 2, there exists a nonempty open subset $U \subset \mathcal{R}$ such that $\dim(\mathcal{A}^{-1}(\mathbf{b})) = 0$ for all $\mathbf{b} \in U$. This implies that $\chi_{\mathbf{b}}$ has finitely many points. Hence $\chi_{\mathbf{b}} \neq \emptyset$.

We now prove the converse. Assume $\chi_{\mathbf{b}} \neq \emptyset$. We will apply Lemma 3 by setting $X = \overline{\mathcal{M}_r}$, $Y = \mathcal{A}(\overline{\mathcal{M}_r})$ and $\pi = \mathcal{A}$. (To apply this lemma, please note that it does not matter whether we take the closure in $\mathbb{P}^m$ or in $\mathbb{C}^m$ since $\mathbb{C}^m$ is an open set in $\mathbb{P}^m$ and the Zariski topology of the affine space $\mathbb{C}^m$ is induced from the Zariski topology of $\mathbb{P}^m$. $\overline{\mathcal{M}_r}$ is an affine variety. In particular, it is a quasi-projective variety.)

By our assumption, $\chi_{\mathbf{b}}$ is not empty. It follows that there is a point $p \in Y$ such that $\pi^{-1}(p)$ is zero dimensional. Since zero is the least dimension possible, we have $\mu = 0$. Hence, using (12) above, we have $\dim(\overline{\mathcal{M}_1}) = \dim(\mathcal{R})$. □

Finally, we need the following definition.

Definition 1 The *degree* of an affine or projective variety with dimension k is the number of intersection points with k hyperplanes in general position.

It has been shown [19, Example 14.4.11] that the degree of the algebraic variety $\overline{\mathcal{M}_r}$ is

$$\prod_{i=0}^{n-r-1} \frac{\binom{n+i}{r}}{\binom{r+i}{r}}.$$

In particular, the degree of $\mathcal{M}_1$ is

$$\prod_{i=0}^{n-2} \frac{n+i}{1+i}. \tag{13}$$

We are now ready to prove another main result in this section.

Theorem 2 *Given a vector* $\mathbf{b} \in \mathbb{R}_+^m$ *lies in the range* $\mathcal{R}_+$. *Assume that* $\chi_{\mathbf{b}} \neq \emptyset$. *Then the number of distinct solutions in* $\chi_{\mathbf{b}}$ *is less than or equals to* $\prod_{i=0}^{n-2} \frac{n+i}{1+i}$.

Proof For any fixed $\mathbf{b}$, the matrices M which satisfy $\mathcal{A}(M) = \mathbf{b}$ and $\text{rank}(M) = 1$ are exactly the intersection points of the variety $\overline{\mathcal{M}_1}$ with m hyperplanes, namely the hyperplanes defined by equations $\langle A_i, M \rangle = b_i, i = 1, \cdots, m$. Since $m > \dim(\overline{\mathcal{M}_r}) = 2n - 1$, the number of intersection points would be less than degree of $\overline{\mathcal{M}_1}$ generically. So, the number of positive semidefinite matrices M which satisfy $\mathcal{A}(M) = \mathbf{b}$ and $\text{rank}(M) = 1$ would be no more than the degree of $\overline{\mathcal{M}_1}$. Finally, using the exact formula for the degree from (13), the result follows. □

3 A DC Based Algorithm for Phase Retrieval

For convenience, we simply discuss the case where $\mathbf{x}$ and $\mathbf{a}_j$, $j = 1, \ldots, m$ are real. The complex case can be treated in the same way from the algorithmic perspective. Indeed, when $\mathbf{x} \in \mathbb{C}^n$ and $\mathbf{a}_j \in \mathbb{C}^n$, $j = 1, \ldots, m$, we only need to write $\mathbf{x} = \mathbf{x}_R + \sqrt{-1}\mathbf{x}_I$ and similar for $\mathbf{a}_j$. Letting $\mathbf{u} = [\mathbf{x}_R^\top\ \mathbf{x}_I^\top]^\top \in \mathbb{R}^{2n}$, we view $F_1(\mathbf{x})$ as the function $G_1(\mathbf{u}) = F_1(\mathbf{x}_R + \sqrt{-1}\mathbf{x}_I)$. Then $G_1(\mathbf{u})$ is a convex function of real variable $\mathbf{u}$. Similarly, $G_2(\mathbf{u}) = F_2(\mathbf{x}_R + \sqrt{-1}\mathbf{x}_I)$ is a convex function of $\mathbf{u}$. The difference is that we can no longer recover $\mathbf{u}$ up to a sign but up to an orthogonal matrix. That is, for any orthogonal matrix $O \in \mathbb{R}^{2n\times 2n}$, the vector $O\mathbf{u}$ is also the true solution.

Recall that we aim to recover $\mathbf{x}$ by minimizing $F(\mathbf{x}) = F_1(\mathbf{x}) - F_2(\mathbf{x})$ in (4). It is easy to see that the minimization can happen in a bounded region $\mathcal{R}$ due to the coercive condition $f(x) \to \infty$ when $x \to \infty$. Our DC based method is given as follows. Start from any iterative solution $\mathbf{x}^{(k)}$, we solve the following convex minimization problem:

$$\mathbf{x}^{(k+1)} = \arg\min_{\mathbf{x}\in\mathbb{R}^n} F_1(\mathbf{x}) - \nabla F_2(\mathbf{x}^{(k)})^\top(\mathbf{x} - \mathbf{x}^{(k)}) \tag{14}$$

for $k \geq 1$, where $\mathbf{x}^{(1)}$ is an initial guess. The choice of $\mathbf{x}^{(1)}$ will be discussed later. Without loss of generality, we always assume $\mathbf{x}^{(1)}$ is located in a bounded region $\mathcal{R}$.

Our goal in this section is to show the sequence $\{\mathbf{x}^{(k)}\}_{k=1}^{\infty}$ converges to a critical point. Later, we will discuss how to find a global minimization by choosing a good initial guess $\mathbf{x}^{(1)}$ appropriately. For a fixed $\mathbf{x}^{(k)}$, there are many standard iterative methods to solve the convex minimization problem (14), such as the gradient descent method with various acceleration techniques. After getting $\mathbf{x}^{(k+1)}$ from solving (14), we update $\mathbf{x}^{(k)}$ with $\mathbf{x}^{(k+1)}$ and then solve (14) again. Hence, there are two iterative procedures. The iterative procedure for solving (14) is an inner iteration which will be discussed in the next section. In this section, we mainly discuss the outer iteration assuming $\mathbf{x}^{(k+1)}$ has been found.

We will state the following assumptions on functions F_1 and F_2:

1. The gradient function ∇F_1 has Lipschitz constant L_1 in bounded region $\mathcal{R}$. That is, $\|\nabla F_1(\mathbf{x}) - \nabla F_1(\mathbf{y})\|_2 \leq L_1\|\mathbf{x} - \mathbf{y}\|_2$ for all vectors $\mathbf{x}, \mathbf{y} \in \mathcal{R}$.
2. F_2 is a strongly convex function with parameter ℓ in $\mathcal{R}$. That is, $F_2(\mathbf{y}) \geq F_2(\mathbf{x}) + \nabla F_2(\mathbf{x})^\top(\mathbf{y} - \mathbf{x}) + \frac{\ell}{2}\|\mathbf{y} - \mathbf{x}\|^2$ for all vectors $\mathbf{x}, \mathbf{y} \in \mathcal{R}$.

Observe that the function $F_2 = 2\sum_{i=1}^m b_i f(\mathbf{a}_i^\top\mathbf{x})$. Through a simple calculation, the Hessian matrix of function F_2 is

$$H_{F_2} = \sum_{i=1}^m 2b_i f''(\mathbf{a}_i^\top\mathbf{x})\mathbf{a}_i\mathbf{a}_i^\top,$$

where $f''(x) \ge 0$ since the convexity of f. Thus the parameter ℓ of strong convexity corresponds to the minimal eigenvalue of H_{F_2}. In fact, we can prove that in the case of the standard phase retrieval problem with $f(x) = x^2$, the function F_2 is strongly convex under a standard assumption that the measurement vectors are generic and measurement values are essential, see Theorem 8 in the Appendix. Similar result also holds for the general phase retrieval problem with $f(x) = |x|^4$.

We first start with a standard result for our DC based algorithm:

Theorem 3 *Assume F_2 is a strongly convex function with parameter ℓ. Starting from any initial guess $\mathbf{x}^{(1)}$, let $\mathbf{x}^{(k+1)}$ be the solution of (14) for all $k \ge 1$. Then*

$$F(\mathbf{x}^{(k+1)}) \le F(\mathbf{x}^{(k)}) - \frac{\ell}{2}\|\mathbf{x}^{(k+1)} - \mathbf{x}^{(k)}\|^2, \quad \forall k \ge 1. \tag{15}$$

Furthermore, it holds $\nabla F_1(\mathbf{x}^{(k+1)}) - \nabla F_2(\mathbf{x}^{(k)}) = 0$.

Proof By the strong convexity of F_2, we have

$$F_2(\mathbf{x}^{(k+1)}) \ge F_2(\mathbf{x}^{(k)}) + \nabla F_2(\mathbf{x}^{(k)})^\top (\mathbf{x}^{(k+1)} - \mathbf{x}^{(k)}) + \frac{\ell}{2}\|\mathbf{x}^{(k+1)} - \mathbf{x}^{(k)}\|^2.$$

Recall that $F(\mathbf{x}) = F_1(\mathbf{x}) - F_2(\mathbf{x})$. Combining with (14), we obtain that

$$\begin{aligned} F(\mathbf{x}^{(k+1)}) &= F_1(\mathbf{x}^{(k+1)}) - F_2(\mathbf{x}^{(k+1)}) \\ &\le F_1(\mathbf{x}^{(k+1)}) - \nabla F_2(\mathbf{x}^{(k)})^\top (\mathbf{x}^{(k+1)} - \mathbf{x}^{(k)}) - F_2(\mathbf{x}^{(k)}) - \frac{\ell}{2}\|\mathbf{x}^{(k+1)} - \mathbf{x}^{(k)}\|^2 \\ &\le F_1(\mathbf{x}^{(k)}) - F_2(\mathbf{x}^{(k)}) - \frac{\ell}{2}\|\mathbf{x}^{(k+1)} - \mathbf{x}^{(k)}\|^2 = F(\mathbf{x}^{(k)}) - \frac{\ell}{2}\|\mathbf{x}^{(k+1)} - \mathbf{x}^{(k)}\|^2. \end{aligned}$$

Since $\mathbf{x}^{(k+1)}$ is the minima of (14), the property $\nabla F_1(\mathbf{x}^{(k+1)}) - \nabla F_2(\mathbf{x}^{(k)}) = 0$ follows from the first order optimality condition directly. □

Next, we use the Kurdyka-Łojasiewicz (KL) inequality to establish the convergence rate of $\mathbf{x}^{(k)}$. The applications which use the KL inequality to solve various minimization problems can be found in [1, 3, 41, 43]. The following is our major theorem in this section.

Theorem 4 *Suppose that $F(\mathbf{x}) = F_1(\mathbf{x}) - F_2(\mathbf{x})$ is a real analytic function. Assume the gradient function ∇F_1 has Lipschitz constant $L_1 > 0$ and F_2 is a strongly convex function with parameter $\ell > 0$ in bounded region $\mathcal{R}$. Starting from any initial guess $\mathbf{x}^{(1)}$, let $\mathbf{x}^{(k+1)}$ be the solution in (14) for all $k \ge 1$. Then $\mathbf{x}^{(k)}, k \ge 1$ converges to a critical point of F. Furthermore, if we let $\mathbf{x}^*$ be the limit, then*

$$\|\mathbf{x}^{(k+1)} - \mathbf{x}^*\| \le C\tau^k \tag{16}$$

for a positive constant C and $\tau \in (0, 1)$.

To prove the theorem, we need the following KL inequality which is central to the global convergence analysis.

Definition 2 (Łojasiewicz [27]) We say a function $f(\mathbf{x})$ satisfies the Kurdyka-Lojasiewicz (KL) property at point $\bar{\mathbf{x}}$ if there exists $\theta \in [0, 1)$ such that

$$|f(\mathbf{x}) - f(\bar{\mathbf{x}})|^{\theta} \leq C\text{dist}(0, \partial f(\mathbf{x}))$$

in a neighborhood $B(\bar{\mathbf{x}}, \delta)$ for some $\delta > 0$, where $C > 0$ is a constant independent of $\mathbf{x}$. In other words, there exists a function $\varphi(s) = cs^{1-\theta}$ with $\theta \in [0, 1)$ such that it holds

$$\varphi'(|f(\mathbf{x}) - f(\bar{\mathbf{x}})|)\text{dist}(0, \partial f(\mathbf{x})) \geq 1 \tag{17}$$

for any $\mathbf{x} \in B(\bar{\mathbf{x}}, \delta)$ with $f(\mathbf{x}) \neq f(\bar{\mathbf{x}})$.

This property is introduced by Lojasiewicz on the real analytic functions, which the inequality (17) holds in a critical point with $\theta \in [1/2, 1)$. Later, many extensions of the above inequality are proposed. Typically, the extension to the setting of o-minimal structure in [26] is a general version. Recently, the KL inequality is extended to nonsmooth subanalytic functions. For our proof in the setting of phase retrieval, we need to specify $\theta = 1/2$. Indeed, we shall include an elementary proof to justify that our choice of $\theta = 1/2$ can be achieved. To show this, we need the following proposition.

Proposition 1 *Suppose that $f : \mathbb{R}^n \mapsto \mathbb{R}$ is a continuously twice differentiable function whose Hessian $H_f(\mathbf{x})$ is invertible at a critical point $\mathbf{x}^*$ of f. Then there exists a positive constant C, an exponent $\theta = 1/2$ and a positive number δ such that*

$$|f(\mathbf{x}) - f(\mathbf{x}^*)|^{\theta} \leq C\|\nabla f(\mathbf{x})\|, \quad \forall \mathbf{x} \in B(\mathbf{x}^*, \delta), \tag{18}$$

where $B(\mathbf{x}^*, \delta)$ is a ball at $\mathbf{x}^*$ with radius δ.

Proof Since f is continuous and twice differentiable, using Taylor formula and noting $f(\mathbf{x}^*) = 0$, we have

$$|f(\mathbf{x}) - f(\mathbf{x}^*)| \leq c_1\|\mathbf{x} - \mathbf{x}^*\|^2, \quad \forall \mathbf{x} \in B(\mathbf{x}^*, r)$$

for some $r > 0$. On the other hand, due to the fact the Hessian is invertible, we have

$$\|\nabla f(\mathbf{x})\| = \|\nabla f(\mathbf{x}) - \nabla f(\mathbf{x}^*)\| \geq c_2\|\mathbf{x} - \mathbf{x}^*\|.$$

Combining the above two estimates, we obtain (18) with $\theta = 1/2$ and $C = \sqrt{c_1}/c_2$. □

The importance of the Łajosiewicz inequality is to establish the inequality (18) under the case where f may not have an invertible Hessian at the critical point $\mathbf{x}^*$.

However, for our phase retrieval problem, the Hessian matrix is restricted strong convex at the global minimizer (cf. [36]). In the real case, we can even show that the Hessian is positive definite at a global minimizer, see Theorem 9 in the Appendix. We are now ready to establish Theorem 4.

Proof of Theorem 4 As we have shown in Theorem 3,

$$\frac{\ell}{2}\|\mathbf{x}^{(k+1)} - \mathbf{x}^{(k)}\|^2 \leq F(\mathbf{x}^{(k)}) - F(\mathbf{x}^{(k+1)}). \tag{19}$$

That is, $F(\mathbf{x}^{(k)}), k \geq 1$ is strictly decreasing sequence. Without loss of generality, we assume

$$\mathcal{R} := \{\mathbf{x} \in \mathbb{R}^n, F(\mathbf{x}) \leq F(\mathbf{x}^{(1)})\}.$$

Then the sequence $\{\mathbf{x}^{(k)}\}_{k=1}^{\infty} \subset \mathcal{R}$ is a bounded sequence. It means that there exists a cluster point $\mathbf{x}^*$ and a subsequence $\mathbf{x}^{(k_i)}$ such that $\mathbf{x}^{(k_i)} \to \mathbf{x}^*$. Note that $\{F(\mathbf{x}^{(k)})\}_{k=1}^{\infty}$ is a bounded monotonic descending sequence. Then $F(\mathbf{x}^{(k)}) \to F(\mathbf{x}^*)$ for all $k \geq 1$. We claim that there exists a positive constant C_1 such that

$$C_1\|\mathbf{x}^{(k+1)} - \mathbf{x}^{(k)}\| \leq \sqrt{F(\mathbf{x}^{(k)}) - F(\mathbf{x}^*)} - \sqrt{F(\mathbf{x}^{(k+1)}) - F(\mathbf{x}^*)} \tag{20}$$

holds for all $k \geq k_0$ where k_0 is large enough. To establish this claim, we shall use the Proposition 1. Firstly, we prove that the condition $\nabla F(\mathbf{x}^*) = 0$ holds. Indeed, from Theorem 3 we have

$$\begin{aligned}\|\nabla F(\mathbf{x}^{(k)})\| = \|\nabla F_1(\mathbf{x}^{(k)}) - \nabla F_2(\mathbf{x}^{(k)})\| &= \|\nabla F_1(\mathbf{x}^{(k)}) - \nabla F_1(\mathbf{x}^{(k+1)})\| \\ &\leq L_1\|\mathbf{x}^{(k)} - \mathbf{x}^{(k+1)}\|.\end{aligned}$$

Combining with (19), it gives that $\|\nabla F(\mathbf{x}^{(k_i)})\| \to 0$. By the continuity of gradient function, we have $\|\nabla F(\mathbf{x}^*)\| = 0$ since $\mathbf{x}^{(k_i)} \to \mathbf{x}^*$.

Next, Theorem 9 shows that F has a positive definite Hessian near $\mathbf{x}^*$. Thus the Kurdyka-Lojasiewicz inequality holds for $\theta = 1/2$ by Proposition 1. Consider the function $g(t) = \sqrt{t}$ which is concave over [0, 1] and hence, $g(t) - g(s) \geq g'(t)(t - s)$. From the Kurdyka-Lojasiewicz inequality, there exists a positive constant $c_0 > 0$ and $\delta > 0$ such that

$$\|g'(F(\mathbf{x}) - F(\mathbf{x}^*))\nabla F(\mathbf{x})\| \geq c_0 > 0 \tag{21}$$

for all $\mathbf{x}$ in the neighborhood $B(\mathbf{x}^*, \delta)$ of $\mathbf{x}^*$. Since $F(\mathbf{x}^{(k)}) - F(\mathbf{x}^*) \to 0$ as $k \to \infty$, there is an integer k_0 such that for all $k \geq k_0$ it holds

$$\max\left(\sqrt{2/\ell}, L_1/(\ell c_0)\right) \cdot \sqrt{F(\mathbf{x}^{(k)}) - F(\mathbf{x}^*)} \leq \delta/2. \tag{22}$$

Recall that $\mathbf{x}^{(k_i)} \to \mathbf{x}^*$ as $k_i \to \infty$. Without loss of generality, we may assume that $k_0 = 1$ and $\mathbf{x}^{(1)} \in B(\mathbf{x}^*, \delta/2)$. We next show that $\mathbf{x}^{(k)} \in B(\mathbf{x}^*, \delta)$ for all $k \geq 1$ and prove it by induction. By (22) we have

$$\|\mathbf{x}^{(2)}-\mathbf{x}^*\| \leq \|\mathbf{x}^{(2)}-\mathbf{x}^{(1)}\|+\|\mathbf{x}^{(1)}-\mathbf{x}^*\| \leq \sqrt{2(F(\mathbf{x}^{(1)}) - F(\mathbf{x}^*)/\ell}+\|\mathbf{x}^{(1)}-\mathbf{x}^*\| \leq \delta.$$

Assume that $\mathbf{x}^{(k)} \in B(\mathbf{x}^*, \delta)$ for $k \leq K$. Multiplying $g'(F(\mathbf{x}^{(k)}) - F(\mathbf{x}^*))$ on both sides of (19), we have

$$\begin{aligned}
&\frac{\ell}{2}\|\mathbf{x}^{(k+1)} - \mathbf{x}^{(k)}\|^2 g'(F(\mathbf{x}^{(k)}) - F(\mathbf{x}^*)) \\
&\leq g'(F(\mathbf{x}^{(k)}) - F(\mathbf{x}^*))\left(F(\mathbf{x}^{(k)}) - F(\mathbf{x}^{(k+1)})\right) \\
&\leq \sqrt{F(\mathbf{x}^{(k)}) - F(\mathbf{x}^*)} - \sqrt{F(\mathbf{x}^{(k+1)}) - F(\mathbf{x}^*)},
\end{aligned} \tag{23}$$

where the last inequality follows from the concavity of g. On the other hand, combining the KL inequality (21) with Theorem 3, we have

$$\begin{aligned}
|g'(F(\mathbf{x}^{(k)}) - F(\mathbf{x}^*))| \geq \frac{c_0}{\|\nabla F(\mathbf{x}^{(k)})\|} &= \frac{c_0}{\|\nabla F_1(\mathbf{x}^{(k)}) - \nabla F_2(\mathbf{x}^{(k)})\|} \\
&= \frac{c_0}{\|\nabla F_1(\mathbf{x}^{(k)}) - \nabla F_1(\mathbf{x}^{(k+1)})\|} \\
&\geq \frac{c_0}{L_1\|\mathbf{x}^{(k+1)} - \mathbf{x}^{(k)}\|}.
\end{aligned}$$

Putting it in (23), we obtain

$$\sqrt{F(\mathbf{x}^{(k)}) - F(\mathbf{x}^*)} - \sqrt{F(\mathbf{x}^{(k+1)}) - F(\mathbf{x}^*)} \geq \frac{\ell c_0}{2L_1}\|\mathbf{x}^{(k)} - \mathbf{x}^{(k+1)}\| \tag{24}$$

for all $2 \leq k \leq K$. Taking the sum, it follows that

$$\frac{2L_1}{\ell c_0}\sqrt{F(\mathbf{x}^{(1)}) - F(\mathbf{x}^*)} \geq \sum_{j=1}^{K} \|\mathbf{x}^{(j+1)} - \mathbf{x}^{(j)}\|.$$

Finally, observe that

$$\begin{aligned}
\|\mathbf{x}^{(K+1)} - \mathbf{x}^*\| &\leq \|\mathbf{x}^{(K+1)} - \mathbf{x}^{(1)}\| + \|\mathbf{x}^{(1)} - \mathbf{x}^*\| \\
&\leq \sum_{j=1}^{K} \|\mathbf{x}^{(j+1)} - \mathbf{x}^{(j)}\| + \|\mathbf{x}^{(1)} - \mathbf{x}^*\|
\end{aligned}$$

$$\leq \frac{2L_1}{\ell c_0}\sqrt{F(\mathbf{x}^{(1)}) - F(\mathbf{x}^*)} + \|\mathbf{x}^{(1)} - \mathbf{x}^*\| \leq \delta,$$

where the last inequality follows from (22). Thus $\mathbf{x}^{(K+1)} \in B(\mathbf{x}^*, \delta)$, which means that all $\mathbf{x}^{(k)}$ are in $B(\mathbf{x}^*, \delta)$ and inequality (24) holds for all k. Hence, we arrive at the claim (20) with $C_1 = \ell c_0/(2L_1)$. By summing the inequality (20) above, it follows

$$\sum_{k\geq 1} \|\mathbf{x}^{(k+1)} - \mathbf{x}^{(k)}\| \leq \frac{1}{C_1}\sqrt{F(\mathbf{x}^{(1)}) - F(\mathbf{x}^*)}.$$

That is, $\mathbf{x}^{(k)}$ is a Cauchy sequence and hence, it is convergent with $\mathbf{x}^{(k)} \to \mathbf{x}^\star$. Note that $\nabla F(\mathbf{x}^\star) = 0$, which implies $\mathbf{x}^{(k)}$ converges to a critical point of F.

We next turn to prove the second part. Let $S_k = \sum_{i=k}^{\infty} \|\mathbf{x}^{(i+1)} - \mathbf{x}^{(i)}\|$. It follows from (24) that

$$C_1 S_k = \sum_{i=k}^{\infty} C_1 \|\mathbf{x}^{(i+1)} - \mathbf{x}^{(i)}\|$$

$$\leq \sum_{i=k}^{\infty} (\sqrt{F(\mathbf{x}^{(i)}) - F(\mathbf{x}^*)} - \sqrt{F(\mathbf{x}^{(i+1)}) - F(\mathbf{x}^*)}) \leq \sqrt{F(\mathbf{x}^{(k)}) - F(\mathbf{x}^*)}.$$

Recall from (24) that

$$\sqrt{F(\mathbf{x}^{(k)}) - F(\mathbf{x}^*)} \leq \frac{L_1}{2c_0}\|\mathbf{x}^{(k)} - \mathbf{x}^{(k+1)}\| = C_2(S_k - S_{k+1})$$

where $C_2 = L_1/(2c_0)$. Combining the two above inequality, we obtain

$$S_{k+1} \leq \frac{C_2 - C_1}{C_2} S_k \leq \cdots \leq \tau^k S_0$$

for $\tau = (C_2 - C_1)/(C_2)$. Since $\|\mathbf{x}^{(k)} - \mathbf{x}^*\| \leq S_k$, we complete the proof. □

Remark 1 We should point out that the assumptions on F, F_1 and F_2 in Theorem 4 are easy to satisfy. For example, in standard phase retrieval all these assumptions are satisfied, especially when the region $\mathcal{R}$ is sufficiently small and near the global minimization by a technical initialization. More details can be found in Theorems 8 and 9.

In summary, two obvious consequences are:

1. For any given initial point $\mathbf{x}^{(1)}$, let $D = F(\mathbf{x}^{(1)}) - F(\mathbf{x}^\star) > 0$, where $\mathbf{x}^\star$ is one of the global minimizer of (5). Then

$$F(\mathbf{x}^{(k)}) - F(\mathbf{x}^\star) \le D - \frac{\ell}{2}\sum_{j=1}^{k-1} \|\mathbf{x}^{(j+1)} - \mathbf{x}^{(j)}\|^2.$$

That is, $\mathbf{x}^{(k)}$ is closer to one of global minimizer than the initial guess point.

2. As our approach can find a critical point, if a global minimizer $\mathbf{x}^\star$ is a local minimizer over a neighborhood $N(\mathbf{x}^\star)$ and an initial vector $\mathbf{x}^{(1)}$ is in $N(\mathbf{x}^\star)$, then our approach finds $\mathbf{x}^\star$.

Example 1 In this example, we consider the standard phase retrieval problem where $f(x) = |x|^2$. Assume the measurements are Gaussian random vectors. It has been shown that one can use the initialization from [13, 20] to find an excellent initial vector. More specifically, to recover a vector $x \in \mathbb{R}^n$ (or $x \in \mathbb{C}^n$), if the number of measurements $m \ge O(n)$, then with high probability we have

$$\|\mathbf{x}^{(1)} - \mathbf{x}^*\|_2 \le \delta\|\mathbf{x}^*\|_2,$$

where $\mathbf{x}^*$ is a global minimizer and δ is a sufficient positive constant. Furthermore, in a small neighborhood $N(\mathbf{x}^\star, \delta) := \{\mathbf{x} : \|\mathbf{x} - \mathbf{x}^*\|_2 \le \delta\|\mathbf{x}^*\|_2\}$, the minimizing function $F(x)$ is strongly convex [36]. Thus, our algorithm can converge to the global minimizer by using a good initialization.

4 Computation of the Inner Minimization (14)

We now discuss how to compute the minimization (14). For convenience, we rewrite the minimization in the following form

$$\min_{\mathbf{x}\in\mathbb{R}^n} G(\mathbf{x}) \tag{25}$$

for a differentiable convex function $G(\mathbf{x}) := F_1(\mathbf{x}) - \langle \nabla F_2(\mathbf{x}^{(k)}), \mathbf{x} - \mathbf{x}^{(k)}\rangle$. The first approach is to use the gradient descent method:

$$\mathbf{z}^{(j+1)} = \mathbf{z}^{(j)} - h\nabla G(\mathbf{z}^{(j)}) \tag{26}$$

for $j \in \mathbb{N}$ with $\mathbf{z}^{(1)} = \mathbf{x}^{(k)}$, where $h > 0$ is the step size. It is well-known that if we choose $h \approx 1/(2L)$ where L is the Lipschitz constant of $G(\mathbf{x})$, the gradient descent method (26) will have a linear convergence. It has also been shown that if we choose $h = \nu/L$ with Lipschitz constant L and the strong convex parameter ν, the Nesterov acceleration technique will speed up the convergence rate. Some results are given as follows.

Lemma 5 (Nesterov's Acceleration [30]) *Let $f : \mathbb{R}^n \to \mathbb{R}$ be a ν-strong convex function and the gradient function has L-Lipschitz constant. Start from an arbitrary initial point $\mathbf{u}_1 = \mathbf{z}_1$, the following Nesterov's accelerated gradient descent*

$$\mathbf{z}^{(j+1)} := \mathbf{u}^{(j)} - \frac{\nu}{L}\nabla f(\mathbf{u}^{(j)}),$$
$$\mathbf{u}^{(j+1)} = \mathbf{z}^{(j+1)} - q(\mathbf{z}^{(j+1)} - \mathbf{z}^{(j)}) \tag{27}$$

satisfies

$$f(\mathbf{z}^{(j+1)}) - f(\mathbf{z}^*) \leq \frac{\nu + L}{2}\|\mathbf{z}^{(1)} - \mathbf{z}^*\|^2 \exp(-\frac{j}{\sqrt{L/\nu}}), \tag{28}$$

where $\mathbf{z}^*$ *is the optimal solution and* $q = (\sqrt{L/\nu} - 1)/(\sqrt{L/\nu} + 1)$ *is a constant.*

The role of Nesterov's acceleration is to reduce the number of iterations in (26) significantly. That is, for any tolerance ϵ, we need $O(1/\epsilon)$ number of iterations for the gradient descent method due to the linear convergence, but $O(1/\sqrt{\epsilon})$ number of iterations if Nesterov's acceleration (27) is used.

Since G is twice differentiable, we can certainly use the Newton method to solve (14) because of its quadratic convergence. However, we will not pursue it here due to the fact that when the dimension of $\mathbf{z}$ is large, the Newton method will be extremely slow. Instead, we apply the Barzilai-Borwein (BB) method to choose a good h, which is an excellent approach for the large scale minimization problem (cf. [5]). The iteration of the BB method can be described as

$$\mathbf{z}^{(j+1)} = \mathbf{z}^{(j)} - \beta_j^{-1}\nabla G(\mathbf{z}^{(j)}), \tag{29}$$

where the step size

$$\beta_j = (\mathbf{z}^{(j)} - \mathbf{z}^{(j-1)})^\top(\nabla G(\mathbf{z}^{(j)}) - \nabla G(\mathbf{z}^{(j-1)}))/\|\mathbf{z}^{(j)} - \mathbf{z}^{(j-1)}\|^2. \tag{30}$$

Algorithm 1 The BB Algorithm for the Inner Minimization

Let $\mathbf{u}^{(1)} = \mathbf{z}^{(1)}$ be an initial guess.
For $j \geq 1$, we solve the minimization of (25) by computing β_j according to (30).
Update

$$\mathbf{z}^{(j+1)} := \mathbf{u}^{(j)} - \beta_j^{-1}\nabla G(\mathbf{u}^{(j)})$$
$$\mathbf{u}^{(j+1)} = \mathbf{z}^{(j+1)} - q(\mathbf{z}^{(j+1)} - \mathbf{z}^{(j)}) \tag{31}$$

until a maximum number T of iteration is achieved.
return $\mathbf{u}^T$

Our computation of inner minimization is described in Algorithm 1, which is a combination of the BB technique with Nesterov's acceleration technique. The intuition behind it based on Lemma 5. Since BB method has a good performance in numerical experiment, we can hope our Algorithm 1 has better performance.

There are several modified versions of the BB method available with their convergence analysis in the literature, see, e.g. [15, 46] and the references therein. Although a large number of numerical experiments show that the BB method has excellent performance, however, the convergence rate is still not established yet for general minimizing function F. We next give some necessary and sufficient conditions to show why the Algorithm 1 has a better convergence rate. To this end, we say a algorithm is convergent superlinearly if

$$\sigma_k = \frac{\|\mathbf{u}^{(k+1)} - \mathbf{u}^*\|}{\|\mathbf{u}^{(k)} - \mathbf{u}^*\|} \to 0, \text{ when } k \to \infty.$$

To analyze the convergence of the BB method in our setting, let $\mathbf{s}_{k+1} = \mathbf{u}^{(k+1)} - \mathbf{u}^{(k)}$ and $\mathbf{y}_{k+1} = \nabla G(\mathbf{u}^{(k+1)}) - \nabla G(\mathbf{u}^{(k)})$.

Lemma 6 *Suppose that the function $G(\mathbf{x})$ in (25) is α-strongly convex and the gradient has Lipschitz constant L in a domain D. Let $\mathbf{u}^* \in D$. Assume the sequence $\{\mathbf{u}^{(k)}, k \geq 1\}$ is obtained from the BB method and remains in D. Then $\{\mathbf{u}^{(k)}, k \geq 1\}$ converges super-linearly to $\mathbf{u}^*$ if and only if $(\beta_k - H_G(\mathbf{u}^*))\mathbf{s}_{k+1} = o(\|\mathbf{s}_{k+1}\|)$.*

Proof From BB update rule (29), we have

$$\begin{aligned}\left(\beta_k - H_G(\mathbf{x}^*)\right)\mathbf{s}_{k+1} &= -\nabla G(\mathbf{u}^{(k)}) - H_G(\mathbf{u}^*)\mathbf{s}_{k+1}\\ &= \nabla G(\mathbf{u}^{(k+1)}) - \nabla G(\mathbf{u}^{(k)}) - H_G(\mathbf{u}^*)\mathbf{s}_{k+1} - \nabla G(\mathbf{u}^{(k+1)}).\end{aligned} \tag{32}$$

Since the Hessian matrix $H_G(\mathbf{u})$ is continuous at $\mathbf{u}^*$ and all $\mathbf{u}^{(k)} \in D$, then we have

$$\nabla G(\mathbf{u}^{(k+1)}) - \nabla G(\mathbf{u}^{(k)}) - H_G(\mathbf{u}^*)\mathbf{s}_{k+1} \to 0, \quad k \to \infty.$$

By the assumption that $(\beta_k - H_G(\mathbf{u}^*))\mathbf{s}_{k+1} = o(\|\mathbf{s}_{k+1}\|)$, it implies that

$$\lim_{k\to\infty} \frac{||\nabla G(\mathbf{u}^{(k+1)})||}{||\mathbf{s}_{k+1}||} = 0. \tag{33}$$

Note that

$$\|\nabla G(\mathbf{u}^{(k+1)}) - G(\mathbf{u}^{(k)})\| \leq L\|\mathbf{u}^{(k+1)} - \mathbf{u}^{(k)}\|$$

and

$$||\nabla G(\mathbf{u}^{(k+1)})|| = ||\nabla G(\mathbf{u}^{(k+1)}) - \nabla G(\mathbf{u}^*)|| = ||H_G(\xi_k)(\mathbf{u}^{(k+1)} - \mathbf{u}^*)|| \geq \alpha\|\mathbf{u}^{(k+1)} - \mathbf{u}^*||$$

for $\mathbf{u}^{(k+1)} \in D$, where ξ_k in D. Then, we have

$$\frac{||\nabla G(\mathbf{u}^{(k+1)})||}{||\mathbf{y}_{k+1}||} \geq \frac{\alpha||\mathbf{u}^{(k+1)} - \mathbf{u}^*||}{L||\mathbf{u}^{(k+1)} - \mathbf{u}^*|| + L||\mathbf{u}^{(k)} - \mathbf{u}^*||} = \frac{\alpha\sigma_k}{L(1+\sigma_k)},$$

where $\sigma_k = \frac{||\mathbf{u}^{(k+1)} - \mathbf{u}^*||}{||\mathbf{u}^{(k)} - \mathbf{u}^*||}$. It follows that $\frac{\sigma_k}{1+\sigma_k} \to 0$ and hence, $\sigma_k \to 0$. That is, the BB method converges super-linearly.

On the other hand, if $\sigma_k \to 0$, we can show that $(\beta_k - H_G(\mathbf{u}^*))\mathbf{s}_{k+1} = o(\|\mathbf{s}_{k+1}\|)$. In fact, if $\mathbf{u}^{(k)} \to \mathbf{u}^*$ super-linearly, then we have

$$\lim_{k\to+\infty} \frac{\|\mathbf{u}^{(k+1)} - \mathbf{u}^{(k)}\|}{||\mathbf{u}^{(k)} - \mathbf{u}^*||} = 1. \tag{34}$$

Indeed, since

$$\Big|||\mathbf{u}^{(k+1)} - \mathbf{u}^{(k)}|| - ||\mathbf{u}^{(k)} - \mathbf{u}^*||\Big| \leq ||\mathbf{u}^{(k+1)} - \mathbf{u}^*||,$$

it is clear that

$$\left|\frac{||\mathbf{u}^{(k+1)} - \mathbf{u}^{(k)}||}{||\mathbf{u}^{(k)} - \mathbf{u}^*||} - 1\right| \leq \frac{||\mathbf{u}^{(k+1)} - \mathbf{u}^*||}{||\mathbf{u}^{(k)} - \mathbf{u}^*||} \to 0.$$

Hence, from (34) it follows

$$\begin{aligned} \frac{||\nabla G(\mathbf{u}^{(k+1)})||}{||\mathbf{s}_{k+1}||} &\leq \frac{||\nabla G(\mathbf{u}^{(k+1)}) - \nabla G(\mathbf{u}^*)||}{||\mathbf{s}_{k+1}||} \leq \frac{L\|\mathbf{u}^{(k+1)} - \mathbf{u}^*\|}{\|\mathbf{u}^{(k+1)} - \mathbf{u}^{(k)}\|} \\ &= \frac{\sigma_{k+1}}{\|\mathbf{u}^{(k+1)} - \mathbf{u}^{(k)}\|/\|\mathbf{u}^{(k)} - \mathbf{u}^*\|} \to 0 \end{aligned}$$

because of the denominator is bounded by the property (34). Using the argument at the beginning of the proof, we can see that $(\beta_k - H_G(\mathbf{u}^*))\mathbf{s}_{k+1} = o(\|\mathbf{s}_{k+1}\|)$. This completes the proof. □

5 Sparse Phase Retrieval

In previous sections, several computational algorithms have been developed for the phase retrieval problem based on measurements (1). We now extend the approaches to study the sparse phase retrieval. Suppose that $\mathbf{x}_\flat$ is a sparse solution to the given measurements (1). We want to recover $\mathbf{x}_\flat$ using the DC based algorithm. Firstly, we consider the following optimization

$$\min_{\mathbf{x}\in\mathbb{R}^n \text{ or } \mathbb{C}^n} \lambda\|\mathbf{x}\|_1 + \sum_{i=1}^{m}(f(\langle \mathbf{a}_i, \mathbf{x}\rangle) - b_i)^2, \tag{35}$$

which is a standard approach in compressive sensing by adding $\lambda\|\mathbf{x}\|_1$ to (2). If we take $f(\langle \mathbf{a}_i, \mathbf{x}\rangle) = |\langle \mathbf{a}_i, \mathbf{x}\rangle|^2$, then (35) becomes as the sparse phase retrieval. See [6] and [34] for recent literature on sparse phase retrieval problem.

We now discuss how to solve it numerically. We approach it by using a similar method as in the previous section. Indeed, for the case $\mathbf{x} \in \mathbb{R}^n$ and $\mathbf{a}_i \in \mathbb{R}^n$, we rewrite $\sum_{i=1}^m (f(\langle \mathbf{a}_i, \mathbf{x}\rangle) - b_i)^2$ to be the difference of $F_1(\mathbf{x}) - F_2(\mathbf{x})$ as in (5). The minimization (35) will be approximated by

$$\mathbf{x}^{(k+1)} := \arg\min \lambda\|\mathbf{x}\|_1 + F_1(\mathbf{x}) - \nabla F_2(\mathbf{x}^{(k)})^\top(\mathbf{x} - \mathbf{x}^{(k)}) \tag{36}$$

for any given $\mathbf{x}^{(k)}$. We call this algorithm as sparse DC based method. For the general convex function f, we can also obtain the minimization problem as in (36) with the similar formulation. For convenience, we only consider the case when $\mathbf{x}, \ \mathbf{a}_j, \ j = 1, \ldots, m$ are real. The complex case can be treated in the same way.

To solve (36), we use the proximal gradient method: for any given $\mathbf{y}^{(k)}$, we update it by

$$\begin{aligned}\mathbf{y}^{(k+1)} := \operatorname{argmin} \lambda\|\mathbf{y}\|_1 + F_1(\mathbf{y}^{(k)}) + (\nabla F_1(\mathbf{y}^{(k)}) - \nabla F_2(\mathbf{y}^{(k)}))^\top(\mathbf{y} - \mathbf{y}^{(k)}) \\ + \tfrac{L_1}{2}\|\mathbf{y} - \mathbf{y}^{(k)}\|^2\end{aligned} \tag{37}$$

for $k \geq 1$, where L_1 is the Lipschitz differentiability of F_1. This is a typical DC algorithm discussed in [41]. The above minimization can be easily solved by using shrinkage-thresholding technique as in [7]. Note that Beck and Teboulle in [7] use a Nesterov's acceleration technique to speed up the iteration to form the well-known FISTA. However, we shall use the acceleration technique from [2] which is slightly better than Nesterov's technique. The discussion above furnishes a computational method for sparse phase retrieval problem (35). Let us point out one significant difference between update rule (37) and (14) is that one can find $\mathbf{y}^{(k+1)}$ by using a explicit formula while the solution $\mathbf{x}^{(k+1)}$ of (14) has to be computed using an iterative method as explained before. Thus the sparse phase retrieval is more efficient in this sense.

Let us study the convergence of our sparse phase retrieval method. To the best of the authors' knowledge, the convergence is not available in the literature so far. We first start with a standard result for the ℓ_1-DC based algorithm.

Theorem 5 *Assume F_2 is a strongly convex function with parameter ℓ. Starting from any initial guess $\mathbf{y}^{(1)}$, let $\mathbf{y}^{(k+1)}$ be the solution of (37) for all $k \geq 1$. Then it holds*

$$\lambda\|\mathbf{y}^{(k+1)}\|_1 + F(\mathbf{y}^{(k+1)}) \leq \lambda\|\mathbf{y}^{(k)}\|_1 + F(\mathbf{y}^{(k)}) - \frac{\ell}{2}\|\mathbf{y}^{(k+1)} - \mathbf{y}^{(k)}\|^2, \quad \forall k \geq 1 \tag{38}$$

and

$$\partial g(\mathbf{y}^{(k+1)}) + \nabla F_1(\mathbf{y}^{(k)}) - \nabla F_2(\mathbf{y}^{(k)}) + L_1(\mathbf{y}^{(k+1)} - \mathbf{y}^{(k)}) = 0,$$

where $g(\mathbf{x}) = \lambda\|\mathbf{x}\|_1$ *and* ∂g *denotes the subgradient of* g.

Proof The Lipschitz differentiability of F_1 gives

$$F_1(\mathbf{y}^{(k+1)}) \leq F_1(\mathbf{y}^{(k)}) + \nabla F_2(\mathbf{y}^{(k)})^\top(\mathbf{y}^{(k+1)} - \mathbf{y}^{(k)}) + \frac{L_1}{2}\|\mathbf{y}^{(k+1)} - \mathbf{y}^{(k)}\|^2,$$

where L_1 is the Lipschitz differentiability of F_1. By the strongly convexity of F_2, we have

$$F_2(\mathbf{y}^{(k+1)}) \geq F_2(\mathbf{y}^{(k)}) + \nabla F_2(\mathbf{y}^{(k)})^\top(\mathbf{y}^{(k+1)} - \mathbf{y}^{(k)}) + \frac{\ell}{2}\|\mathbf{y}^{(k+1)} - \mathbf{y}^{(k)}\|^2.$$

Combing the above two inequalities and using used the first order optimality condition for (37), we obtain that

$$\begin{aligned}
&\lambda\|\mathbf{y}^{(k+1)}\|_1 + F(\mathbf{x}^{(k+1)}) = \lambda\|\mathbf{y}^{(k+1)}\|_1 + F_1(\mathbf{y}^{(k+1)}) - F_2(\mathbf{y}^{(k+1)})\\
&\leq \lambda\|\mathbf{y}^{(k+1)}\|_1 + F_1(\mathbf{y}^{(k)}) + \nabla F_1(\mathbf{y}^{(k)})^\top(\mathbf{y}^{(k+1)} - \mathbf{y}^{(k)}) + \frac{L}{2}\|\mathbf{y}^{(k+1)} - \mathbf{y}^{(k)}\|^2\\
&\qquad - F_2(\mathbf{y}^{(k)}) - \nabla F_2(\mathbf{y}^{(k)})^\top(\mathbf{y}^{(k+1)} - \mathbf{y}^{(k)}) - \frac{\ell}{2}\|\mathbf{y}^{(k+1)} - \mathbf{y}^{(k)}\|^2\\
&= F_1(\mathbf{y}^{(k)}) - F_2(\mathbf{y}^{(k)}) - \frac{\ell}{2}\|\mathbf{y}^{(k+1)} - \mathbf{y}^{(k)}\|^2\\
&\quad + \lambda\|\mathbf{y}^{(k+1)}\|_1 + (\nabla F_1(\mathbf{y}^{(k)}) - \nabla F_2(\mathbf{y}^{(k)})^\top(\mathbf{y}^{(k+1)} - \mathbf{y}^{(k)}) + \frac{L}{2}\|\mathbf{y}^{(k+1)} - \mathbf{y}^{(k)}\|^2\\
&\leq F_1(\mathbf{y}^{(k)}) - F_2(\mathbf{y}^{(k)}) - \frac{\ell}{2}\|\mathbf{y}^{(k+1)} - \mathbf{y}^{(k)}\|^2 + \lambda\|\mathbf{y}^{(k)}\|_1\\
&= \lambda\|\mathbf{y}^{(k)}\|_1 + F(\mathbf{y}^{(k)}) - \frac{\ell}{2}\|\mathbf{y}^{(k+1)} - \mathbf{y}^{(k)}\|^2.
\end{aligned}$$

Letting $g(\mathbf{x}) = \lambda\|\mathbf{x}\|_1$, the second property $\partial g(\mathbf{y}^{(k+1)}) + \nabla F_1(\mathbf{y}^{(k)}) - \nabla F_2(\mathbf{y}^{(k)}) + L_1(\mathbf{y}^{(k+1)} - \mathbf{y}^{(k)}) = 0$ follows from the minimization (37). □

Next we show that the sequence $\mathbf{y}^{(k)}, k \geq 1$ from (37) converges to a critical point $\mathbf{y}^*$.

Theorem 6 *Write* $\mathcal{F}(\mathbf{x}) = \lambda\|\mathbf{x}\|_1 + \sum_{i=1}^m (f(\langle \mathbf{a}_i, \mathbf{x}\rangle) - b_i)^2$. *Suppose that* $f(\mathbf{x})$ *is a real analytic function and the gradient* $\nabla f(\mathbf{x})$ *has Lipschitz constant L. Let* $\mathbf{y}^{(k)}, k \geq 1$ *be the sequence obtained from (37). Then it converges to a critical point* $\mathbf{y}^*$ *of* F.

Proof From Theorem 5, we have

$$\frac{\ell}{2}\|\mathbf{y}^{(k+1)} - \mathbf{y}^{(k)}\|^2 \le \mathcal{F}(\mathbf{y}^{(k)}) - \mathcal{F}(\mathbf{y}^{(k+1)}). \tag{39}$$

That is, $\mathcal{F}(\mathbf{y}^{(k)}), k \ge 1$ is a strictly decreasing sequence. Due to the coerciveness, we know that

$$\mathcal{R} := \{\mathbf{x} \in \mathbb{R}^n, \mathcal{F}(\mathbf{y}) \le \mathcal{F}(\mathbf{y}^{(1)})\}$$

is a bounded set. It follows that the sequence $\{\mathbf{y}^{(k)}\}_{k=1}^{\infty} \subset \mathcal{R}$ is a bounded sequence and there exists a cluster point $\mathbf{y}^*$ and a subsequence $\mathbf{y}^{(k_i)}$ such that $\mathbf{y}^{(k_i)} \to \mathbf{y}^*$. Note that $\{\mathcal{F}(\mathbf{y}^{(k)})\}_{k=1}^{\infty}$ is a bounded monotonic descending sequence, and hence $\mathcal{F}(\mathbf{y}^{(k)}) \to \mathcal{F}(\mathbf{y}^*)$ for all $k \ge 1$. We claim that the sequence $\{\mathbf{y}^{(k)}\}_{k=1}^{\infty}$ has finite length, that is,

$$\sum_{k=1}^{\infty} \|\mathbf{y}^{(k+1)} - \mathbf{y}^{(k)}\| < \infty. \tag{40}$$

To establish the claim, we need to use the Kurdyka-Łojasiewicz inequality (cf. [26]). Note that the ℓ_1 norm $\|\mathbf{x}\|_1$ is semialgebraic function and the function $f(\mathbf{x})$ is analytic, so the objective function $\mathcal{F}(\mathbf{x})$ satisfies the KL property at any critical point (cf. [1, 3, 43]). Let us prove that $\|\nabla\mathcal{F}(\mathbf{y}^*)\| = 0$ holds, that is, $\mathbf{y}^*$ is a critical point of $\mathcal{F}$. Indeed, using the second property of (5), we have

$$\begin{aligned}\|\partial\mathcal{F}(\mathbf{y}^{(k)})\| &= \|\partial g(\mathbf{y}^{(k)}) + \nabla F_1(\mathbf{y}^{(k)}) - \nabla F_2(\mathbf{y}^{(k)})\| \\ &\le \|\nabla F(\mathbf{y}^{(k)}) - \nabla F(\mathbf{y}^{(k-1)})\| + L_1\|\mathbf{y}^{(k)} - \mathbf{y}^{(k-1)}\|. \end{aligned} \tag{41}$$

Combining (39) with (41) and using the Lipschitz differentiation of F_1 and F_2, we obtain that $\|\partial F(\mathbf{y}^{(k_i)})\| \to 0$. By the property of subgradient of g and the continuity of the gradients F_1 and F_2, we have $\|\partial\mathcal{F}(\mathbf{y}^*)\| = 0$ when $\mathbf{y}^{(k_i)} \to \mathbf{y}^*$. Thus, $\mathbf{y}^* \in$ domain(∂F), the set of all critical points of $\mathcal{F}$.

Therefore, we can use KL inequality to obtain that

$$\varphi'(\mathcal{F}(\mathbf{y}) - \mathcal{F}(\mathbf{y}^*))\|\partial\mathcal{F}(\mathbf{y})\| \ge 1 \tag{42}$$

for all $\mathbf{y}$ in the neighborhood $B(\mathbf{y}^*, \delta)$. As $\mathcal{F}(\mathbf{y}^{(k)}) - \mathcal{F}(\mathbf{y}^*) \to 0,\ k \to \infty$, there is an integer k_0 such that for all $k \ge k_0$ it holds

$$\max\left(\sqrt{2/\ell}\sqrt{\mathcal{F}(\mathbf{y}^{(k)}) - \mathcal{F}(\mathbf{y}^*)}, 2C/\ell \cdot \varphi(\mathcal{F}(\mathbf{y}^{(k)}) - \mathcal{F}(\mathbf{y}^*))\right) \le \delta/2. \tag{43}$$

Without loss of generality, we may assume that $k_0 = 1$ and $\mathbf{y}^{(1)} \in B(\mathbf{y}^*, \delta/2)$. Let us show that $\mathbf{y}^{(k)}, k \ge 1$ will be in the neighborhood $B(\mathbf{y}^*, \delta)$. We shall use an induction to do so. By (43) we have

$$\|\mathbf{y}^{(2)}-\mathbf{y}^*\| \leq \|\mathbf{y}^{(2)}-\mathbf{y}^{(1)}\|+\|\mathbf{y}^{(1)}-\mathbf{y}^*\| \leq \sqrt{2(\mathcal{F}(\mathbf{y}^{(1)}) - \mathcal{F}(\mathbf{y}^*)/\ell}+\|\mathbf{y}^{(1)}-\mathbf{y}^*\| \leq \delta.$$

Assume that $\mathbf{y}^{(k)} \in B(\mathbf{y}^*, \delta)$ for $k \leq K$. From (5), we have

$$\begin{aligned}\|\partial\mathcal{F}(\mathbf{y}^{k+1})\| &= \|\partial g(\mathbf{y}^{k+1}) + \nabla F(\mathbf{y}^{k+1})\| \\ &= \|\nabla F(\mathbf{y}^{k+1}) - \nabla F(\mathbf{y}^{k}) - L_1(\mathbf{y}^{k+1} - \mathbf{y}^{k})\| \leq C\|\mathbf{y}^{k+1} - \mathbf{y}^{k}\|,\end{aligned}$$

where constant $C := L + L_1/2$. Putting it into (42), it gives that

$$\varphi'(\mathcal{F}(\mathbf{y}^k) - \mathcal{F}(\mathbf{y}^*)) \geq \frac{1}{C\|\mathbf{y}^k - \mathbf{y}^{k-1}\|}. \tag{44}$$

On the other hand, from the concavity of φ we get that

$$\varphi(\mathcal{F}(\mathbf{y}^k)-\mathcal{F}(\mathbf{y}^*))-\varphi(\mathcal{F}(\mathbf{y}^{k+1})-\mathcal{F}(\mathbf{y}^*)) \geq \varphi'(F(\mathbf{y}^k)-F(\mathbf{y}^*))(F(\mathbf{y}^k)-F(\mathbf{y}^{k+1})).$$

Combining (39) with (44), we obtain

$$\varphi(\mathcal{F}(\mathbf{y}^k) - \mathcal{F}(\mathbf{y}^*)) - \varphi(\mathcal{F}(\mathbf{y}^{k+1}) - \mathcal{F}(\mathbf{y}^*)) \geq \frac{\ell}{2C} \cdot \frac{\|\mathbf{y}^{k+1} - \mathbf{y}^k\|^2}{\|\mathbf{y}^k - \mathbf{y}^{k-1}\|}.$$

Multiplying $\|\mathbf{y}^{(k)} - \mathbf{y}^{(k-1)}\|$ on both sides of the above inequality and using a standard inequality $2ab \leq a^2 + b^2$ on the left side, we have

$$\|\mathbf{y}^{(k)}-\mathbf{y}^{(k-1)}\|+\frac{2C}{\ell}(\varphi(\mathcal{F}(\mathbf{y}^k)-\mathcal{F}(\mathbf{y}^*))-\varphi(\mathcal{F}(\mathbf{y}^{k+1})-\mathcal{F}(\mathbf{y}^*))) \geq 2\|\mathbf{y}^{(k)}-\mathbf{y}^{(k+1)}\|$$

for all $2 \leq k \leq K$. It follows that

$$\frac{2C}{\ell}\varphi(\mathcal{F}(\mathbf{y}^{(1)}) - \mathcal{F}(\mathbf{y}^*)) \geq \sum_{j=1}^{K} \|\mathbf{y}^{(j+1)} - \mathbf{y}^{(j)}\| + \|\mathbf{y}^{(K+1)} - \mathbf{y}^{(K)}\|. \tag{45}$$

That is, we have

$$\begin{aligned}\|\mathbf{y}^{(K+1)}-\mathbf{y}^*\| &\leq \|\mathbf{y}^{(K+1)}-\mathbf{y}^{(1)}\|+\|\mathbf{y}^{(1)}-\mathbf{y}^*\| \leq \sum_{j=1}^{K}\|\mathbf{y}^{(j+1)}-\mathbf{y}^{(j)}\|+\|\mathbf{y}^{(1)}-\mathbf{y}^*\| \\ &\leq \frac{2C}{\ell}\varphi(\mathcal{F}(\mathbf{y}^{(1)}) - \mathcal{F}(\mathbf{y}^*)) + \|\mathbf{y}^{(1)} - \mathbf{y}^*\| \leq \delta.\end{aligned}$$

That is, $\mathbf{y}^{(K+1)} \in B(\mathbf{y}^*, \delta)$ which implies that all $\mathbf{y}^{(k)}$ are in $B(\mathbf{y}^*, \delta)$. From the above arguments, we know that the inequality (45) holds for all k, which implies the claim (40) holds. It is clear that (40) implies that $\{\mathbf{y}^{(k)}\}_{k=1}^{\infty}$ is a Cauchy sequence and hence, it is convergent with $\mathbf{y}^{(k)} \to \mathbf{y}^\star$. Note that $\nabla F(\mathbf{y}^\star) = 0$, which implies $\mathbf{y}^{(k)}$ converges to a critical point of F. □

Finally, we show that the convergence rate is linear. To beginning, we give the following technical lemma.

Lemma 7 *Let $g(\mathbf{x}) = \lambda\|\mathbf{x}\|_1$ for $\lambda > 0$. Then for any $\mathbf{x}$, there exists a $\delta > 0$ such that for any $\mathbf{y} \in B(\mathbf{x}, \delta)$, there exists a subgradient ∂g at $\mathbf{y}$ such that*

$$(\partial g(\mathbf{y}) - \partial g(\mathbf{x}))^\top (\mathbf{y} - \mathbf{x}) = 0. \tag{46}$$

Proof For simplicity, we only consider $\mathbf{x} \in \mathbb{R}^1$. Then if $\mathbf{x} \neq 0$, we can find $\delta = |\mathbf{x}| > 0$ such that when $\mathbf{y} \in B(\mathbf{x}, \delta)$, we have $\partial g(\mathbf{y}) = \partial g(\mathbf{x})$ and hence, we have (46). If $\mathbf{x} = 0$, for any $y \neq 0$, we choose $\partial g(0)$ according to $\mathbf{y}$, i.e. $\partial g(0) = 1$ if $\mathbf{y} > 0$ and $\partial g(0) = -1$ if $\mathbf{y} < 0$. Then we have (46). □

In the following lemma, we need the sparse set $\mathcal{R}_s$

$$\mathcal{R}_s := \{\mathbf{x} \in \mathbb{R}^n : \|\mathbf{x}\|_0 \leq s\} = \bigcup_{\substack{I \subset \{1,\cdots,n\} \\ |I|=s}} \mathbb{R}^s_I, \tag{47}$$

which is the union of all canonical subspaces $\mathbb{R}^s_I = \text{span}\{\mathbf{e}_{i_1}, \cdots, \mathbf{e}_{i_s}\}$ if $I = \{i_1, i_2, \cdots, i_s\}$.

Lemma 8 *Let $\mathcal{F}(\mathbf{x}) = g(\mathbf{x}) + F(\mathbf{x})$ with $g(\mathbf{x}) = \lambda\|\mathbf{x}\|_1$. Suppose that F is L-Lipschitz differentiable. Let $\mathbf{x}^*$ be a critical point of $\mathcal{F}$ as explained in (6). Suppose that either all entries of $\mathbf{x}^*$ are nonzero or $\mathbf{x}^* \in \mathbb{R}^s_I$ for some $s \in \{1, \cdots, n\}$. Then there exists $\delta > 0$ such that for all $\mathbf{x} \in B(\mathbf{x}^*, \delta)$,*

$$|\mathcal{F}(\mathbf{x}) - \mathcal{F}(\mathbf{x}^*)| \leq C\|\mathbf{x} - \mathbf{x}^*\|^2. \tag{48}$$

Proof Under the assumption that either all entries of $\mathbf{x}^*$ are nonzero or $\mathbf{x}^* \in \mathbb{R}^s_I$ for some $s \in \{1, \cdots, n\}$, we know that $\mathcal{F}(\mathbf{x})$ is differentiable at $\mathbf{x}^*$. Since $\mathbf{x}^*$ is a critical point, we have

$$\partial\mathcal{F}(\mathbf{x}^*) = \partial g(\mathbf{x}^*) + \nabla F(\mathbf{x}^*) = 0.$$

Combing it with (36), we obtain

$$\begin{aligned}
\mathcal{F}(\mathbf{x}) - \mathcal{F}(\mathbf{x}^*) &= g(\mathbf{x}) - g(\mathbf{x}^*) + F(\mathbf{x}) - F(\mathbf{x}^*) \\
&\leq \partial g(\mathbf{x})^\top (\mathbf{x} - \mathbf{x}^*) + \nabla F(\mathbf{x}^*)(\mathbf{x} - \mathbf{x}^*) + \frac{1}{2}(\mathbf{x} - \mathbf{x}^*)^\top \nabla^2 F(\xi)(\mathbf{x} - \mathbf{x}^*) \\
&= (\partial g(\mathbf{x}) - \partial g(\mathbf{x}^*))^\top (\mathbf{x} - \mathbf{x}^*) + \frac{1}{2}(\mathbf{x} - \mathbf{x}^*)^\top \nabla^2 F(\xi)(\mathbf{x} - \mathbf{x}^*) \\
&= \frac{1}{2}(\mathbf{x} - \mathbf{x}^*)^\top \nabla^2 F(\xi)(\mathbf{x} - \mathbf{x}^*),
\end{aligned}$$

where ξ is a point in between $\mathbf{x}^*$ and $\mathbf{x}$. That is, $|\mathcal{F}(\mathbf{x}) - \mathcal{F}(\mathbf{x}^*)| \leq C\|\mathbf{x} - \mathbf{x}^*\|^2$ for a positive constant C. □

We are now ready to establish the following result on the rate of convergence.

Theorem 7 *Suppose that F_2 is strongly convex. Starting from any initial guess $\mathbf{x}^{(1)}$, let $\mathbf{x}^{(k+1)}$ be the solution of (14) for all $k \geq 1$. Then for any $\epsilon > 0$, there exists a point $\mathbf{x}^*$ such that either $\mathbf{x}^{(k+1)} \in B(\mathbf{x}^*, \epsilon)$ or*

$$\|\mathbf{x}^{(k+1)} - \mathbf{x}^*\| \leq C_\epsilon \tau^k \tag{49}$$

for a positive constant C_ϵ dependent on ϵ and $\tau \in (0, 1)$.

Proof As we have shown in Theorems 5 and 6, the sequence $\mathbf{x}^{(k)}, k \geq 1$ converges to a critical point $\mathbf{x}^*$ of $\mathcal{F}$. Furthermore, we have

$$C_0\|\mathbf{x}^{(k+1)} - \mathbf{x}^{(k)}\|^2 \leq (\mathcal{F}(\mathbf{x}^{(k)}) - \mathcal{F}(\mathbf{x}^*)) - (\mathcal{F}(\mathbf{x}^{(k+1)}) - \mathcal{F}(\mathbf{x}^*)) \tag{50}$$

for a positive constant C_0. We now claim that

$$C_1\|\mathbf{x}^{(k+1)} - \mathbf{x}^{(k)}\| \leq \sqrt{\mathcal{F}(\mathbf{x}^{(k)}) - \mathcal{F}(\mathbf{x}^*)} - \sqrt{\mathcal{F}(\mathbf{x}^{(k+1)}) - \mathcal{F}(\mathbf{x}^*)} \tag{51}$$

holds for a positive constant C_1. To establish this claim, we first note that Lemma 8 gives

$$\frac{1}{\sqrt{\mathcal{F}(\mathbf{x}^{(k)}) - \mathcal{F}(\mathbf{x}^*)}} \geq \frac{C}{\|\mathbf{x}^{(k)} - \mathbf{x}^*\|}.$$

Multiplying the above inequality to (50), we have

$$C_0 C \frac{\|\mathbf{x}^{(k+1)} - \mathbf{x}^{(k)}\|^2}{\|\mathbf{x}^{(k)} - \mathbf{x}^*\|} \leq \frac{(\mathcal{F}(\mathbf{x}^{(k)}) - \mathcal{F}(\mathbf{x}^*)) - (\mathcal{F}(\mathbf{x}^{(k+1)}) - \mathcal{F}(\mathbf{x}^*))}{\sqrt{\mathcal{F}(\mathbf{x}^{(k)}) - \mathcal{F}(\mathbf{x}^*)}}. \tag{52}$$

Consider $h(t) = \sqrt{t}$ which is concave over $[0, 1]$ and we know $h(t) - h(s) \geq h'(t)(t - s)$. Thus, the right-hand side above is less than or equal to the right-hand side of (51). We next show that the left-hand side of the inequality satisfies

$$\frac{\|\mathbf{x}^{(k+1)} - \mathbf{x}^{(k)}\|}{\|\mathbf{x}^{(k)} - \mathbf{x}^*\|} \geq \frac{\|\mathbf{x}^{(k+1)} - \mathbf{x}^{(k)}\|}{\|\mathbf{x}^{(k+1)} - \mathbf{x}^*\| + \|\mathbf{x}^{(k+1)} - \mathbf{x}^{(k)}\|}.$$

Note that F is strongly convex outside the ball $B(\mathbf{x}^*, \epsilon)$. If $\mathbf{x}^{(k+1)}$ is within the $B(\mathbf{x}^*, \epsilon)$, then we complete the proof. Otherwise, the strong convexity of F outside $B(\mathbf{x}^*, \epsilon)$ (see Theorem 9 for the real case and Theorem 10 for the complex case) gives

$$C_\epsilon \|\mathbf{x}^{(k+1)} - \mathbf{x}^*\| \le \|\nabla F(\mathbf{x}^{(k+1)}) - \nabla F(\mathbf{x}^*)\|$$

for a positive constant dependent on ϵ. As $\partial g(\mathbf{x}^*) + \nabla F(\mathbf{x}^*) = 0$, Lemma 7 implies that $-\nabla F(\mathbf{x}^*) = \partial g(\mathbf{x}^*) = \partial g(\mathbf{x}^{(k+1)})$ when $\mathbf{x}^{(k+1)}$ is close to $\mathbf{x}^*$ enough (i.e., the support of $\mathbf{x}^{(k+1)}$ is the same as the support of $\mathbf{x}^*$ and the sign of each entry in $\mathbf{x}^{(k+1)}$ is the same to $\mathbf{x}^*$). By Theorem 5, we have

$$\partial F(\mathbf{x}^{(k+1)}) - \partial F(\mathbf{x}^*) = \nabla F(\mathbf{x}^{(k+1)}) - \nabla F(\mathbf{x}^{(k)}) - L_1(\mathbf{x}^{(k+1)} - \mathbf{x}^{(k)}).$$

In other words, using the Lipschitz differentiability of F, it holds

$$C_\epsilon \|\mathbf{x}^{(k+1)} - \mathbf{x}^*\| \le (L + L_1)\|\mathbf{x}^{(k+1)} - \mathbf{x}^{(k)}\|$$

and

$$\frac{\|\mathbf{x}^{(k+1)} - \mathbf{x}^{(k)}\|}{\|\mathbf{x}^{(k)} - \mathbf{x}^*\|} \ge \frac{\|\mathbf{x}^{(k+1)} - \mathbf{x}^{(k)}\|}{((L + L_1)/C_\epsilon + 1)\|\mathbf{x}^{(k+1)} - \mathbf{x}^{(k)}\|} = \frac{C_\epsilon}{L + L_1 + C_\epsilon}.$$

The left-hand side of (52) can be simplified to be

$$C_0 C \frac{C_\epsilon}{L_1 + L + 1}\|\mathbf{x}^{(k+1)} - \mathbf{x}^{(k)}\|,$$

which implies the claim (51) holds. By summing the inequality (51), it follows

$$\sum_{k\ge 1} \|\mathbf{x}^{(k+1)} - \mathbf{x}^{(k)}\| \le \frac{1}{C_1}\sqrt{f(\mathbf{x}^{(1)}) - f(\mathbf{x}^*)}.$$

Then the remaining part of the proof is similar to the proof of Theorem 4 and we leave the details to the interested readers. □

6 Numerical Results

In this section, we give some numerical experiments for our DC based algorithm and the ℓ_1-DC based algorithm. We compare the empirical success rate of our DC based algorithms with WF [9] and Gauss-Newton [20] methods. All experiments are carried out with 1000 repeated trials. The results show that the DC based algorithm is able to recover real signals with probability around 80% with $m \approx 2n$ measurements, where n is the dimension of signals. As demonstrated in [13], more precisely the Figures 8 and 9 in [13], it needs $m \approx 3n$ measurements to recover real signals for truncated Wirtinger flow and Wirtinger flow algorithms. Similarly, the DC based algorithm can recover the complex signals with $m \approx 3n$ measurements (cf. Table 2) while WF requires $m \approx 4n$. Finally, for the sparse signals, the ℓ_1-DC based algorithm with thresholding technique needs only $m \approx n$ measurements.

6.1 Phase Retrieval for Real and Complex Signals

Example 2 In this example, we consider to recover a real signal $\mathbf{x_b}$ from the given measurements (1) using Gaussian random measurement vectors $\mathbf{a}_j$, $j = 1, \cdots, m$. We fix $n = 128$ and consider the number of measurements m is around twice the dimension of $\mathbf{x_b}$, i.e., $m = kn/16$ for $k = 24, 25, \cdots, 35$. For the initialization, we first obtain a initial guess by the initialization algorithm in [20] and then improve the initial guess by applying alternating projection method discussed in Algorithm 3. We say a trail is successful if the relative error is less than 10^{-5}. Table 1 gives the empirical success rate of recovering $\mathbf{x_b}$ for DC, WF and Gauss-Newton methods. From Table 1, we can see that the DC based algorithm can recover the solutions with probability large than 60% under $m \geq 2n$. According to the result in [4], one needs $m \geq 2n - 1$ measurements to guarantee the recovery of all real signals. Thus the DC based algorithm almost reaches the theoretical low bound.

Example 3 We next repeat Example 2 using a litter more number of measurements. The numbers of successes for the Wirtinger Flow algorithm, Gauss-Newton algorithm and the DC based algorithm are shown in Table 2. One can see that the performance of the DC based algorithm is the best and can achieve 95% success rate with $m = 2.5n$.

Example 4 This example is to show the robustness of the DC based algorithm. We repeat the computation in Example 2 for noisy measurements. There are two ways to generate the noisy measurements. One way is to add the noises η_j to b_j directly and obtain

Table 1 The numbers of successes over 1000 repeated trials versus the number of measurements m/n listed above

m/n	1.5	1.5625	1.6250	1.6875	1.75	1.8125	1.875	1.9375	2	2.0625	2.125	2.1875
WF successes	0	0	0	0	0	1	11	10	24	27	41	64
GN successes	0	0	0	18	13	36	71	114	167	251	315	415
DC successes	50	78	119	182	266	318	406	542	600	681	744	807

Table 2 The numbers of successes over 1000 repeated trails versus the number of measurements m/n listed above, where $n = 128$

m/n	2.4375	2.5	2.5625	2.625	2.6875	2.75	2.8125	2.875	2.937	3
WF successes	168	220	254	352	372	459	513	612	641	706
GN successes	728	749	844	886	908	934	931	963	968	982
DC successes	944	952	975	982	984	989	994	993	995	998

$$\hat{b}_j = |\langle \mathbf{a}_j, \mathbf{x_b} \rangle|^2 + \eta_j, \quad j = 1, \cdots, m. \tag{53}$$

Another way is

$$\tilde{b}_j = |\langle \mathbf{a}_j, \mathbf{x_b} \rangle + \delta_j|^2 + \eta_j, \quad j = 1, \cdots, m, \tag{54}$$

where δ_j and η_j are noises. For noisy measurements (53), we assume that η_j are subject to uniform random distribution between $[-\beta, \beta]$ with mean zero, where $\beta = 10^{-1}, 10^{-3}$ and 10^{-5}. If the stopping tolerance ϵ satisfies $\epsilon \geq \beta$, then the Gauss-Newton method and DC based method produce the same empirical success rate as in Table 2. For noisy measurements (54), we assume that both ϵ_j and δ_j are subject to uniform distribution between $[-\beta, \beta]$ with mean zero. Similarly, if the stopping tolerance ϵ satisfies $\epsilon \geq \beta$, then both algorithms can recover the true solution.

Example 5 In this example, we use the DC based algorithm and the Gauss-Newton method to recover the complex signals. We choose $n = 128$ and the number of measurements m is around $3n$, i.e., $m \approx 3n$. For Gaussian random measurements $\mathbf{a}_j = \mathbf{a}_{j,R} + \mathbf{i}\mathbf{a}_{j,I}$, $j = 1, \cdots, m$, we aim to recover $\mathbf{z} \in \mathbb{C}^n$ with $\mathbf{z} = \mathbf{x} + \mathbf{i}\mathbf{y}$ from $|\langle \mathbf{a}_j, \mathbf{z} \rangle|^2$, $j = 1, \cdots, m$. The maximum iteration numbers for WF, GN, DC are 3000, 100 and 1000, respectively. We say a trial is successful if the relative error is less than 10^{-5}. Table 3 gives the numbers of successes for WF, Gauss-Newton and the DC based methods with 1000 repeated trials. From the Table 3, we can see that the DC based algorithm can recover the complex signals very well when $m \geq 3n$, which is slightly better than the GN algorithm and much better than the WF algorithm.

We next present some numerical experiments to demonstrate that the ℓ_1-DC based algorithm works well. The procedure is presented in Algorithm 2, where a modified Attouch-Peypouquet technique [2] is used. We use the step size $\beta_k = k/(k+\alpha)$ for the first few k iterations, say $k \leq K$, and then a fixed step size β_K for the remaining iterations.

Example 6 In this example, we test the performance of Algorithm 2 for recovering the real signals without sparsity. We choose $n = 128$ and the number of measurements $m = 1.1n, 1.2n, \ldots, 2.5n$. The target signal $\mathbf{x_b}$ and the measurement vectors $\mathbf{a}_j$, $j = 1, \ldots, m$ are Gaussian random vectors. We choose the parameter $\lambda = 10^{-5}$ in Algorithm 2. The numbers of successes are summarized in Table 4.

Table 3 The numbers of successes over 1000 repeated trials based on m/n listed above for complex case

m/n	2.938	3	3.062	3.125	3.187	3.25	3.312	3.375	3.437	3.5	3.562	3.625	3.687	3.75
WF	0	0	0	0	0	0	0	0	0	56	192	204	322	401
GN	191	338	304	416	452	536	594	739	744	762	801	815	910	912
DC	422	563	537	565	623	730	829	887	881	894	954	946	981	986

Algorithm 2 ℓ_1-DC Based Algorithm

We use the same initialization as in the previous examples.
while $k \geq 1$ **do**
1° Solve (37) to get $\mathbf{y}^{(k+1)}$.
2° Apply a modified Attouch-Peypouquet technique to get a new $\mathbf{y}^{(k+1)}$
until the maximal number of iterations is reached.
end while
return $\mathbf{y}^T$

Table 4 The numbers of successes over 1000 repeated trials based on Algorithm 2

m/n	1.5	1.6	1.7	1.8	1.9	2	2.1	2.2	2.3	2.4	2.5
ℓ_1-DC alg.	0	0	206	317	352	557	724	913	938	947	994

The results show that the ℓ_1-DC based algorithm can recover the real signals with high probability if $m \geq 2.2n$.

6.2 *Phase Retrieval of Sparse Signals*

We next turn to consider how to use our ℓ_1-DC based algorithm to recover the sparse signals. We know that if the number of measurements $m \leq 2n$, then many existing algorithms will fail to recover the signals. For our DC based algorithm, it can recover any signal when $m \approx 2n$, no matter sparse or not. However, when $m \approx 1.5n$, we are not able to recover the general signals. The purpose of our numerical experiments is to see if we are able to recover the sparse signals when $m \approx n$. By using the sparsity, we will enhance the ℓ_1-DC based algorithm with the projection technique. More specifically, we use the hard thresholding technique to project $\mathbf{y}^{(k+1)}$ from (7) into the set of all s-sparse vectors. This leads to an ℓ_1-DC based algorithm with hard thresholding which given below.

Algorithm 3 ℓ_1-DC Based Algorithm with Hard Thresholding

Obtain a initial guess with the initialization in [9].
while $k \geq 1$ **do**
Solve (37) using the shrinkage-thresholding technique to get $\mathbf{y}^{(k+1)}$.
Apply a modified Attouch-Peypouquet technique to get a new $\mathbf{y}^{(k+1)}$.
Project $\mathbf{y}^{(k+1)}$ into the collection of s-sparse set $\mathcal{R}_s$. That is, let $\mathbf{z}^{(k+1)}$ be the solution of the following minimization problem:

$$\sigma_s(\mathbf{x}^k) = \min_{\mathbf{z} \in \mathcal{R}_s} \|\mathbf{y}^{(k+1)} - \mathbf{z}\|_1. \tag{55}$$

Update $\mathbf{y}^{(k+1)} = \mathbf{z}^{(k+1)}$.
end while
return the maximal number of iterations $\mathbf{y}^T$

Table 5 The numbers of successes with sparsities $s = 2, 4, 5, 10, 20$ over 1000 repeated trials

m/n		0.5	0.6	0.7	0.8	0.9	1.0	1.1	1.2	1.3	1.4	1.5	1.6	1.7	1.8	1.9	2
Alg. 3	$s = 2$	422	517	566	619	662	710	745	829	846	837	833	849	862	870	877	946
Alg. 3	$s = 4$	187	333	429	474	590	673	693	747	778	811	819	832	840	862	873	919
Alg. 3	$s = 5$	71	116	264	383	452	594	618	674	726	771	802	821	837	858	869	894
Alg. 3	$s = 10$	0	0	0	52	151	247	416	385	537	482	590	680	701	737	796	812
Alg. 3	$s = 20$	0	0	0	0	0	55	156	180	227	271	368	422	451	527	574	599

Example 7 In this example, we show that our ℓ_1-DC based algorithm with hard thresholding works well. We choose $n = 128$ and the number of measurements $m = 0.5n, 0.6n, \ldots, 2n$. For each m, we test the performance of Algorithm 3 with sparsities $s = 2, 4, 5, 10, 20$. The experiments are implemented under 1000 repeated trials. The results on the numbers of successes are presented in Table 5. From Table 5, we can see that Algorithm 3 is able to recover sparse solutions with high probability.

Acknowledgments The authors "Ming-Jun Lai and Abraham Varghese" are partly supported by the National Science Foundation under grant DMS-1521537.

Zhiqiang Xu was supported by NSFC grant (11422113, 91630203, 11331012) and by National Basic Research Program of China (973 Program 2015CB856000).

Appendix

In this section we give some deterministic description of the minimizing function F as well as strong convexity of F_2. We will show that at any global minimizer, the Hessian matrix of F is positive definite in the real case and is nonnegative positive definite in the complex case. These results are used when we apply the KL inequality. For convenience, let $A_\ell = \mathbf{a}_\ell \bar{\mathbf{a}}_\ell^\top$ be the Hermitian matrix of rank one for $\ell = 1, \cdots, m$.

Definition 3 We say $\mathbf{a}_j, j = 1, \cdots, m$ are a_0-generic if there exists a positive constant $a_0 \in (0, 1)$ such that

$$\|(\mathbf{a}_{j_1}^* \mathbf{y}, \ldots, \mathbf{a}_{j_n}^* \mathbf{y})\| \geq a_0 \|\mathbf{y}\|, \quad \forall \mathbf{y} \in \mathbb{C}^n$$

holds for any $1 \leq j_1 < j_2 < \cdots < j_n \leq m$.

Theorem 8 *Let $m \geq n$. Assume $\mathbf{a}_j, j = 1, \cdots, m$ are a_0-generic for some constant a_0. If there exist n nonzero elements among the measurements $b_j, j = 1, \cdots, m$, then for the phase retrieval problem with $f(x) = |x|^2$, F_2 is positive definite.*

Proof Recall that $F_2 = 2\sum_{i=1}^m b_i f(\mathbf{a}_k^\top \mathbf{x})$. Then the Hessian matrix of F_2 is

$$H_{F_2}(\mathbf{x}) = 2\sum_{i=1}^{m} b_i f''(\mathbf{a}_k^\top \mathbf{x})\mathbf{a}_i\mathbf{a}_i^\top.$$

Note that $f''(x) = 2$. Thus we have

$$H_{F_2}(\mathbf{x}) = 4\sum_{\ell=1}^{m} b_\ell A_\ell.$$

Let $b_0 = \min\{b_\ell \neq 0\}$. Then

$$\mathbf{y}^\top H_{F_2}(\mathbf{x})\mathbf{y} \geq 4b_0\|(\mathbf{a}_{j_1}^*\mathbf{y}, \ldots, \mathbf{a}_{j_n}^*\mathbf{y})\|^2 \geq 4b_0 a_0^2\|\mathbf{y}\|^2.$$

Thus, F_2 is strongly convex. □

Theorem 9 *Let $H_F(\mathbf{x})$ be the Hessian matrix of the function $F(\mathbf{x})$ and let $\mathbf{x}^\star$ be a global minimizer of (2). Suppose that $\mathbf{a}_j$, $j = 1, \cdots, m$ are a_0-generic. Then $H_F(\mathbf{x}^\star)$ is positive definite in a neighborhood of $\mathbf{x}^*$.*

Proof Recall that $A_\ell = \mathbf{a}_\ell\bar{\mathbf{a}}_\ell^\top$ for $\ell = 1, \cdots, m$. It is easy to see

$$\nabla F(\mathbf{x}) = 2\sum_{\ell=1}^{m}(\mathbf{x}^\top A_\ell\mathbf{x} - b_\ell)A_\ell\mathbf{x}$$

and the entries h_{ij} of the Hessian $H_F(\mathbf{x})$ is

$$h_{ij} = \frac{\partial}{\partial x_i}\frac{\partial}{\partial x_j}f(\mathbf{x}) = 2\sum_{\ell=1}^{m}(\mathbf{x}^\top A_\ell\mathbf{x} - b_\ell)a_{ij}(\ell) + 4\sum_{p=1}^{n} a_{i,p}(\ell)x_p\sum_{q=1}^{n} a_{j,q}(\ell)x_q,$$

where $A_\ell = [a_{ij}(\ell)]_{ij=1}^n$. Since $(\mathbf{x}^*)^\top A_\ell\mathbf{x}^* = b_\ell$ for all $\ell = 1, \cdots, m$, the first summation term of h_{ij} above is zero at $\mathbf{x}^*$. Let $M(\mathbf{y}) = \mathbf{y}^\top H_f(\mathbf{x}^*)\mathbf{y}$ be a quadratic function of $\mathbf{y}$. Then we have

$$\begin{aligned} M(\mathbf{y}) &= 4\sum_{\ell=1}^{m}(\mathbf{y}^\top A_\ell\mathbf{x}^*(\mathbf{x}^*)^\top A_\ell\mathbf{y} = 4\sum_{\ell=1}^{m}|\mathbf{y}^\top A_\ell\mathbf{x}^*|^2 \\ &= 4\sum_{\ell=1}^{m}|\mathbf{y}^\top\mathbf{a}_\ell|^2|\bar{\mathbf{a}}_\ell^\top\mathbf{x}^*|^2 \geq 4a_0\|\mathbf{x}^*\|^2\|\mathbf{y}\|^2, \end{aligned}$$

where the inequality follows from the fact that $\mathbf{a}_j$, $j = 1, \cdots, m$ are a_0-generic. It implies that $H_F(\mathbf{x}^*)$ is positive definite. □

Next, we show that the Hessian $H_F(\mathbf{x}^*)$ is nonnegative definite at the global minimizer $\mathbf{x}^\star$ in the complex case. To this end, we first introduce some notations.

Write $\mathbf{a}_\ell = a_\ell + \mathbf{i}c_\ell$ for $\ell = 1, \cdots, m$. For $\mathbf{z} = \mathbf{x} + \mathbf{i}\mathbf{y}$, we have $\mathbf{a}_\ell^\top \mathbf{z}^* = b_\ell$ for the global minimizer $\mathbf{z}^*$. Writing $f_\ell(\mathbf{x}, \mathbf{y}) = |\mathbf{a}_\ell^\top \mathbf{z}|^2 - b_\ell = (a_\ell^\top \mathbf{x} - c_\ell^\top \mathbf{y})^2 + (c_\ell^\top \mathbf{x} + a_\ell^\top \mathbf{y})^2 - b_\ell$, we consider

$$f(\mathbf{x}, \mathbf{y}) = \frac{1}{m} \sum_{\ell=1}^{m} f_\ell^2.$$

The gradient of f can be easily computed as follows: $\nabla f = [\nabla_\mathbf{x} f, \nabla_\mathbf{y} f]$ with

$$\nabla_\mathbf{x} f(\mathbf{x}, \mathbf{y}) = \frac{4}{m} \sum_{\ell=1}^{m} f_\ell(\mathbf{x}, \mathbf{y})[(a_\ell^\top \mathbf{x} - c_\ell^\top \mathbf{y})a_\ell + (c_\ell^\top \mathbf{x} + a_\ell^\top \mathbf{y})c_\ell]$$

and

$$\nabla_\mathbf{y} f(\mathbf{x}, \mathbf{y}) = \frac{4}{m} \sum_{\ell=1}^{m} f_\ell(\mathbf{x}, \mathbf{y})[(a_\ell^\top \mathbf{x} - c_\ell^\top \mathbf{y})(-c_\ell) + (c_\ell^\top \mathbf{x} + a_\ell^\top \mathbf{y})a_\ell].$$

Furthermore, the Hessian of F is given by

$$H_F(\mathbf{x}, \mathbf{y}) = \left[\nabla_\mathbf{x}\nabla_\mathbf{x} f(\mathbf{x}, \mathbf{y})\ \nabla_\mathbf{x}\nabla_\mathbf{y} f(\mathbf{x}, \mathbf{y});\ \nabla_\mathbf{y}\nabla_\mathbf{x} f(\mathbf{x}, \mathbf{y})\ \nabla_\mathbf{y}\nabla_\mathbf{y} f(\mathbf{x}, \mathbf{y})\right],$$

where the terms $\nabla_\mathbf{x}\nabla_\mathbf{x} f(\mathbf{x}, \mathbf{y}), \cdots, \nabla_\mathbf{y}\nabla_\mathbf{y} f(\mathbf{x}, \mathbf{y})$ are given below.

$$\nabla_\mathbf{x}\nabla_\mathbf{x} f(\mathbf{x}, \mathbf{y}) = \frac{4}{m} \sum_{\ell=1}^{m} f_\ell(\mathbf{x}, \mathbf{y})[a_\ell a_\ell^\top + c_\ell c_\ell^\top]$$
$$+\frac{8}{m} \sum_{\ell=1}^{m} [(a_\ell^\top \mathbf{x} - c_\ell^\top \mathbf{y})a_\ell + (c_\ell^\top \mathbf{x} + a_\ell^\top \mathbf{y})c_\ell][(a_\ell^\top \mathbf{x} - c_\ell^\top \mathbf{y})a_\ell^\top + (c_\ell^\top \mathbf{x} + a_\ell^\top \mathbf{y})c_\ell^\top]$$

and

$$\nabla_\mathbf{x}\nabla_\mathbf{y} f(\mathbf{x}, \mathbf{y}) = \frac{4}{m} \sum_{\ell=1}^{m} f_\ell(\mathbf{x}, \mathbf{y})[a_\ell(-c_\ell)^\top + c_\ell a_\ell^\top]$$
$$+\frac{8}{m} \sum_{\ell=1}^{m} [(a_\ell^\top \mathbf{x} - c_\ell^\top \mathbf{y})a_\ell + (c_\ell^\top \mathbf{x} + a_\ell^\top \mathbf{y})c_\ell][(a_\ell^\top \mathbf{x} - c_\ell^\top \mathbf{y})(-c_\ell)^\top + (c_\ell^\top \mathbf{x} + a_\ell^\top \mathbf{y})a_\ell^\top].$$

The terms $\nabla_\mathbf{y}\nabla_\mathbf{x} f(\mathbf{x}, \mathbf{y})$ and $\nabla_\mathbf{y}\nabla_\mathbf{y} f(\mathbf{x}, \mathbf{y})$ can be obtained similarly.

Theorem 10 *For phase retrieval problem in the complex case, the Hessian matrix $H_f(\mathbf{x}^*, \mathbf{y}^*)$ at any global minimizer $\mathbf{z}^* := (\mathbf{x}^*, \mathbf{y}^*)$ satisfies $H_f(\mathbf{x}^*, \mathbf{y}^*) \geq 0$. Furthermore, $H_f(\mathbf{x}^*, \mathbf{y}^*) = 0$ along the direction $[-(\mathbf{y}^*)^\top, (\mathbf{x}^*)^\top]^\top$.*

Proof At the global minimizer $\mathbf{z}^* = \mathbf{x}^* + \mathbf{i}\mathbf{y}^*$, we have

$$\nabla_{\mathbf{x}}\nabla_{\mathbf{x}} f(\mathbf{x}^*, \mathbf{y}^*) = \frac{8}{m}\sum_{\ell=1}^{m} [(a_\ell^\top \mathbf{x}^* - c_\ell^\top \mathbf{y}^*)a_\ell + (c_\ell^\top \mathbf{x}^* + a_\ell^\top \mathbf{y}^*)c_\ell]\times$$
$$[(a_\ell^\top \mathbf{x}^* - c_\ell^\top \mathbf{y}^*)a_\ell^\top + (c_\ell^\top \mathbf{x}^* + a_\ell^\top \mathbf{y}^*)c_\ell^\top],$$

$$\nabla_{\mathbf{x}}\nabla_{\mathbf{y}} f(\mathbf{x}^*, \mathbf{y}^*) = \frac{8}{m}\sum_{\ell=1}^{m} [(a_\ell^\top \mathbf{x}^* - c_\ell^\top \mathbf{y}^*)a_\ell + (c_\ell^\top \mathbf{x}^* + a_\ell^\top \mathbf{y}^*)c_\ell]\times$$
$$[(a_\ell^\top \mathbf{x}^* - c_\ell^\top \mathbf{y}^*)(-c_\ell)^\top + (c_\ell^\top \mathbf{x}^* + a_\ell^\top \mathbf{y}^*)a_\ell^\top]$$

and similar for the other two terms. It is easy to check that for any $\mathbf{w} = \mathbf{u} + \mathbf{i}\mathbf{v}$ with $\mathbf{u}, \mathbf{v} \in \mathbb{R}^n$, we have

$$\begin{aligned}
&[\mathbf{u}^\top\ \mathbf{v}^\top]^\top H_F(\mathbf{x}^*, \mathbf{y}^*)\,[\mathbf{u}\,;\,\mathbf{v}]\\
&= \frac{8}{m}\sum_{\ell=1}^{m}[(a_\ell^\top \mathbf{x}^* - c_\ell^\top \mathbf{y}^*)a_\ell^\top \mathbf{u} + (c_\ell^\top \mathbf{x}^* + a_\ell^\top \mathbf{y}^*)c_\ell^\top \mathbf{u}]^2\\
&+\frac{8}{m}\sum_{\ell=1}^{m}[(a_\ell^\top \mathbf{x}^* - c_\ell^\top \mathbf{y}^*)(-c_\ell)^\top \mathbf{v} + (c_\ell^\top \mathbf{x}^* + a_\ell^\top \mathbf{y}^*)a_\ell^\top \mathbf{v}]^2\\
&+\frac{8}{m}\sum_{\ell=1}^{m}2[(a_\ell^\top \mathbf{x}^* - c_\ell^\top \mathbf{y}^*)a_\ell^\top \mathbf{u} + (c_\ell^\top \mathbf{x}^* + a_\ell^\top \mathbf{y}^*)c_\ell^\top \mathbf{u}]\times\\
&[(a_\ell^\top \mathbf{x}^* - c_\ell^\top \mathbf{y}^*)(-c_\ell)^\top \mathbf{v} + (c_\ell^\top \mathbf{x}^* + a_\ell^\top \mathbf{y}^*)a_\ell^\top \mathbf{v}]\\
&= \frac{8}{m}\sum_{\ell=1}^{m}[(a_\ell^\top \mathbf{x}^* - c_\ell^\top \mathbf{y}^*)a_\ell^\top \mathbf{u} + (c_\ell^\top \mathbf{x}^* + a_\ell^\top \mathbf{y}^*)c_\ell^\top \mathbf{u} + (a_\ell^\top \mathbf{x}^* - c_\ell^\top \mathbf{y}^*)(-c_\ell)^\top \mathbf{v}\\
&+(c_\ell^\top \mathbf{x}^* + a_\ell^\top \mathbf{y}^*)a_\ell^\top \mathbf{v}]^2\\
&\geq 0.
\end{aligned}$$

It means that $H_f(\mathbf{x}^*, \mathbf{y}^*) \geq 0$. Furthermore, if we choose $\mathbf{u} = -\mathbf{y}^*$ and $\mathbf{v} = \mathbf{x}^*$, then it is easy to show that

$$[-(\mathbf{y}^*)^\top\ (\mathbf{x}^*)^\top]^\top H_f(\mathbf{x}^*, \mathbf{y}^*)\left[-\mathbf{y}^*\,;\,\mathbf{x}^*\right] = 0,$$

which gives that the Hessian H_F along this direction is zero. □

References

1. Attouch, H., Bolte, J.: On the convergence of the proximal algorithm for nonsmooth functions involving analytic features. Math. Program. **116**, 5–16 (2009)
2. Attouch, H., Peypouquet, J.: The rate of convergence of Nesterov's accelerated forward-backward method is actually faster than $1/k^2$. SIAM J. Optim. **26**, 1824–1834 (2016)
3. Attouch, H., Bolte, J., Redont, P., Soubeyran, A.: Proximal alternating minimization and projection methods for nonconvex problems: an approach based on the Kurdyka-Łojasiewicz inequality. Math. Oper. Res. **35**(2), 438–457 (2010)
4. Balan, R., Casazza, P., Edidin, D.: On signal reconstruction without phase. Appl. Comput. Harmon. Anal. **20**(3), 345–356 (2006)
5. Barzilai, J., Borwein, J.M.: Two-point step size gradient methods. IMA J. Numer. Anal. **8**(1), 141–148 (1988)
6. Beck, A., Eldar, Y.C.: Sparsity constrained nonlinear optimization: optimality conditions and algorithms. SIAM J. Optim. **23**(3), 1480–1509 (2013)
7. Beck, A., Teboulle, M.: A fast iterative shrinkage-thresholding algorithm for linearinverse problems. SIAM J. Imaging Sci. **2**, 183–202 (2009)
8. Bostan, E., Soltanolkotabi, M., Ren, D., Waller, L.: Accelerated Wirtinger flow for multiplexed Fourier ptychographic microscopy. In: 2018 25th IEEE International Conference on Image Processing, pp. 3823–3827 (2018)
9. Cai, T.T., Li, X., Ma, Z.: Optimal rates of convergence for noisy sparse phase retrieval via thresholded Wirtinger flow. Ann. Stat. **44**(5), 2221–2251 (2016)
10. Candes, E.J., Eldar, Y.C., Strohmer, T., Voroninski, V.: Phase retrieval via matrix completion. SIAM Rev. **57**(2), 225–251 (2015)
11. Candes, E.J., Li, X., Soltanolkotabi, M.: Phase retrieval from coded diffraction patterns. Appl. Comput. Harmon. Anal. **39**(2), 277–299 (2015)
12. Candes, E.J., Li, X., Soltanolkotabi, M.: Phase retrieval via Wirtinger flow: theory and algorithms. IEEE Trans. Inf. Theory **61**(4), 1985–2007 (2015)
13. Chen, Y., Candes, E.J.: Solving random quadratic systems of equations is nearly as easy as solving linear systems. Commun. Pure Appl. Math. **70**(5), 822–883 (2017)
14. Conca, A., Edidin, D., Hering, M., Vinzant, C.: An algebraic characterization of injectivity in phase retrieval. Appl. Comput. Harmon. Anal. **38**(2), 346–356 (2015)
15. Dai, Y.H., Liao, L.Z.: R-linear convergence of the Barzilai and Borwein gradient method. IMA J. Numer. Anal. **22**, 1–10 (2002)
16. Dainty, C., Fienup, J.R.: Phase retrieval and image reconstruction for astronomy. In: Image Recovery: Theory and Application, pp. 231–275. Academic, New York (1987)
17. Deutsch, F.R.: Best Approximation in Inner Product Spaces. Springer Science and Business Media, New York (2012)
18. Fienup, J.R.: Phase retrieval algorithms: a comparison. Appl. Opt. **21**(15), 2758–2769 (1982)
19. Fulton, W.: Intersection Theory, vol. 2. Springer Science and Business Media, New York (2013)
20. Gao, B., Xu, Z.: Phaseless recovery using the Gauss–Newton method. IEEE Trans. Signal Process. **65**(22), 5885–5896 (2017)
21. Gong, P., Zhang, C., Lu, Z., Huang, J., Ye, J.: A general iterative shrinkage and thresholding algorithm for non-convex regularized optimization problems. In: The 30th International Conference on Machine Learning, pp. 37–45 (2013)
22. Harris, J.: Algebraic Geometry: A First Course, vol. 133. Springer Science and Business Media, New York (2013)
23. Hartshorne, R.: Algebraic Geometry, vol. 52. Springer Science and Business Media, New York (2013)
24. Huang, M., Xu, Z.: Phase retrieval from the norms of affine transformations (2018). arXiv:1805.07899

25. Jaganathan, K., Eldar, Y.C., Hassibi, B.: Phase retrieval: an overview of recent developments (2015). arXiv:1510.07713v1
26. Kurdyka, K.: On gradients of functions definable in o-minimal structures. Ann. Inst. Fourier **48**(3), 769–784 (1998)
27. Łojasiewicz, S.: Une propriété topologique des sous-ensembles analytiques réels. Les équations aux dérivées partielles **117**, 87–89 (1963)
28. Miao, J., Ishikawa, T., Johnson, B., Anderson, E.H., Lai, B., Hodgson, K.O.: High resolution 3D X-ray diffraction microscopy. Phys. Rev. Lett. **89**(8), 088303 (2002)
29. Millane, R.P.: Phase retrieval in crystallography and optics. J. Opt. Soc. Am. A **7**(3), 394–411 (1990)
30. Nesterov, Y.: Introductory Lectures on Convex Optimization: A Basic Course, vol. 87. Springer Science and Business Media, New York (2013)
31. Robert, W.H.: Phase problem in crystallography. J. Opt. Soc. Am. A **10**(5), 1046–1055 (1993)
32. Rumelhart, D.E., Hinton, G.E., Williams, R.J.: Learning representations by back-propagating errors. Nature **323**(6088), 533–536 (1986)
33. Shafarevich, I.R.: Basic Algebraic Geometry 1. Springer, Berlin (2013)
34. Shechtman, Y., Beck, A., Eldar, Y.C.: GESPAR: efficient phase retrieval of sparse signals. IEEE Trans. Signal Process. **62**(4), 928–938 (2014)
35. Shechtman, Y., Eldar, Y.C., Cohen, O., Chapman, H.N., Miao, J., Segev, M.: Phase retrieval with application to optical imaging: a contemporary overview. IEEE Signal Process. Mag. **32**(3), 87–109 (2015)
36. Sun, J., Qu, Q., Wright, J.: A geometric analysis of phase retrieval. Found. Comput. Math. **18**(5), 1131–1198 (2018)
37. Tao, P.D.: The DC (difference of convex functions) programming and DCA revisited with DC models of real world nonconvex optimization problems. Ann. Oper. Res. **133**, 23–46 (2005)
38. Vinzant, C.: A small frame and a certificate of its injectivity. In: 2015 International Conference on Sampling Theory and Applications, pp. 197–200 (2015)
39. Wang, Y., Xu, Z.: Phase retrieval for sparse signals. Appl. Comput. Harmon. Anal. **37**, 531–544 (2014)
40. Wang, Y., Xu, Z.: Generalized phase retrieval: measurement number, matrix recovery and beyond. Appl. Comput. Harmon. Anal. **47**(2), 423–446 (2019)
41. Wen, B., Chen, X., Pong, T.K. A proximal difference-of-convex algorithm with extrapolation. Comput. Optim. Appl. **69**(2), 297–324 (2018)
42. Wu, C., Li, C., Long, Q.: A DC programming approach for sensor network localization with uncertainties in anchor positions. J. Ind. Manag. Optim. **10**(3), 817–826 (2014)
43. Xu, Y., Yin, W.: A block coordinate descent method for regularized multiconvex optimization with applications to nonnegative tensor factorization and completion. SIAM J. Imaging Sci. **6**(3), 1758–1789 (2013)
44. Zariski, O.: On the purity of the branch locus of algebraic functions. Proc. Natl. Acad. Sci. **44**(8), 791–796 (1958)
45. Zhang, H., Liang, Y.: Reshaped Wirtinger flow for solving quadratic systems of equations. In: Advances in Neural Information Processing Systems, pp. 2622–2630 (2016)
46. Zheng, Y., Zheng, B.: A new modified Barzilai–Borwein gradient method for the quadratic minimization problem. J. Optim. Theory Appl. **172**(1), 179–186 (2017)

Modifications of Prony's Method for the Recovery and Sparse Approximation with Generalized Exponential Sums

Ingeborg Keller and Gerlind Plonka

Abstract In this survey we describe some modifications of Prony's method. In particular, we consider the recovery of general expansions into eigenfunctions of linear differential operators of first order. We show, how these expansions can be reconstructed from function samples using generalized shift operators. We derive an ESPRIT-like algorithm for the generalized recovery method and illustrate, how the method can be simplified if some frequency parameters are known beforehand. Furthermore, we present a modification of Prony's method for sparse approximation with exponential sums which leads to a non-linear least-squares problem.

Keywords Generalized Prony method · Generalized exponential sums · Shifted Gaussians · Eigenfunctions of linear operators · Sparse signal approximation · Nonstationary signals

1 Introduction

The recovery and sparse approximation of structured functions is a fundamental problem in many areas of signal processing and engineering. In particular, exponential sums and their generalizations play an important role in time series analysis and in system theory [13, 15], in the theory of annihilating filters, and for the recovery of signals with finite rate of innovation [3, 10, 26, 35, 37], as well as for linear prediction methods [17, 34]. For system reduction, Prony's method is related to the problem of low-rank approximation of structured matrices (particularly Hankel matrices) and corresponding nonlinear least-squares problems [18, 36]. There is a close relation between Prony's method and Padé approximation [4, 9]. Extended models have also been studied in [16]. Exponential sums started to become more important for sparse approximation of smooth functions, see [5, 6, 12, 23], and

I. Keller · G. Plonka (✉)
University of Göttingen, Institute for Numerical and Applied Mathematics, Göttingen, Germany
e-mail: i.keller@math.uni-goettingen.de; plonka@math.uni-goettingen.de
http://na.math.uni-goettingen.de

G. E. Fasshauer et al. (eds.), *Approximation Theory XVI*, Springer Proceedings in Mathematics & Statistics 336, https://doi.org/10.1007/978-3-030-57464-2_7

this question is closely related to approximation in Hardy spaces and the theory of Adamjan, Arov and Krein, see [1, 2, 22].

1.1 The Classical Prony Method

A fundamental problem discussed in many papers is the recovery of exponential sums of the form

$$f(x) := \sum_{j=1}^{M} c_j \, \mathrm{e}^{\alpha_j x} = \sum_{j=1}^{M} c_j \, z_j^x, \qquad \text{with} \quad z_j := \mathrm{e}^{\alpha_j}, \tag{1}$$

where the coefficients $c_j \in \mathbb{C} \setminus \{0\}$ as well as the pairwise different frequency parameters $\alpha_j \in \mathbb{C}$ or equivalently, $z_j \in \mathbb{C} \setminus \{0\}$ are unknown. For simplicity we assume that the number of terms M is given beforehand. One important question appears: What information about f is needed in order to solve this recovery problem uniquely?

The classical Prony method uses the equidistant samples $f(0), f(1), \ldots, f(2M-1)$. Indeed, if we suppose that $\operatorname{Im} \alpha_j$, $j = 1, \ldots, M$, lies in a predefined interval of length 2π, as e.g. $[-\pi, \pi)$, these $2M$ samples are sufficient. This can be seen as follows.

We can view $f(x)$ as the solution of a homogeneous linear difference equation of order M with constant coefficients and try to identify these constant coefficients in a first step. We define the characteristic polynomial (Prony polynomial) with the help of its (yet unknown) zeros $z_j = \mathrm{e}^{\alpha_j}$, $j = 1, \ldots, M$, and consider its monomial representation,

$$p(z) := \prod_{j=1}^{M} (z - \mathrm{e}^{\alpha_j}) = z^M + \sum_{k=0}^{M-1} p_k \, z^k.$$

Then the coefficients p_k, $k = 0, \ldots, M-1$, and $p_M = 1$ satisfy

$$\sum_{k=0}^{M} p_k f(k+m) = \sum_{k=0}^{M} p_k \sum_{j=1}^{M} c_j z_j^{k+m} = \sum_{j=1}^{M} c_j z_j^m \sum_{k=0}^{M} p_k z_j^k = \sum_{j=1}^{M} c_j z_j^m \, p(z_j) = 0$$

for all $m \in \mathbb{Z}$. Thus the coefficients p_k of the linear difference equation can be computed by solving the linear system

$$\sum_{k=0}^{M-1} p_k \, f(k+m) = -f(M+m), \qquad m = 0, \ldots, M-1.$$

Knowing $p(z)$, we can simply compute its zeros $z_j = e^{\alpha_j}$, and in a further step the coefficients c_j, $j = 1, \ldots, M$, by solving the (overdetermined) system

$$f(\ell) = \sum_{j=1}^{M} c_j \, z_j^{\ell}, \qquad \ell = 0, \ldots, 2M-1.$$

In practice there are different numerical algorithms available for this method, which take care for the inherit numerical instability of this approach, see e.g. [14, 24, 27, 29, 31]. Note that for a given arbitrary vector $(f_k)_{k=0}^{2M-1}$, the interpolation problem

$$f_k = \sum_{j=1}^{M} c_j \, z_j^{k}, \qquad k = 0, \ldots, 2M-1,$$

may not be solvable, see e.g. [8]. The characteristic polynomial $p(z)$ of the homogeneous difference equation $\sum_{k=0}^{M} p_k f_{k+m} = 0$, $m = 0, \ldots, M-1$, may have zeros with multiplicity greater than 1, whereas the exponential sum in (1) is only defined for pairwise different zeros. In this paper, we will exclude the case of zeros with multiplicity greater than 1. However, the zeros e^{α_j} of the characteristic polynomial $p(z)$ resp. the parameters α_j, $j = 1, \ldots, M$, may be arbitrarily close. This may lead to highly ill-conditioned system matrices $(f(k+m))_{k,m=0}^{M-1}$.

1.2 Content of This Paper

In this paper, we will particularly consider the following questions.

1. How can we generalize Prony's method in order to recover other expansions than (1)?
2. What kind of information is needed in order to recover the considered expansion?
3. How can we modify Prony's method such that we are able to optimally approximate a given (large) vector of function values in the Euclidean norm by a sparse exponential sum?

To tackle the first question, we introduce the operator based general Prony method in [33] and apply it to study expansions of the form

$$f(x) = \sum_{j=1}^{M} c_j \, H(x) \, e^{\alpha_j G(x)}, \qquad x \in [a, b] \subset \mathbb{R}, \tag{2}$$

where $c_j, \alpha_j \in \mathbb{C}$, $c_j \neq 0$, α_j pairwise different, $G, H \in C^{\infty}(\mathbb{R})$ are predefined functions, where G is strictly monotone on $[a, b]$, and H is nonzero on $[a, b]$. This model covers many interesting examples as e.g. shifted Gaussians, generalized monomial sums and others. For the expansions (2) we will derive different sets of samples which are sufficient for the recovery of all model parameters, thus answering the second question.

In regard to question 3 we will focus on the case of f as in (1) and (2) and show how the methods can be modified for optimal approximation, and how to treat the case of noisy measurements.

The outline of the paper is as follows. First we will introduce the idea of the operator based Prony method by looking at the recovery problem of the classical exponential sum from different angles. In Sect. 3, we study the recovery of the more general expansion f of the form (2). We will show that (2) can be viewed as an expansion into eigenfunctions of a differential operator of first order and thus, according to the generalized Prony method in [21], can be recovered using higher order derivative values of f. We will show, how to find a new generalized shift operator possessing the same eigenfunctions. This leads to a recovery method that requires only function values of f instead of derivative values. The idea will be further illustrated with several examples in Sect. 3.3. Section 4 is devoted to the numerical treatment of the generalized recovery method. We will derive an ESPRIT-like algorithm for the computation of all unknown parameters in the expansion (2). This algorithm also applies if the number of terms M in the expansion (2) is not given beforehand. Furthermore, we show in Sect. 4.3, how the recovery problem can be simplified if some frequencies α_j are known beforehand (while the corresponding coefficients c_j are unknown). Finally, in Sect. 5 we study the optimal approximation with exponential sums in the Euclidean norm. This leads to a nonlinear least squares problem which we tackle directly using a Levenberg-Marquardt iteration. Our approach is essentially different from earlier algorithms, as e.g. [7, 19, 20, 38].

2 Operator Based View to Prony's Method

In order to tackle the questions 1 and 2 in Sect. 1.2, we start by reconsidering Prony's method. As an introductory example, we study the exponential sum in (1) from a slightly different viewpoint. For $h \in \mathbb{R} \setminus \{0\}$ let $S_h : C^\infty(\mathbb{R}) \to C^\infty(\mathbb{R})$ be the shift operator given by $S_h f := f(\cdot + h)$. Then, for any $\alpha \in \mathbb{C}$, the function $\mathrm{e}^{\alpha x}$ is an eigenfunction of S_h with corresponding eigenvalue $\mathrm{e}^{\alpha h}$, i.e.,

$$(S_h \mathrm{e}^{\alpha \cdot})(x) = \mathrm{e}^{\alpha(x+h)} = \mathrm{e}^{\alpha h}\, \mathrm{e}^{\alpha x}.$$

Therefore, the exponential sum in (1) can be seen as a sparse expansion into eigenfunctions of the shift operator S_h. The eigenvalues $\mathrm{e}^{\alpha_j h}$ are pairwise different, if we assume that $\operatorname{Im} \alpha_j \in [-\pi/h, \pi/h)$. Now we consider the Prony polynomial

$$p(z) := \prod_{j=1}^{M} (z - e^{\alpha_j h}) = \sum_{k=0}^{M} p_k \, z^k$$

defined by the (unknown) eigenvalues $e^{\alpha_j h}$ corresponding to the active eigenfunctions in the expansion f in (1). Then, for any predefined $x_0 \in \mathbb{R}$ we have

$$\begin{aligned}\sum_{k=0}^{M} p_k f(x_0 + h(k+m)) &= \sum_{k=0}^{M} p_k (S_h^{k+m} f)(x_0) = \sum_{k=0}^{M} p_k \sum_{j=1}^{M} c_j (S_h^{k+m} e^{\alpha_j \cdot})(x_0) \\ &= \sum_{j=1}^{M} c_j \sum_{k=0}^{M} p_k e^{\alpha_j (hm+hk)} \, e^{\alpha_j x_0} \\ &= \sum_{j=1}^{M} c_j e^{\alpha_j hm} p(e^{\alpha_j h}) e^{\alpha_j x_0} = 0, \end{aligned} \tag{3}$$

i.e., we can reconstruct $p(z)$ by solving this homogeneous system for $m = 0, \ldots, M-1$. We conclude that the exponential sum in (1) can be recovered from the samples $f(h\ell + x_0)$, $\ell = 0, \ldots, 2M-1$. This is a slight generalization of the original Prony method in Sect. 1.1 as we introduced an arbitrary sampling distance $h \in \mathbb{R} \setminus \{0\}$ and a starting point $x_0 \in \mathbb{R}$.

Moreover, we can also replace the samples $(S_h^{k+m} f)(x_0) = f(h(k+m) + x_0)$ in the above computation (3) by any other representation of the form $F(S_h^{k+m} f)$, where $F : C^\infty(\mathbb{R}) \to \mathbb{C}$ is a linear functional satisfying $F(e^{\alpha \cdot}) \neq 0$ for all $\alpha \in \mathbb{C}$, since

$$\sum_{k=0}^{M} p_k F(S_h^{k+m} f) = \sum_{k=0}^{M} p_k \sum_{j=1}^{M} c_j F(S_h^{k+m} e^{\alpha_j \cdot}) = \sum_{j=1}^{M} c_j e^{\alpha_j hm} p(e^{\alpha_j h}) \, F(e^{\alpha_j \cdot}) = 0.$$

Any set of samples of the form $F(S_h^\ell f)$, $\ell = 0, \ldots, 2M-1$, is sufficient to recover f in (1), and the above set is obtained using the point evaluation functional $F = F_{x_0}$ with $F_{x_0} f := f(x_0)$ with $x_0 \in \mathbb{R}$. For further generalizations of the sampling scheme we refer to [33].

This operator-based view leads us to the generalized Prony method introduced in [21], which can be applied to recover any sparse expansion into eigenfunctions of a linear operator.

To illustrate this idea further, let us consider now the differential operator $D : C^\infty(\mathbb{R}) \to C^\infty(\mathbb{R})$ given by $(Df)(x) := f'(x)$ with f' denoting the first derivative of f. Due to

$$(De^{\alpha \cdot})(x) = \alpha \, e^{\alpha x}$$

we observe that exponentials $e^{\alpha x}$ are eigenfunctions of D corresponding to the eigenvalues $\alpha \in \mathbb{C}$. Thus, the sum of exponentials in (1) can also be seen as a sparse expansion into eigenfunctions of the differential operator D. Similarly as before let now

$$\widetilde{p}(z) := \prod_{j=1}^{M}(z-\alpha_j) = \sum_{k=0}^{M} \widetilde{p}_k\, z^k$$

be the characteristic polynomial being defined by the eigenvalues α_j corresponding to the active eigenfunctions of D in (1), where again $\widetilde{p}_M = 1$ holds. Choosing the functional $Ff := f(x_0)$ for some fixed $x_0 \in \mathbb{R}$, we find for any integer $m \geq 0$

$$\sum_{k=0}^{M} \widetilde{p}_k F(D^{k+m} f) = \sum_{k=0}^{M} \widetilde{p}_k f^{(k+m)}(x_0) = \sum_{k=0}^{M} \widetilde{p}_k \sum_{j=1}^{M} c_j \alpha_j^{k+m}\, e^{\alpha_j x_0}$$
$$= \sum_{j=1}^{M} c_j \alpha_j^m\, \widetilde{p}(\alpha_j)\, e^{\alpha_j x_0} = 0.$$

Thus we can determine $\widetilde{p}_k$, $k = 0, \ldots, M-1$, from $\sum_{k=0}^{M} \widetilde{p}_k f^{(k+m)}(x_0) = 0$ for $m = 0, \ldots, M-1$ and $\widetilde{p}_M = 1$, and recover the zeros α_j of $\widetilde{p}$ in a first step. The c_j are computed in a second step the same way as in the classical case. We conclude that also the sample set $f^{(\ell)}(x_0)$, $\ell = 0, \ldots, 2M-1$, for any fixed value $x_0 \in \mathbb{R}$, is sufficient to recover f. Note that here we do not have any restrictions regarding $\operatorname{Im} \alpha_j$.

This example already shows, that there exist many different sample sets that may be used to recover the exponential sum. In particular, each sample set of the form $F(A^\ell h)$, $\ell = 0, \ldots, 2M-1$, where $A : C^\infty(\mathbb{R}) \to C^\infty(\mathbb{R})$ is a linear operator with eigenfunctions $e^{\alpha x}$ corresponding to pairwise different eigenvalues α (covering the range of α_j in (1)), and where F is an arbitrary (fixed) linear functional satisfying $F(e^{\alpha \cdot}) \neq 0$ for all $\alpha \in \mathbb{C}$, can be employed for recovery.

However, in practice it is usually much easier to provide function samples of the form $f(x_0 + h\ell)$ than higher order derivative values $f^\ell(x_0)$ for $\ell = 0, \ldots, 2M-1$. Therefore, for more general expansions, for example of the form (2), we will raise the following question which has also been investigated in [33]: Suppose we already found a set of samples which is (theoretically) sufficient to recover the expansion at hand. Is it possible to find other sets of samples which can be more easily acquired and also admit a unique recovery of the sparse expansion? In terms of linear operators, we can reformulate this idea: Suppose that we have already found an operator A, such that a considered expansion f is a sparse expansion into M eigenfunctions of A (corresponding to pairwise different eigenvalues). Is it possible to find another operator B that possesses the same eigenfunctions, such that the samples $\widetilde{F}(B^\ell)f$ (with some suitable linear functional $\widetilde{F}$) can be easier obtained than $F(A^\ell)f$ for $\ell = 0, \ldots, 2M-1$?

Back to our introductory example for the exponential sum (1). Let the linear functional F be given as $Ff := f(0)$. Assume that we have found the recovery of (1) from the samples $f^{(\ell)}(0)$, $\ell = 0, \ldots, 2M-1$ first. This sampling set corresponds to the linear differential operator $A = D$ with $Df = f'$. How can we find the shift operator $B = S_h$, knowing just the fact, that (1) can be viewed as a sparse expansion into eigenfunctions of D? Is there a simple link between the linear differential operator D and the shift operator S_h?

This is indeed the case. Taking $\varphi \in C^{\infty}(\mathbb{R})$ with $\varphi(x) = \mathrm{e}^{hx}$, and applying φ (formally) to D, we observe for each exponential $\mathrm{e}^{\alpha x}$, $\alpha \in \mathbb{C}$,

$$\varphi(D)\mathrm{e}^{\alpha\cdot} = \mathrm{e}^{hD}\mathrm{e}^{\alpha\cdot} = \sum_{\ell=0}^{\infty} \frac{h^{\ell}}{\ell!} D^{\ell}\mathrm{e}^{\alpha\cdot} = \left(\sum_{\ell=0}^{\infty} \frac{h^{\ell}}{\ell!}\alpha^{\ell}\right)\mathrm{e}^{\alpha\cdot} = \mathrm{e}^{\alpha h}\,\mathrm{e}^{\alpha\cdot} = S_h\mathrm{e}^{\alpha\cdot}.$$

Therefore, we have $\varphi(D)f = S_h f$ for f in (1). We note that φ also maps the eigenvalues of the differential operator onto the eigenvalues of the shift operator. This idea to switch from differential operators to other more suitable operators will be also applied to general sparse expansions in the next section.

3 Recovery of Generalized Exponential Sums

In this section we focus on the recovery of more general sparse expansions. Let $G : \mathbb{R} \to \mathbb{R}$ be a given function in $C^{\infty}(\mathbb{R})$, which is strictly monotone in a given interval $[a, b] \subset \mathbb{R}$, and let $H : \mathbb{R} \to \mathbb{R}$ be in $C^{\infty}(\mathbb{R})$ and nonzero in $[a, b]$. We consider expansions of the form

$$f(x) = \sum_{j=1}^{M} c_j\, H(x)\, \mathrm{e}^{\alpha_j G(x)}, \qquad x \in [a, b] \subset \mathbb{R}, \tag{4}$$

with $c_j \in \mathbb{C} \setminus \{0\}$ and pairwise different $\alpha_j \in \mathbb{C}$. Obviously, (1) is a special case of (4) with $G(x) = x$ and $H(x) \equiv 1$. In order to recover f, we need to identify the parameters c_j and α_j, $j = 1, \ldots, M$.

3.1 Expansion into Eigenfunctions of a Linear Differential Operator

According to our previous considerations in Sect. 2, we want to apply the so-called generalized Prony method introduced in [21], where we view (4) as an expansion into eigenfunctions of a linear operator.

Step 1 First we need to find a linear operator A that possesses the functions $H(x)\mathrm{e}^{\alpha_j G(x)}$ as eigenfunctions for any $\alpha_j \in \mathbb{C}$. For this purpose, let us define the functions

$$g(x) := \frac{1}{G'(x)}, \qquad \eta(x) := -g(x)\frac{H'(x)}{H(x)} = -\frac{H'(x)}{G'(x)H(x)}, \tag{5}$$

which are well defined on $[a, b]$, since G' and H have no zeros in $[a, b]$. Then the differential operator $A : C^{\infty}(\mathbb{R}) \to C^{\infty}(\mathbb{R})$ with

$$Af(x) := g(x)f'(x) + \eta(x)f(x) \tag{6}$$

satisfies

$$\begin{aligned} A\left(H(\cdot)\mathrm{e}^{\alpha_j G(\cdot)}\right)(x) &= g(x)\left(\alpha_j G'(x)H(x) + H'(x)\right)\mathrm{e}^{\alpha_j G(x)} + \eta(x)\, H(x)\,\mathrm{e}^{\alpha_j G(x)} \\ &= \alpha_j\, H(x)\,\mathrm{e}^{\alpha_j G(x)}, \qquad \alpha_j \in \mathbb{C}, \end{aligned}$$

i.e., the differential operator A indeed possesses the eigenfunctions $H(x)\,\mathrm{e}^{\alpha_j G(x)}$ with corresponding eigenvalues $\alpha_j \in \mathbb{C}$.

Step 2 To reconstruct f in (4), we can apply a similar procedure as in Sect. 2. Let

$$\widetilde{p}(z) := \prod_{j=1}^{M}(z - \alpha_j) = \sum_{k=0}^{M} \widetilde{p}_k\, z^k, \qquad \widetilde{p}_M = 1, \tag{7}$$

be the characteristic polynomial defined by the (unknown) eigenvalues α_j that correspond to the active eigenfunctions of the operator A in the expansion (4). Let $F\colon C^{\infty}(\mathbb{R}) \to \mathbb{C}$ be the point evaluation functional $Ff := f(x_0)$ with $x_0 \in [a, b]$, such that $H(x_0) \neq 0$ and $G'(x_0) \neq 0$. Then, for f as in (4) we observe that

$$\begin{aligned} \sum_{k=0}^{M} \widetilde{p}_k\, F(A^{m+k} f) &= \sum_{k=0}^{M} \widetilde{p}_k \sum_{j=1}^{M} c_j\, F\left(A^{k+m}\left(H(\cdot)\,\mathrm{e}^{\alpha_j G(\cdot)}\right)\right) \\ &= \sum_{k=0}^{M} \widetilde{p}_k \sum_{j=1}^{M} c_j\, \alpha_j^{k+m}\, F\left(H(\cdot)\,\mathrm{e}^{\alpha_j G(\cdot)}\right) \\ &= \sum_{j=1}^{M} c_j\, \alpha_j^m \left(\sum_{k=0}^{M} \widetilde{p}_k\, \alpha_j^k\right)\left(H(x_0)\,\mathrm{e}^{\alpha_j G(x_0)}\right) = 0 \end{aligned} \tag{8}$$

for all integers $m \geq 0$. Thus we can compute the coefficients $\widetilde{p}_k$, $k = 0, \ldots, M-1$, using the values $F(A^{\ell} f)$, $\ell = 0, \ldots, 2M-1$. Having determined the polynomial

$\widetilde{p}(z)$ in (7), we can compute its zeros α_j, and afterwards solve a linear equation system to reconstruct the complex coefficients c_j in (4).

However, the question remains, how to obtain the needed data $F(A^\ell f)$, $\ell = 0, \ldots, 2M-1$. We obtain

$$\begin{aligned} F(A^0 f) &= f(x_0), \\ F(A^1 f) &= g(x_0) f'(x_0) + \eta(x_0) f(x_0), \\ F(A^2 f) &= g(x_0)^2 f''(x_0) + [g(x_0) g'(x_0) + 2g(x_0)\eta(x_0)] f'(x_0) \\ &\quad + [g(x_0)\eta'(x_0) + \eta(x_0)^2] f(x_0). \end{aligned} \tag{9}$$

Since g and η (and their derivatives) are known beforehand, it is sufficient to provide the first $2M$ derivative values of f at one point $x_0 \in [a, b]$ in order to reconstruct f. Therefore we can conclude.

Theorem 1 *Let $G, H \in C^\infty([a, b])$, such that G' and H have no zeros on $[a, b]$, and let $x_0 \in [a, b]$ be fixed. Then f in (4) can be viewed as an expansion into eigenfunctions of the differential operator A as in (6), and can be uniquely reconstructed from the derivative samples $f^{(\ell)}(x_0)$, $\ell = 0, \ldots, 2M-1$.*

Proof As seen from the above computations, the operator A of the form (6) indeed possesses the eigenfunctions $H(x)\,\mathrm{e}^{\alpha_j G(x)}$. In order to reconstruct the parameters α_j, we first have to compute the required values $F(A^\ell f) = (A^\ell f)(x_0)$, $\ell = 0, \ldots, 2M-1$. For this purpose, we need to determine the lower triangular matrix $\mathbf{L} = (\lambda_{m,\ell})_{m,\ell=0}^{2M-1} \in \mathbb{R}^{2M\times 2M}$ such that

$$\left(F(A^\ell f)\right)_{\ell=0}^{2M-1} = \left((A^\ell f)(x_0)\right)_{\ell=0}^{2M-1} = \mathbf{L}\,\left(f^{(\ell)}(x_0)\right)_{\ell=0}^{2M-1}.$$

As seen in (9), we have already $\lambda_{0,0} := 1$, $\lambda_{1,0} := g(x_0)$, $\lambda_{1,1} := \eta(x_0)$. Generally, to obtain the entries of $\mathbf{L}$, we have to consider the elements $\lambda_{m,\ell}$ as functions in x, starting with $\lambda_{0,0}(x) \equiv 1$. By induction, it follows from

$$A^\ell f(x) = \sum_{r=0}^{\ell} \lambda_{\ell,r}(x)\, f^{(r)}(x)$$

that

$$\begin{aligned} A^{\ell+1} f(x) &= \sum_{r=0}^{\ell} g(x)\left(\lambda'_{\ell,r}(x)\, f^{(r)}(x) + \lambda_{\ell,r}(x) f^{(r+1)}(x)\right) + \eta(x)\,\lambda_{\ell,r}(x)\, f^{(r)}(x) \\ &= \sum_{r=0}^{\ell} \left(g(x)\,\lambda'_{\ell,r}(x) + \eta(x)\,\lambda_{\ell,r}(x)\right)\, f^{(r)}(x) + g(x)\,\lambda_{\ell,r}(x) f^{(r+1)}(x). \end{aligned}$$

We conclude the recursion

$$\lambda_{\ell+1,r}(x) := \begin{cases} g(x)\,\lambda'_{\ell,r}(x) + \eta(x)\,\lambda_{\ell,r}(x) & r = 0, \\ g(x)\,(\lambda'_{\ell,r}(x) + \lambda_{\ell,r-1}(x)) + \eta(x)\,\lambda_{\ell,r}(x) & r = 1, \ldots, \ell, \\ g(x)\,\lambda_{\ell,r-1}(x) & r = \ell+1. \end{cases}$$

The matrix entries $\lambda_{\ell,k} := \lambda_{\ell,k}(x_0)$ are well-defined by assumption on H and G. In a second step, we solve the homogeneous equation system (8),

$$\sum_{k=0}^{M} \widetilde{p}_k\, F(A^{k+m} f) = 0, \qquad m = 0, \ldots, M-1.$$

Then we can determine the characteristic polynomial $\widetilde{p}$ in (7) and extract its zeros α_j. Finally, the coefficients c_j can be computed from the linear system

$$F(A^\ell f) = (A^\ell f)(x_0) = \sum_{j=1}^{M} c_j\,(A^\ell(H(\cdot)\,\mathrm{e}^{\alpha_j G(\cdot)}))(x_0) = H(x_0) \sum_{j=1}^{M} c_j \alpha_j^\ell \mathrm{e}^{\alpha_j G(x_0)}$$

for $\ell = 0, \ldots, 2M-1$. □

However, the values $f^{(r)}(x_0)$, $r = 0, \ldots, 2M-1$, may not be easily accessible, and we need some extra effort to compute $F(A^\ell f)$ from the derivatives of f.

3.2 Expansion into Eigenfunctions of a Generalized Shift Operator

Our goal is to find a different set of sample values for the recovery of f in (4), which is easier to obtain but also sufficient for a unique reconstruction. Thus we need to find an operator B which has the same eigenfunctions as A in (6). In addition, we require that $F(B^\ell f)$ (with some point evaluation functions F) can be easily obtained from function values of f. Similarly as in Sect. 2, we consider the linear operator $B = \varphi(A) = \exp(hA)$ with A in (6) and $h \in \mathbb{R} \setminus \{0\}$. We observe for f in (4),

$$\begin{aligned} \exp(hA) f &= \sum_{\ell=0}^{\infty} \frac{h^\ell}{\ell!} A^\ell f = \sum_{\ell=0}^{\infty} \frac{h^\ell}{\ell!} \sum_{j=1}^{M} c_j\, A^\ell \left(H(\cdot)\mathrm{e}^{\alpha_j G(\cdot)}\right) \\ &= \sum_{\ell=0}^{\infty} \frac{h^\ell}{\ell!} \sum_{j=1}^{M} c_j \alpha_j^\ell \left(H(\cdot)\,\mathrm{e}^{\alpha_j G(\cdot)}\right) = \sum_{j=1}^{M} c_j \left(\sum_{\ell=0}^{\infty} \frac{h^\ell}{\ell!} \alpha_j^\ell\right) \left(H(\cdot)\,\mathrm{e}^{\alpha_j G(\cdot)}\right) \end{aligned}$$

$$= \sum_{j=1}^{M} c_j \, \mathrm{e}^{\alpha_j h} \left(H(\cdot) \, \mathrm{e}^{\alpha_j G(\cdot)} \right) = H(\cdot) \sum_{j=1}^{M} c_j \, \mathrm{e}^{\alpha_j (h+G(\cdot))}$$

$$= H(\cdot) \sum_{j=1}^{M} c_j \, \mathrm{e}^{\alpha_j G(G^{-1}(h+G(\cdot)))}$$

$$= \frac{H(\cdot)}{H(G^{-1}(h+G(\cdot)))} \sum_{j=1}^{M} c_j H(G^{-1}(h+G(\cdot))) \, \mathrm{e}^{\alpha_j G(G^{-1}(h+G(\cdot)))}$$

$$= \frac{H(\cdot)}{H(G^{-1}(h+G(\cdot)))} \, f\left(G^{-1}(h+G(\cdot)) \right).$$

Therefore, we define the generalized shift operator

$$S_{H,G,h} f(x) := \frac{H(x)}{H(G^{-1}(h+G(x)))} f\left(G^{-1}(h+G(x)) \right), \tag{10}$$

which depends on the functions H, G, and the step size $h \in \mathbb{R} \setminus \{0\}$. This shift operator has also been introduced in [25] and satisfies the properties

$$S_{H,G,h_2}\left(S_{H,G,h_1} f\right) = S_{H,G,h_1}\left(S_{H,G,h_2} f\right) = S_{H,G,h_1+h_2} f$$

for all $h_1, h_2 \in \mathbb{R}$, and

$$S^k_{H,G,h} f = S_{H,G,kh} f \tag{11}$$

for $k \in \mathbb{Z}$, see Theorem 2.1 in [25]. Observe that the generalized shift operator in (10) is already well defined for continuous functions H, G, and we don't need to assume that G and H are in $C^\infty(\mathbb{R})$. We only need to ensure that G^{-1} and $1/H$ are well defined within the considered sampling interval. We summarize this in the following theorem.

Theorem 2 *Let G, H be continuous functions on an interval $[a, b]$, such that G is strictly monotone in $[a, b]$ and H has no zeros in $[a, b]$. Assume that the pairwise different parameters α_j in the expansion*

$$f(x) = \sum_{j=1}^{M} c_j \, H(x) \, \mathrm{e}^{\alpha_j G(x)}, \qquad x \in [a, b] \subset \mathbb{R}, \tag{12}$$

satisfy $\mathrm{Im}\, \alpha_j \in (-T, T]$ and that $c_j \in \mathbb{C} \setminus \{0\}$. Then f can be uniquely reconstructed from the sample values $f(G^{-1}(h\ell + G(x_0) + h\ell))$, $\ell = 0, \ldots, 2M-1$, where x_0, h are taken such that $0 < |h| < \frac{\pi}{T}$ and $G(x_0 + h\ell) \in [G(a), G(b)]$ for $G(a) < G(b)$ or $G(x_0) + h\ell \in [G(b), G(a)]$ for $G(a) > G(b)$.

Proof From the arguments above, we can conclude that $H(x)\,\mathrm{e}^{\alpha_j G(x)}$ is an eigenfunction of the generalized shift operator $S_{H,G,h}$ in (10) associated with the eigenvalue $\mathrm{e}^{\alpha_j h}$, since

$$\begin{aligned} S_{H,G,h}(H(\cdot)\mathrm{e}^{\alpha_j G(\cdot)}) &= \frac{H(\cdot)}{H(G^{-1}(h+G(\cdot)))}\left(H(G^{-1}(h+G(\cdot)))\mathrm{e}^{\alpha_j G(G^{-1}(h+G(\cdot)))}\right) \\ &= H(\cdot)\,\mathrm{e}^{\alpha_j(h+G(\cdot))} = \mathrm{e}^{\alpha_j h}\,H(\cdot)\,\mathrm{e}^{\alpha_j G(\cdot)}. \end{aligned}$$

Further, for $\mathrm{Im}\,\alpha_j \in (-T, T]$, and $0 < |h| < \frac{\pi}{T}$, the eigenvalues $\mathrm{e}^{\alpha_j h}$ corresponding to active eigenfunctions in (4) are pairwise different, such that we can uniquely derive the active eigenfunctions $H(x)\mathrm{e}^{\alpha_j G(x)}$ in (12) from the corresponding active eigenvalues. We define the Prony polynomial

$$p(z) := \prod_{j=1}^{M}(z - \mathrm{e}^{\alpha_j h}) = \sum_{k=0}^{M} p_k\, z^k \quad \text{with} \quad p_M = 1, \tag{13}$$

using the (unknown) eigenvalues $\mathrm{e}^{\alpha_j h}$, where p_k, $k = 0, \ldots, M-1$, are the (unknown) coefficients of the monomial representation of $p(z)$. Then, we conclude

$$\begin{aligned} \sum_{k=0}^{M} p_k\,(S_{H,G,h}^{k+m} f)(x_0) &= \sum_{k=0}^{M} p_k \sum_{j=1}^{M} c_j\,(S_{H,G,h}^{k+m} H(\cdot)\,\mathrm{e}^{\alpha_j G(\cdot)})(x_0) \\ &= \sum_{k=0}^{M} p_k \sum_{j=1}^{M} c_j\,\mathrm{e}^{\alpha_j h(k+m)}\,H(x_0)\,\mathrm{e}^{\alpha_j G(x_0)} \\ &= H(x_0)\sum_{j=1}^{M} c_j\,\mathrm{e}^{\alpha_j hm}\,\mathrm{e}^{\alpha_j G(x_0)} \sum_{k=0}^{M} p_k\,(\mathrm{e}^{\alpha_j h})^k \\ &= H(x_0)\sum_{j=1}^{M} c_j\,\mathrm{e}^{\alpha_j hm}\,\mathrm{e}^{\alpha_j G(x_0)}\,p(\mathrm{e}^{\alpha_j h}) = 0 \end{aligned} \tag{14}$$

for all integers m, where by definition

$$(S_{H,G,h}^{k+m} f)(x_0) = \frac{H(x_0)}{H(G^{-1}(h(k+m)+G(x_0)))} f(G^{-1}(h(k+m)+G(x_0))).$$

Thus, we can compute the coefficients p_k, $k = 0, \ldots, M-1$, from the homogeneous linear system

$$\sum_{k=0}^{M} p_k \, (S_{H,G,h}^{k+m} f)(x_0) = H(x_0) \sum_{k=0}^{M} p_k \frac{f(G^{-1}(h(k+m)+G(x_0)))}{H(G^{-1}(h(k+m)+G(x_0)))} = 0, \tag{15}$$

for $m = 0, \ldots, M-1$, and $p_M = 1$, or equivalently from

$$\sum_{k=0}^{M-1} p_k \frac{f(G^{-1}(h(k+m)+G(x_0)))}{H(G^{-1}(h(k+m)+G(x_0)))} = -\frac{f(G^{-1}(h(M+m)+G(x_0)))}{H(G^{-1}(h(M+m)+G(x_0)))}, \tag{16}$$

for $m = 0, \ldots, M-1$. The conditions on h and x_0 in the theorem ensure that we only use samples of f in $[a, b]$. The equation system (16) is always uniquely solvable, since the coefficient matrix is invertible. This can be deduced as follows. For f in (12),

$$\left(\frac{f(G^{-1}(h(k+m)+G(x_0)))}{H(G^{-1}(h(k+m)+G(x_0)))} \right)_{m,k=0}^{M-1} = \left(\sum_{j=1}^{M} c_j \, \mathrm{e}^{\alpha_j (h(k+m)+G(x_0)))} \right)_{m,k=0}^{M-1}$$

$$= \left(\mathrm{e}^{\alpha_j hm} \right)_{m=0, j=1}^{M-1, M} \operatorname{diag}\left(c_1 \mathrm{e}^{\alpha_1 G(x_0)}, \ldots, c_M \mathrm{e}^{\alpha_M G(x_0)} \right) \left(\mathrm{e}^{\alpha_j hk} \right)_{j=1, m=0}^{M, M-1}. \tag{17}$$

The first and the last matrix factor are invertible Vandermonde matrices with pairwise different nodes $\mathrm{e}^{\alpha_j h}$, and the diagonal matrix is invertible, since $c_j \neq 0$.

Having solved (16), we can reconstruct $p(z)$ and extract all its zeros $z_j = \mathrm{e}^{\alpha_j h}$. In a second step we can compute the coefficients c_j from the overdetermined system

$$f(G^{-1}(h\ell + G(x_0)) = \sum_{j=1}^{M} c_j \, H(G^{-1}(h\ell + G(x_0))) \, \mathrm{e}^{\alpha_j (h\ell + G(x_0))}, \tag{18}$$

for $\ell = 0, \ldots, 2M-1$. □

3.3 Application to Special Expansions

The model (4) covers many special expansions, and we want to illustrate some of them.

3.3.1 Classical Exponential Sums

Obviously, the model (1) is a special case of (4) with $G(x) := x$ and $H(x) := 1$. In this case, we have

$$g(x) \equiv 1, \quad \eta(x) \equiv 0$$

in (5) such that A in (6) reduces to $Af = f'$. The generalized shift operator in (10) with $G^{-1}(x) = x$ is of the form $S_{1,x,h} f(x) = f(h + x)$ and is therefore just the usual shift operator S_h in Sect. 2. By Theorem 1, the sample values $f^{(\ell)}(x_0)$, $\ell = 0, \ldots, 2M - 1$ are sufficient for recovery of f, where in this case the interval $[a, b]$ can be chosen arbitrarily in $\mathbb{R}$ and thus also x_0. Theorem 2 provides the set of sample values $f(x_0 + h\ell)$ similarly as we had seen already in Sect. 2.

3.3.2 Expansions into Shifted Gaussians

We want to reconstruct expansions of the form

$$f(x) = \sum_{j=1}^{M} c_j \, \mathrm{e}^{-\beta(x-\alpha_j)^2}, \tag{19}$$

where $\beta \in \mathbb{R} \setminus \{0\}$ is known beforehand, and we need to find $c_j \in \mathbb{C} \setminus \{0\}$ and pairwise different $\alpha_j \in \mathbb{C}$, see also [25, 37].

First, we observe that the functions

$$\mathrm{e}^{-\beta(x-\alpha_j)^2} = \mathrm{e}^{-\beta\alpha_j^2} \, \mathrm{e}^{-\beta x^2} \, \mathrm{e}^{2\beta\alpha_j x},$$

are of the form $H(x) \, \mathrm{e}^{\alpha_j G(x)}$, with

$$H(x) := \mathrm{e}^{-\beta\alpha_j^2} \mathrm{e}^{-\beta x^2}, \qquad G(x) := 2\beta x.$$

Using the results in Sects. 3.1 and 3.2, (5) yields

$$g(x) = \frac{1}{G'(x)} = \frac{1}{2\beta}, \qquad \eta(x) = -g(x) \frac{H'(x)}{H(x)} = -\frac{1}{2\beta}(-2\beta x) = x.$$

Therefore, the operator A defined in (6) simplifies to $Af(x) := \frac{1}{2\beta} f'(x) + x\, f(x)$ and

$$A\left(\mathrm{e}^{-\beta(\cdot-\alpha_j)^2}\right)(x) = \left(\frac{1}{2\beta}(-2\beta(x - \alpha_j)) + x\right) \mathrm{e}^{-\beta(x-\alpha_j)^2} = \alpha_j \, \mathrm{e}^{-\beta(x-\alpha_j)^2}.$$

Thus, we can reconstruct f in (19) according to Theorem 1 from the derivative samples $f^{(\ell)}(x_0)$, $\ell = 0, \ldots, 2M - 1$. Here, x_0 can be chosen arbitrarily in $\mathbb{R}$, since $G'(x) = 2\beta \neq 0$ and $H(x) \neq 0$ for all $x \in \mathbb{R}$, which means that the interval $[a, b]$ can be chosen arbitrarily in Theorem 1.

Another sampling set is obtained by Theorem 2. The generalized shift operator $S_{H,G,h}$ in (10) reduces to

$$S_{H,G,h} f(x) = \frac{\mathrm{e}^{-\beta x^2}}{\mathrm{e}^{-\beta((h+2\beta x)/2\beta)^2}} f\left(\frac{h+2\beta x}{2\beta}\right) = \mathrm{e}^{h(x+h/4\beta)} f\left(x + \frac{h}{2\beta}\right). \tag{20}$$

Then

$$\begin{aligned} S_{H,G,h}(\mathrm{e}^{-\beta(\cdot-\alpha_j)^2})(x) &= \mathrm{e}^{h(x+h/4\beta)}\, \mathrm{e}^{-\beta(x+\frac{h}{2\beta}-\alpha_j)^2} \\ &= \mathrm{e}^{h\alpha_j}\, \mathrm{e}^{-\beta(x-\alpha_j)^2}. \end{aligned}$$

Therefore, the expansion in (19) is an expansion into eigenfunctions of the generalized shift operator in (20) and can be reconstructed from the equidistant samples

$$f\left(x_0 + \frac{h\ell}{2\beta}\right), \qquad \ell = 0, \ldots, 2M-1,$$

where $x_0 \in \mathbb{R}$ can be chosen arbitrarily and $0 < |h| < \frac{\pi}{T}$, where T is the a priori known bound satisfying $|\alpha_j| < T$ for all $j = 1, \ldots, M$. Since the interval $[a, b]$ occurring in Theorem 2 can be taken arbitrarily large, we can always take it such that

$$\frac{|G(b) - G(a)|}{2M} = \frac{2|\beta|(b-a)}{2M} > \frac{\pi}{T},$$

and therefore, there is no further condition on the choice of h. We note that the procedure also applies for $\beta \in \mathbb{C} \setminus \{0\}$. In this case we can use the substitution $\widetilde{\alpha}_j = \alpha_j 2\beta$ and take $G(x) = x$.

Remark 1 In particular, the model (19) includes expansions into modulated shifted Gaussians

$$f(x) = \sum_{j=1}^{M} c_j\, \mathrm{e}^{2\pi \mathrm{i} x \kappa_j}\, \mathrm{e}^{-\beta(x-s_j)^2}$$

with $\kappa_j \in [0, 1)$ and $s_j \in \mathbb{R}$ which have been considered in [25]. Since

$$\mathrm{e}^{2\pi \mathrm{i} x \kappa_j}\, \mathrm{e}^{-\beta(x-s_j)^2} = \mathrm{e}^{-\beta s_j^2}\, \mathrm{e}^{-\beta x^2}\, \mathrm{e}^{-x(2\beta s_j + 2\pi \mathrm{i} \kappa_j)},$$

we choose $\alpha_j := 2\beta s_j + 2\pi \mathrm{i}\kappa_j$, $j = 1, \ldots, M$. Then the reconstruction of the α_j is sufficient to find the parameters s_j and κ_j from the real and the imaginary part of α_j, respectively.

Example 1 We illustrate the recovery of expansions into shifted Gaussians and consider f of the form (19) with $M = 10$ and $\beta = \mathrm{i}$. The original parameters in Table 1 have been obtained by applying a uniform random choice from the intervals

Table 1 Parameters c_j and α_j for $f(x)$ in (19) with $M = 10$, see Fig. 1

	$j=1$	$j=2$	$j=3$	$j=4$	$j=5$	$j=6$	$j=7$	$j=8$	$j=9$	$j=10$
Re c_j	−1.754	−1.193	0.174	−1.617	2.066	−1.831	−1.644	−1.976	−1.634	−0.386
Im c_j	−0.756	1.694	−0.279	−1.261	1.620	1.919	−0.245	−1.556	−0.968	−0.365
α_j	0.380	−0.951	0.411	0.845	−1.113	−1.530	−0.813	−0.725	−0.303	−0.031

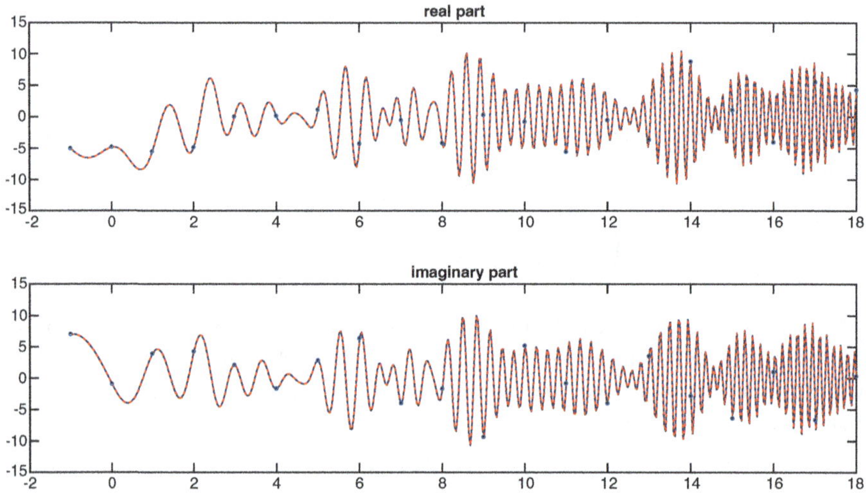

Fig. 1 Real and imaginary part of the signal $f(x)$ consisting of shifted Gaussians as given in Example 1. The black dots indicate the used signal values. Here the reconstructed signal is shown in red and cannot be distinguished from the original signal $f(x)$

$(-3, 3) + \mathrm{i}(-2, 2)$ for c_j and from $(-2, 2)$ for α_j. Since β is complex, we use $G(x) = x$ and the substitution $\widetilde{\alpha}_j = 2\mathrm{i}\alpha_j$. Further, we choose $x_0 = -1$ and $h = 1$.

Figure 1 represents the outcome of such reconstruction. The numerical treatment of the generalized Prony method is studied in more detail in Sect. 4. For the computation of this example we have used Algorithm 1 (see Sect. 4.1) with the minimal number of 20 samples $f(k)$, $k = -1, \ldots, 18$. The samples are represented as black dots in Fig. 1. The obtained maximal reconstruction error for the parameters α_j parameters c_j are

$$err_\alpha = 1.518622755454592 \cdot 10^{-11}, \qquad err_c = 5.286537816367291 \cdot 10^{-10}.$$

3.3.3 Expansions into Functions of the Form $\exp(\alpha_j \sin x)$

We want to reconstruct expansions of the form

$$f(x) = \sum_{j=1}^{M} c_j \, \mathrm{e}^{\alpha_j \sin x}, \tag{21}$$

where we need to find $c_j \in \mathbb{C} \setminus \{0\}$ and pairwise different $\alpha_j \in \mathbb{C}$. Here, $\mathrm{e}^{\alpha_j \sin x}$ is of the form $H(x)\,\mathrm{e}^{\alpha_j G(x)}$ with $H(x) := 1$ and $G(x) := \sin(x)$. To ensure that $G(x)$ is strictly monotone, we choose the interval $[-\frac{\pi}{2}+\delta, \frac{\pi}{2}-\delta]$ with some small $\delta > 0$. With $g(x) = (G'(x))^{-1} = (\cos(x))^{-1}$ and $\eta(x) = 0$ the operator A defined in (6) simplifies to $Af(x) = (\cos(x))^{-1} f'(x)$ and

$$A(\mathrm{e}^{\alpha_j \sin(\cdot)})(x) = \frac{1}{\cos(x)}(\alpha_j \cos(x)\,\mathrm{e}^{\alpha_j \sin(x)}) = \alpha_j \,\mathrm{e}^{\alpha_j \sin(x)}.$$

According to Theorem 1 we can therefore reconstruct f in (21) from the derivative samples $f^{(\ell)}(x_0)$ for some $x_0 \in [-\frac{\pi}{2}+\delta, \frac{\pi}{2}-\delta]$.

Using Theorem 2, we define with $H(x) := 1$ and $G(x) := \sin(x)$ the generalized shift operator

$$S_{H,G,h} f(x) = f(G^{-1}(h + G(x)) = f(\arcsin(h + \sin(x))).$$

We have to choose x_0 and h such that all samples $f(\arcsin(h\ell + \sin(x_0)))$ that we require for the reconstruction are well-defined, i.e., $\sin(x_0) + h\ell \in [-\frac{\pi}{2}+\delta, \frac{\pi}{2}-\delta]$ for $\ell = 0, \ldots, 2M-1$. This is for example ensured for $x_0 = -\frac{\pi}{2} + \frac{h}{2}$ and $0 < h \leq \frac{\pi}{2M+1}$.

Example 2 We illustrate the reconstruction of a function $f(x)$ of the form (21) with $M = 10$ and with real parameters c_j and α_j in Table 2 that have been obtained by applying a uniform random choice from the intervals $(-3, 3)$ for c_j and from $(-\pi, \pi)$ for α_j. We choose a sampling distance $h = \frac{1}{17}$ and a starting point $x_0 = -\frac{\pi}{2} + \frac{h}{2} = -\frac{\pi}{2} + \frac{1}{34}$. The reconstruction is performed using Algorithm 1 in Sect. 4.1.

The reconstruction problem is very ill-posed in this setting, since the measurements all have to be taken from a small interval, see Fig. 2. The possible sampling distance strongly depends on the length of the interval, where $G(x)$ is strictly monotone, as well as on the slope of $G^{-1}(x)$. Therefore, we cannot reconstruct the exact parameters with high precision, however, the reconstructed function is still a very good approximation of f, see Fig. 2.

Table 2 Parameters c_j and α_j for $f(x)$ in (21) with $M = 10$, see Fig. 2

	$j=1$	$j=2$	$j=3$	$j=4$	$j=5$	$j=6$	$j=7$	$j=8$	$j=9$	$j=10$
c_j	2.104	0.363	2.578	1.180	0.497	1.892	2.274	2.933	−2.997	2.192
α_j	1.499	0.540	−1.591	1.046	−2.619	0.791	1.011	1.444	2.455	3.030

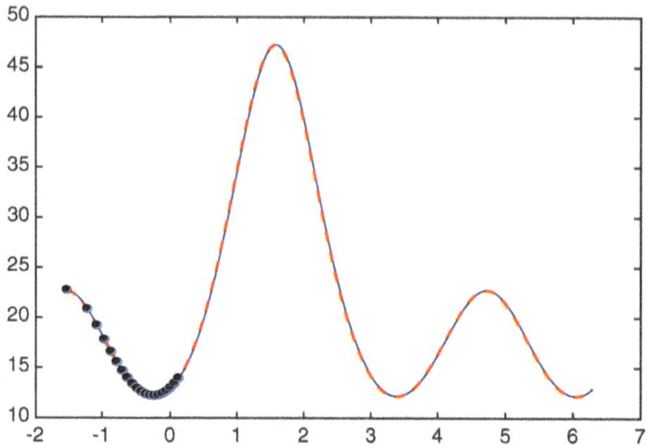

Fig. 2 Signal $f(x)$ in (21) consisting of $M = 10$ terms according to Table 2. The black dots indicate the used signal values and the reconstructed signal is shown in red

4 Numerical Treatment of the Generalized Prony Method

In this section, we consider some numerical procedures to recover the parameters α_j, c_j, $j = 1, \ldots, M$, in (4) resp. (12).

4.1 The Simple Prony Algorithm

First we summarize the direct algorithm for the recovery of f in (12) from the function values $f(G^{-1}(h\ell + G(x_0)))$, $\ell = 0, \ldots, 2M - 1$, according to the proof of Theorem 2.

Algorithm 1
Input: $M \in \mathbb{N}$, $h > 0$, sampled values $f(G^{-1}(h\ell+G(x_0)))$, $\ell = 0, \ldots, 2M-1$.

1. Solve the linear system (16) to find the vector $\mathbf{p} = (p_0, \ldots, p_{M-1})^T$.
2. Compute all zeros $z_j \in \mathbb{C}$, $j = 1, \ldots, M$, of $p(z) = \sum_{k=0}^{M-1} p_k\, z^k + z^M$.
3. Extract the coefficients $\alpha_j := \frac{1}{h} \log z_j$ from $z_j = \mathrm{e}^{\alpha_j h}$, $j = 1, \ldots .M$.
4. Solve the system (18) to compute $c_1, \ldots, c_M \in \mathbb{C}$.

Output: $\alpha_j \in \mathbb{R} + \mathrm{i}[\frac{-\pi}{h}, \frac{\pi}{h})$, $c_j \in \mathbb{C}$, $j = 1, \ldots, M$.

The assumptions of Theorem 2 imply that the coefficient matrix of the linear system (16) is the invertible Hankel matrix,

$$\mathbf{H}_M := \left(\frac{f(G^{-1}(h(k+m)+G(x_0)))}{H(G^{-1}(h(k+m)+G(x_0)))} \right)_{k,m=0}^{M-1}.$$

However, the factorization (17) indicates that $\mathbf{H}_M$ may have very high condition number that particularly depends on the condition number of the Vandermonde matrix $\left(e^{\alpha_j hm}\right)_{m=0,j=1}^{M-1,M}$.

4.2 ESPRIT for the Generalized Prony Method

We are interested in a more stable implementation of the recovery method and present a modification of the ESPRIT method, see [24, 28, 29, 31] for the classical exponential sum. We assume that the number of terms M in (4) is not given beforehand, but L is a known upper bound of M. In the following, we use the notation $\mathbf{A}_{K,N}$ for a rectangular matrix in $\mathbb{C}^{K\times N}$ and $\mathbf{A}_K$ for a square matrix in $\mathbb{C}^{K\times K}$, i.e., the subscripts indicate the matrix dimension.

Let

$$f_\ell := \frac{f(G^{-1}(h\ell+G(x_0)))}{H(G^{-1}(h\ell+G(x_0)))}, \qquad \ell = 0, \ldots, 2N-1, \tag{22}$$

be given and well defined, where $N \geq L \geq M$. We consider first the rectangular Hankel matrix

$$\mathbf{H}_{2N-L,L+1} := (f_{\ell+m})_{\ell,m=0}^{2N-L-1,L} \in \mathbb{C}^{(2N-L)\times(L+1)}.$$

For exact data, (14) implies that rank $\mathbf{H}_{2N-L,L+1} = M$. We therefore compute the singular value decomposition of $\mathbf{H}_{2N-L,L+1}$,

$$\mathbf{H}_{2N-L,L+1} = \mathbf{U}_{2N-L}\, \mathbf{D}_{2N-L,L+1}\, \mathbf{W}_{L+1}, \tag{23}$$

with unitary square matrices $\mathbf{U}_{2N-L}$, $\mathbf{W}_{L+1}$ and a rectangular diagonal matrix $\mathbf{D}_{2N-L,L+1}$ containing the singular values of $\mathbf{H}_{2N-L,L+1}$. We determine the numerical rank M of $\mathbf{H}_{2N-L,L+1}$ by inspecting its singular values $\widetilde{\sigma}_1 \geq \widetilde{\sigma}_2 \geq \ldots \geq \widetilde{\sigma}_{L+1} \geq 0$. We find M as the number of singular values being larger than a predefined bound ϵ. Usually, we can find a clear gap between $\widetilde{\sigma}_M$ and the further singular values $\widetilde{\sigma}_{M+1}, \ldots, \widetilde{\sigma}_{L+1}$, which are close to zero. We redefine the Hankel matrix and consider $\mathbf{H}_{2N-M,M+1} := (f_{\ell+m})_{\ell,m=0}^{2N-M-1,M} \in \mathbb{C}^{(2N-M)\times(M+1)}$ with the corresponding SVD

$$\mathbf{H}_{2N-M,M+1} = \mathbf{U}_{2N-M}\, \mathbf{D}_{2N-M,M+1}\, \mathbf{W}_{M+1}, \tag{24}$$

with unitary matrices $\mathbf{U}_{2N-M}$ and $\mathbf{W}_{M+1}$. For exact data, $\mathbf{H}_{2N-M,M+1}$ has rank M, and $\mathbf{D}_{2N-M,M+1}^T = (\text{diag}(\sigma_1, \ldots, \sigma_M, 0), \mathbf{0}) \in \mathbb{R}^{(M+1)\times(2N-M)}$ with $\sigma_1 \geq \sigma_2 \geq \ldots \geq \sigma_M > 0$.

We introduce the sub-matrices $\mathbf{H}_{2N-M,M}(0)$ and $\mathbf{H}_{2N-M,M}(1)$ given by

$$\mathbf{H}_{2N-M,M+1} = \left(\mathbf{H}_{2N-M,M}(0), (f_{\ell+M})_{\ell=0}^{2N-M-1}\right) = \left((f_\ell)_{\ell=0}^{2N-M-1}, \mathbf{H}_{2N-M,M}(1)\right),$$

i.e., we obtain $\mathbf{H}_{2N-M,M}(0)$ be removing the last column of $\mathbf{H}_{2N-M,M+1}$ and $\mathbf{H}_{2N-M,M}(1)$ by removing the first column of $\mathbf{H}_{2N-M,M+1}$. Recalling (16) we have for exact data

$$\mathbf{H}_{2N-M,M}(0)\,\mathbf{p} = -\,(f_{\ell+M})_{\ell=0}^{2N-M-1}, \qquad (25)$$

where $\mathbf{p} = (p_0, \ldots, p_{M-1})^T$ contains the coefficients of the Prony polynomial in (13). Let now

$$\mathbf{C}_M(\mathbf{p}) := \begin{pmatrix} 0 & 0 & \ldots & 0 & -p_0 \\ 1 & 0 & \ldots & 0 & -p_1 \\ 0 & 1 & \ldots & 0 & -p_2 \\ \vdots & \vdots & & \vdots & \vdots \\ 0 & 0 & \ldots & 1 & -p_{M-1} \end{pmatrix} \in \mathbb{C}^{M\times M}$$

be the (unknown) companion matrix of $\mathbf{p}$ having the M zeros of $p(z)$ in (13) as eigenvalues. By (25) it follows that

$$\mathbf{H}_{2N-M,M}(0)\,\mathbf{C}_M(\mathbf{p}) = \mathbf{H}_{2N-M,M}(1). \qquad (26)$$

This observation leads to the following algorithm. According to (24) we find the factorizations

$$\mathbf{H}_{2N-M,M}(0) = \mathbf{U}_{2N-M}\,\mathbf{D}_{2N-M,M+1}\,\mathbf{W}_{M+1,M}(0),$$
$$\mathbf{H}_{2N-M,M}(1) = \mathbf{U}_{2N-M}\,\mathbf{D}_{2N-M,M+1}\,\mathbf{W}_{M+1,M}(1),$$

where $\mathbf{W}_{M+1,M}(0)$ is obtained by removing the last column of $\mathbf{W}_{M+1}$ and $\mathbf{W}_{M+1,M}(1)$ by removing its first column. Now, (26) implies

$$\mathbf{D}_{2N-M,M+1}\mathbf{W}_{M+1,M}(0)\,\mathbf{C}_M(\mathbf{p}) = \mathbf{D}_{2N-M,M+1}\mathbf{W}_{M+1,M}(1).$$

Multiplication with the generalized inverse

$$\mathbf{D}_{2N-M,M+1}^{\dagger} = \left(\text{diag}\,(\frac{1}{\sigma_1}, \ldots \frac{1}{\sigma_M}, 0), \mathbf{0}\right) \in \mathbb{R}^{(M+1)\times(2N-M)},$$

finally yields

$$\mathbf{W}_M(0)\,\mathbf{C}_M(\mathbf{p}) = \mathbf{W}_M(1),$$

where the square matrices $\mathbf{W}_M(0)$ and $\mathbf{W}_M(1)$ are obtained from $\mathbf{W}_{M+1,M}(0)$ and $\mathbf{W}_{M+1,M}(1)$, respectively, by removing the last row. Thus, the eigenvalues of $\mathbf{C}_M(\mathbf{p})$ are equal to the eigenvalues of

$$\mathbf{W}_M(0)^{-1}\,\mathbf{W}_M(1),$$

where $\mathbf{W}_M(0)$ is invertible since $\mathbf{C}_M(\mathbf{p})$ is invertible. (We can assume here that $z_j \neq 0$ since $z_j = \mathrm{e}^{\alpha_j}$.) We therefore obtain the following new algorithm.

Algorithm 2 (ESPRIT for the generalized Prony method)
Input: $L, N \in \mathbb{N}$, $L \leq N$, L upper bound for the number M of terms in (12), sample values f_ℓ, $\ell = 0, \ldots, 2N-1$ as given in (22), $G(x_0)$.

1. Compute the SVD of the rectangular Hankel matrix $\mathbf{H}_{2N-L,L+1}$ as in (23). Determine the numerical rank M of $\mathbf{H}_{2N-L,L+1}$, and compute the SVD of $\mathbf{H}_{2N-M,M+1} = \mathbf{U}_{2N-M}\,\mathbf{D}_{2N-M,M+1}\,\mathbf{W}_{M+1}$.
2. Build the restricted matrix $\mathbf{W}_M(0)$ by removing the last column and the last row of $\mathbf{W}_{M+1}$ and $\mathbf{W}_M(1)$ by removing the first column and the last row of $\mathbf{W}_{M+1}$. Compute the eigenvalues z_j, $j = 0, \ldots, M$, of $\mathbf{W}_M(0)^{-1}\mathbf{W}_M(1)$.
3. Extract the coefficients $\alpha_j := \frac{1}{h}\log z_j$ from $z_j = \mathrm{e}^{\alpha_j h}$, $j = 1, \ldots, M$.
4. Solve the overdetermined system

$$f_\ell = \sum_{j-1}^{M} c_j\, z_j^{G(x_0)/h}\, z_j^\ell, \qquad \ell = 0, \ldots, 2N-1,$$

to compute $c_1, \ldots, c_M \in \mathbb{C}$.

Output: $M, \alpha_j \in \mathbb{R} + \mathrm{i}[\frac{-\pi}{h}, \frac{\pi}{h})$, $c_j \in \mathbb{C}$, $j = 1, \ldots, M$.

Example 3 We compare the performance of the classical Prony method in Algorithm 1 with the ESPRIT method in Algorithm 2 and focus on the reconstruction of the frequency parameters for f of the form (21). In our numerical example we choose $M = 5$, $x_0 = -\frac{\pi}{2} + \frac{1}{34}$, $h = \frac{1}{17}$ and the parameter vectors $\boldsymbol{\alpha} = (\alpha_j)_{j=1}^M$, $\mathbf{c} = (c_j)_{j=1}^M$ as

$$\boldsymbol{\alpha} = (\frac{\pi}{2}, \frac{\mathrm{i}\pi}{4}, 0.4 + \mathrm{i}, -0.5, -1)^T \text{ and } \mathbf{c} = (0.5, 2, -3, 0.4\mathrm{i}, -0.2)^T.$$

For Algorithm 1 we have only used the first $N = 10$ samples. For the ESPRIT Algorithm 2 we have used $N = 15$, i.e., 30 sample values, and have fixed an upper bound $L = 10$. For the rank approximation we have applied a bound $\epsilon = 10^{-8}$. For comparison we also tested Algorithm 2 with an upper bound of $L = 13$. In Table 3,

Table 3 Reconstructed parameters α_j in Example 3 provided by Algorithms 1 and 2

j	Exact α_j	α_j (Algorithm 1)	α_j (Algorithm 2, $L = 10$)	α_j (Algorithm 2, $L = 13$)
$j = 1$	$\frac{\pi}{2}$	$1.57121 + 6.0886 \cdot 10^{-5}\mathrm{i}$	$1.57079 - 2.3198 \cdot 10^{-8}\mathrm{i}$	$1.57079 - 2.5066\mathrm{i} \cdot 10^{-8}$
$j = 2$	$\frac{\mathrm{i}\pi}{4}$	$0.00231 + 0.7928\mathrm{i}$	$2.00492 \cdot 10^{-6} + 0.7854\mathrm{i}$	$2.00522 \cdot 10^{-6} + 0.7854\mathrm{i}$
$j = 3$	$0.4 + \mathrm{i}$	$0.40168 + 0.9982\mathrm{i}$	$0.4000 + 1\mathrm{i}$	$0.4000 + 1\mathrm{i}$
$j = 4$	-0.5	$-0.49944 - 0.0013\mathrm{i}$	$-0.5 - 4.3008 \cdot 10^{-7}\mathrm{i}$	$-0.5 - 4.5298 \cdot 10^{-7}\mathrm{i}$
$j = 5$	-1	$-1.00019 - 0.0042\mathrm{i}$	$-1.0 - 1.1763^{-6}\mathrm{i}$	$-1.0 - 1.16642 \cdot 10^{-6}\mathrm{i}$

we present the results of parameter reconstruction using Algorithms 1 and 2. The reconstruction of the frequency values using Algorithm 2 is in the case for $L = 10$ as well as in the case $L = 13$ much more accurate than the reconstruction using Algorithm 1. For both upper bounds L the reconstruction error is of the same order. Lemma 3.1 in [30] suggests that a sufficiently large choice of $L \approx N$ is a good choice.

Remark 2 The Hankel matrices occurring in the considered reconstruction problems can have a very high condition number. However, there are stable algorithms available to compute the SVD for such Hankel matrices, particularly for the square case, see e.g. [11].

4.3 Simplification in Case of Partially Known Frequency Parameters

In some applications, one or more of the parameters α_j, or equivalently $z_j = \mathrm{e}^{\alpha_j h}$ in the expansion (12), may be already known beforehand. However, if the corresponding coefficients c_j are unknown, we cannot just eliminate the term $c_j\, H(x)\, \mathrm{e}^{\alpha_j G(x)}$ from the sum in (12) to get new measurements of the simplified sum from the original measurements. However, we can use the following approach. Recall that the vector $\mathbf{p} = (p_0, \ldots, p_M)^T$ of coefficients of the Prony polynomial

$$p(z) = \sum_{k=0}^{M} p_k z^k = \prod_{j=1}^{M} (z - z_j)$$

satisfies by (15) and (16)

$$\mathbf{H}_{2N-M,M+1}\, \mathbf{p} = \mathbf{0},$$

where the Hankel matrix $\mathbf{H}_{2N-M,M+1}$ is constructed from f_ℓ in (22) as in the previous section. Assume that z_1 is already known beforehand, and let

$$q(z) := \prod_{j=2}^{M}(z - z_j) = \sum_{k=0}^{M-1} q_k z^k,$$

with the coefficient vector $\mathbf{q} := (q_0, \ldots, q_{M-1})^T$. Then $p(z) = (z-z_1)q(z)$ implies for the coefficient vectors

$$\mathbf{p} = \begin{pmatrix} 0 \\ q_0 \\ \vdots \\ q_{M-1} \end{pmatrix} - z_1 \begin{pmatrix} q_0 \\ \vdots \\ q_{M-1} \\ 0 \end{pmatrix}$$

and thus

$$\mathbf{H}_{2N-M,M+1}\mathbf{p} = \left(\mathbf{H}_{2N-M,M}(1) - z_1 \mathbf{H}_{2N-M,M}(0)\right) \mathbf{q} = \mathbf{0},$$

with $\mathbf{H}_{2N-M,M}(0)$ and $\mathbf{H}_{2N-M,M}(1)$ denoting the submatrices of $\mathbf{H}_{2N-M,M+1}$, where either the last column or the first column is removed. Therefore, we easily find the new Hankel matrix

$$\widetilde{\mathbf{H}}_{2N-M,M} = \mathbf{H}_{2N-M,M}(1) - z_1 \mathbf{H}_{2N-M,M}(0)$$

for the reduced problem. Observe from (22), that the new components of the matrix $\mathbf{H}_{2N-M,M}(1) - z_1 \mathbf{H}_{2N-M,M}(0)$ are of the form

$$\begin{aligned} \widetilde{f}_\ell = f_{\ell+1} - z_1 f_\ell &= \sum_{j=1}^{M} c_j \mathrm{e}^{\alpha_j(h(\ell+1)+G(x_0))} - \mathrm{e}^{\alpha_1 h} \sum_{j=1}^{M} c_j \mathrm{e}^{\alpha_j(h\ell+G(x_0))} \\ &= \sum_{j=2}^{M} c_j(\mathrm{e}^{\alpha_j h} - \mathrm{e}^{\alpha_1 h})\mathrm{e}^{\alpha_j(h\ell+G(x_0))}, \end{aligned}$$

i.e., the coefficients c_j, $j = 2, \ldots, M$, are changed to $\widetilde{c}_j = c_j(\mathrm{e}^{\alpha_j h} - \mathrm{e}^{\alpha_1 h})$. Thus, we can use the samples $\widetilde{f}_\ell$ to recover the shorter sum $\sum_{j=2}^{M} \widetilde{c}_j \mathrm{e}^{\alpha_j G(x)}$. Once we have computed the remaining α_j, $j = 2, \ldots, M$ we obtain the coefficients c_j, $j = 1, \ldots, M$, by solving the linear system (18).

5 Modified Prony Method for Sparse Approximation

In this section, we want to consider the question, how to approximate a given data vector $\mathbf{y} = (y_k)_{k=0}^N \in \mathbb{C}^{N+1}$ with $N \geq 2M-1$ by a new vector $\mathbf{f} = (f_k)_{k=0}^N \in \mathbb{C}^{N+1}$ whose elements are structured as

$$f_k = \sum_{j=1}^{M} c_j \, z_j^k,$$

i.e., $\mathbf{f}$ only depends on the parameter vectors $\mathbf{c} = (c_j)_{j=1}^M$ and $\mathbf{z} = (z_j)_{j=1}^M$. In this setting, the length N of the data vector $\mathbf{y}$ is usually much larger than M, i.e., $N \gg M$, while M is assumed to be small, say $M < 30$. We assume that for the given data $\mathbf{y}$ the corresponding Hankel matrix $\mathbf{H} := (y_{k+m})_{k=0,m=0}^{N-M-1,M-1}$ has full rank, i.e., that the given data cannot be exactly represented by an exponential sum with less than M terms, as it can be also seen from the factorization (17). Further, we assume that $c_j \in \mathbb{C} \setminus \{0\}$ and that $z_j \in \mathbb{C} \setminus \{0\}$ are pairwise distinct.

5.1 *The Nonlinear Least-Squares Problem*

We want to solve the minimization problem

$$\operatorname*{argmin}_{\mathbf{c},\mathbf{z} \in \mathbb{C}^M} \left\| \mathbf{y} - \left(\sum_{j=1}^{M} c_j \, z_j^k \right)_{k=0}^{N} \right\|_2 . \tag{27}$$

This problem occurs in two different scenarios. The first one is the problem of parameter estimation in case of noisy data. Assume that we have noisy samples $y_k = f(k) + \epsilon_k$, $k = 0, \ldots, N$, of $f(x) = \sum_{j=1}^M c_j z_j^x$, where ϵ_k are i.i.d. random variables with $\epsilon_k \in N(0, \sigma^2)$. In the second scenario we consider the sparse nonlinear approximation problem to find a function $f(x) = \sum_{j=1}^M c_j z_j^x$, which minimizes $\sum_{\ell=0}^N |y_\ell - f(\ell)|^2$. With the Vandermonde matrix

$$\mathbf{V_z} := \begin{pmatrix} 1 & 1 & \ldots & 1 \\ z_1 & z_2 & \ldots & z_M \\ z_1^2 & z_2^2 & \ldots & z_M^2 \\ \vdots & \vdots & & \vdots \\ z_1^N & z_2^N & \ldots & z_M^N \end{pmatrix} \in \mathbb{C}^{(N+1)\times M}$$

we have $\mathbf{f} = \mathbf{V_z}\,\mathbf{c}$, and the problem (27) can be reformulated as

$$\underset{\mathbf{c},\mathbf{z}\in\mathbb{C}^M}{\operatorname{argmin}} \|\mathbf{y} - \mathbf{V_z c}\|_2.$$

For given $\mathbf{z}$, the linear least squares problem $\underset{\mathbf{c}\in\mathbb{C}^M}{\operatorname{argmin}} \|\mathbf{y} - \mathbf{V_z c}\|_2$ can be directly solved, and we obtain $\mathbf{c} = \mathbf{V_z^+ y} = [\mathbf{V_z^* V_z}]^{-1}\mathbf{V_z^* y}$, since $\mathbf{V_z}$ has full rank M. Thus (27) can be simplified to

$$\begin{aligned}\underset{\mathbf{z}\in\mathbb{C}^M}{\operatorname{argmin}} \|\mathbf{y} - \mathbf{V_z V_z^+ y}\|_2^2 &= \underset{\mathbf{z}\in\mathbb{C}^M}{\operatorname{argmin}} \|(\mathbf{I} - \mathbf{P_z})\mathbf{y}\|_2^2 \\ &= \underset{\mathbf{z}\in\mathbb{C}^M}{\operatorname{argmin}} (\mathbf{y}^*\mathbf{y} - \mathbf{y}^*\mathbf{P_z y}) = \underset{\mathbf{z}\in\mathbb{C}^M}{\operatorname{argmax}}\, \mathbf{y}^*\mathbf{P_z y},\end{aligned}$$

where $\mathbf{P_z} := \mathbf{V_z V_z^+}$ is the projection matrix satisfying $\mathbf{P_z} = \mathbf{P_z^*} = \mathbf{P_z^2}$, $\mathbf{P_z V_z} = \mathbf{V_z}$ as well as $\mathbf{V_z^+ P_z} = \mathbf{V_z^+}$. Hence, similarly as for Prony's method, we can concentrate on finding the parameters z_j in $\mathbf{z}$ first.

Let now $\mathbf{r}(\mathbf{z}) := \mathbf{P_z y} \in \mathbb{C}^{N+1}$. Then the optimization problem is equivalent to

$$\underset{\mathbf{z}\in\mathbb{C}^M}{\operatorname{argmax}} \|\mathbf{r}(\mathbf{z})\|_2^2 = \underset{\mathbf{z}\in\mathbb{C}^M}{\operatorname{argmax}} \|\mathbf{P_z y}\|_2^2. \tag{28}$$

To derive an iterative algorithm for solving (28), we first determine the Jacobian $\mathbf{J_z}$ of $\mathbf{r}(\mathbf{z}) = (r_\ell(\mathbf{z}))_{\ell=0}^N$.

Theorem 3 *The Jacobian matrix $\mathbf{J_z} \in \mathbb{C}^{(N+1)\times M}$ of $\mathbf{r}(\mathbf{z})$ in (28) is given by*

$$\begin{aligned}\mathbf{J_z} &:= \left(\frac{\partial r_\ell(\mathbf{z})}{\partial z_j}\right)_{\ell=0,j=1}^{N,M} \\ &= (\mathbf{I}_{N+1} - \mathbf{P_z})\mathbf{V}'_\mathbf{z}\operatorname{diag}(\mathbf{V_z^+ y}) + (\mathbf{V_z^+})^* \operatorname{diag}\left((\mathbf{V}'_\mathbf{z})^*(\mathbf{I}_{N+1} - \mathbf{P}_z)\mathbf{y}\right), \end{aligned}\tag{29}$$

where $\mathbf{I}_{N+1}$ denotes the identity matrix of size $N+1$,

$$\mathbf{V}'_\mathbf{z} := \begin{pmatrix} 0 & 0 & \dots & 0 \\ 1 & 1 & \dots & 1 \\ 2z_1 & 2z_2 & \dots & 2z_M \\ \vdots & \vdots & & \vdots \\ Nz_1^{N-1} & Nz_2^{N-1} & \dots & Nz_M^{N-1} \end{pmatrix} \in \mathbb{C}^{(N+1)\times M},$$

and $\operatorname{diag}(\mathbf{q})$ *denotes the diagonal matrix of size $M \times M$ for a vector $\mathbf{q} \in \mathbb{C}^M$. In particular,*

$$\nabla \|\mathbf{r}(\mathbf{z})\|_2^2 = 2\mathbf{J_z^*}\,\mathbf{r}(\mathbf{z}) = \operatorname{diag}((\mathbf{V}'_\mathbf{z})^T(\mathbf{I}_{N+1} - \mathbf{P}_z)\overline{\mathbf{y}})\,\mathbf{V_z^+ y}. \tag{30}$$

Proof First, observe that $\frac{\partial}{\partial z_j}\mathbf{V_z}$ is a rank-1 matrix of the form

$$\frac{\partial}{\partial z_j}\mathbf{V_z} = \mathbf{z}'_j\, \mathbf{e}_j^* \in \mathbb{C}^{(N+1)\times M}, \qquad j = 1, \ldots, M,$$

where $\mathbf{z}'_j = (0, 1, 2z_j, 3z_j^2, \ldots, Nz_j^{N-1})^T$ and $\mathbf{e}_j$ is the jth unit vector of length M. Then we obtain

$$\begin{aligned}
\frac{\partial}{\partial z_j}\mathbf{r}(\mathbf{z}) &= \frac{\partial}{\partial z_j}(\mathbf{P_z y}) = \frac{\partial}{\partial z_j}\left(\mathbf{V_z}[\mathbf{V_z^* V_z}]^{-1}\mathbf{V_z^* y}\right) \\
&= (\mathbf{z}'_j\, \mathbf{e}_j^*)\mathbf{V_z^+ y} - (\mathbf{V_z^+})^*\left[(\mathbf{z}'_j\, \mathbf{e}_j^*)^*\mathbf{V_z} + \mathbf{V_z^*}(\mathbf{z}'_j\, \mathbf{e}_j^*)\right]\mathbf{V_z^+ y} + (\mathbf{V_z^+})^*(\mathbf{z}'_j\, \mathbf{e}_j^*)^*\mathbf{y} \\
&= (\mathbf{V_z^+ y})_j\, \mathbf{z}'_j - ((\mathbf{z}'_j)^*\mathbf{P_z y})(\mathbf{V_z^+})^*\mathbf{e}_j - (\mathbf{V_z^+ y})_j \mathbf{P_z z}'_j + ((\mathbf{z}'_j)^*\mathbf{y})(\mathbf{V_z^+})^*\mathbf{e}_j \\
&= (\mathbf{V_z^+ y})_j(\mathbf{I}_{N+1} - \mathbf{P_z})\mathbf{z}'_j + ((\mathbf{z}'_j)^*(\mathbf{I}_{N+1} - \mathbf{P_z})\mathbf{y})(\mathbf{V_z^+})^*\mathbf{e}_j \\
&= (\mathbf{V_z^+ y})_j(\mathbf{I}_{N+1} - \mathbf{P_z})\mathbf{V}'_\mathbf{z}\mathbf{e}_j + ((\mathbf{z}'_j)^*(\mathbf{I}_{N+1} - \mathbf{P_z})\mathbf{y})(\mathbf{V_z^+})^*\mathbf{e}_j,
\end{aligned}$$

where $(\mathbf{V_z^+ y})_j$ denotes the jth component of $\mathbf{V_z^+ y}$. From this observation, we immediately find $\mathbf{J_z}$ in (29). Further, this formula implies

$$\begin{aligned}
\mathbf{J_z^*}\mathbf{r}(\mathbf{z}) &= \left(\mathrm{diag}\overline{\mathbf{V_z^+ y}}\right)(\mathbf{V}'_\mathbf{z})^*(\mathbf{I}_{N+1} - \mathbf{P_z})\mathbf{P_z y} + \left(\mathrm{diag}((\mathbf{V}'_\mathbf{z})^*(\mathbf{I}_{N+1} - \mathbf{P_z})\mathbf{y})\right)^*\mathbf{V_z^+ P_z y} \\
&= \mathrm{diag}\left((\mathbf{V}'_\mathbf{z})^T(\mathbf{I}_{N+1} - \mathbf{P_z})\overline{\mathbf{y}}\right)\mathbf{V_z^+ y}.
\end{aligned}$$

□

Corollary 1 *Let* $\mathbf{y} \in \mathbb{C}^{N+1}$ *be given and assume that* $(y_{k+m})_{k=0,m=0}^{N-M+1,M-1}$ *has full rank* M. *Then, a vector* $\mathbf{z} \in \mathbb{C}^M$ *solving (28) necessarily satisfies*

$$(\mathbf{V}'_\mathbf{z})^*(\mathbf{I}_{N+1} - \mathbf{P_z})\mathbf{y} = \mathbf{0}.$$

Proof If $\mathbf{z}$ solves (28), then $\nabla\|\mathbf{r}(\mathbf{z})\|_2^2 = 0$. Now, the assertion follows from (30) using the information that $\mathbf{c} = \mathbf{V_z^+ y}$ has no vanishing components. □

Remark 3

1. The necessary condition in Corollary 1 can be used to build an iterative algorithm for updating the vector $\mathbf{z}$ where we start with $\mathbf{z}^{(0)}$ obtained from the ESPRIT Algorithm 2. We then search for $\mathbf{z}^{(j+1)}$ by solving

$$(\mathbf{V}'_{\mathbf{z}^{(j+1)}})^*(\mathbf{I}_{N+1} - \mathbf{P}_{\mathbf{z}^{(j)}})\mathbf{y} = \mathbf{0},$$

i.e., by computing the zeros of the polynomial with coefficient vector

$$\mathrm{diag}(0, 1, 2, \ldots, N)\,(\mathbf{I}_{N+1} - \mathbf{P}_{\mathbf{z}^{(j)}})\mathbf{y}$$

and taking the subset of M zeros which is closest to the previous set $\mathbf{z}^{(j)}$. We will further elaborate on this approach in the future.

2. This approach is different from most ideas to solve (27) in the literature, see e.g. [7, 19, 20] and the recent survey [38]. In that papers, one first transfers the problem of finding $\mathbf{z} \in \mathbb{C}^M$ into the problem of finding the vector $\mathbf{p} = (p_k)_{k=0}^M \in \mathbb{C}^{M+1}$ with $\|\mathbf{p}\|_2 = 1$, such that $p(z_j) = \sum_{k=0}^M p_k z_j^k = 0$ for all $j = 1, \ldots, M$, thereby imitating the idea of Prony's method. Introducing the matrix

$$\mathbf{X}_\mathbf{p}^T = \begin{pmatrix} p_0 & p_1 & \cdots & p_M & & & \\ & p_0 & p_1 & \cdots & p_M & & \\ & & \ddots & & & \ddots & \\ & & & p_0 & p_1 & \cdots & p_M \end{pmatrix} \in \mathbb{C}^{(N-M+1)\times(N+1)}$$

that satisfies $\mathbf{X}_\mathbf{p}^T \mathbf{V}_\mathbf{z} = \mathbf{0}$, we obtain a projection matrix

$$\overline{\mathbf{P}}_\mathbf{p} := \overline{\mathbf{X}}_\mathbf{p}\overline{\mathbf{X}}_\mathbf{p}^+ = \overline{\mathbf{X}}_\mathbf{p}[\mathbf{X}_\mathbf{p}^T\overline{\mathbf{X}}_\mathbf{p}]^{-1}\mathbf{X}_\mathbf{p}^T = (\mathbf{I}_{N+1} - \mathbf{P}_\mathbf{z}),$$

and (28) can be rephrased as

$$\underset{\substack{\mathbf{p}\in\mathbb{C}^{M+1}\\ \|\mathbf{p}\|_2=1}}{\operatorname{argmin}} \|\overline{\mathbf{P}}_\mathbf{p}\mathbf{y}\|_2^2 = \underset{\substack{\mathbf{p}\in\mathbb{C}^{M+1}\\ \|\mathbf{p}\|_2=1}}{\operatorname{argmin}}\, \mathbf{y}^*\overline{\mathbf{X}}_\mathbf{p}[\mathbf{X}_\mathbf{p}^T\overline{\mathbf{X}}_\mathbf{p}]^{-1}\mathbf{X}_\mathbf{p}^T\mathbf{y}.$$

5.2 Gauß-Newton and Levenberg-Marquardt Iteration

Another approach than given in Remark 3 to solve the non-linear least squares problem (28) is the following. We approximate $\mathbf{r}(\mathbf{z}+\boldsymbol{\delta})$ using its first order Taylor expansion $\mathbf{r}(\mathbf{z}) + \mathbf{J}_\mathbf{z}\boldsymbol{\delta}$. Now, instead of maximizing $\|\mathbf{r}(\mathbf{z}+\boldsymbol{\delta})\|_2^2$ we consider

$$\underset{\boldsymbol{\delta}\in\mathbb{C}^M}{\operatorname{argmax}} \|\mathbf{r}(\mathbf{z}) + \mathbf{J}_\mathbf{z}\boldsymbol{\delta}\|_2^2 = \underset{\boldsymbol{\delta}\in\mathbb{C}^M}{\operatorname{argmax}}(\|\mathbf{r}(\mathbf{z})\|_2^2 + (\mathbf{r}(\mathbf{z})^*\mathbf{J}_\mathbf{z}\,\boldsymbol{\delta} + \boldsymbol{\delta}^*\mathbf{J}_\mathbf{z}^*\,\mathbf{r}(\mathbf{z}) + \boldsymbol{\delta}^*\mathbf{J}_\mathbf{z}^*\,\mathbf{J}_\mathbf{z}\,\boldsymbol{\delta})$$

which yields

$$2\,\mathrm{Re}(\mathbf{J}_\mathbf{z}^*\,\mathbf{r}(\mathbf{z})) + 2\mathbf{J}_\mathbf{z}^*\,\mathbf{J}_\mathbf{z}\,\boldsymbol{\delta} = \mathbf{0}.$$

Thus, starting with the vector $\mathbf{z}^{(0)}$ obtained from Algorithm 2, the jth step of the Gauß-Newton iteration is of the form

$$(\mathbf{J}_{\mathbf{z}^{(j)}}^*\mathbf{J}_{\mathbf{z}^{(j)}})\boldsymbol{\delta}^{(j)} = -\mathrm{Re}\,(\mathbf{J}_{\mathbf{z}^{(j)}}^*\mathbf{r}(\mathbf{z}^{(j)}))$$

to get the improved vector $\mathbf{z}^{(j+1)} = \mathbf{z}^{(j)} + \delta^{(j)}$. Since $(\mathbf{I}_{N+1} - \mathbf{P}_{\mathbf{z}^{(j)}})\mathbf{y}$ may already be close to the zero vector, the matrix $(\mathbf{J}^*_{\mathbf{z}^{(j)}} \mathbf{J}_{\mathbf{z}^{(j)}})$ is usually ill-conditioned. Therefore, we regularize by changing the matrix in each step to $(\mathbf{J}^*_{\mathbf{z}^{(j)}} \mathbf{J}_{\mathbf{z}^{(j)}}) + \lambda_j \mathbf{I}_M$ and obtain the Levenberg-Marquardt iteration

$$((\mathbf{J}^*_{\mathbf{z}^{(j)}} \mathbf{J}_{\mathbf{z}^{(j)}}) + \lambda_j \mathbf{I}_M)\, \delta^{(j)} = -\mathrm{Re}\, (\mathbf{J}^*_{\mathbf{z}^{(j)}} \mathbf{r}(\mathbf{z}^{(j)})).$$

In this algorithm, we need to fix the parameters λ_j, which are usually taken very small. If we arrive at a (local) maximum, then the right-hand side in the Levenberg-Marquardt iteration vanishes, and we obtain $\delta^{(j)} = \mathbf{0}$.

Remark 4

1. The considered non-linear least squares problem is also closely related to structured low-rank approximation, see [18, 36]. Further, instead of the Euclidean norm, one can consider the maximum norm, see [6, 12] or the 1-norm, see [32].
2. Some questions remain. How good is the approximation with exponential sums, if $(y_\ell)_{\ell=0}^N$ is known to be a sampling sequence of a function in a given smoothness space, and what is the convergence rate with respect to the number of terms M? The authors are not aware of a complete answer to this question. However, in [6] it has been shown that the function $1/x$ can be approximated by an M-term exponential sum with an error $\mathcal{O}(\exp(c\sqrt{M})$. Also the results in [5] and [23] indicate that we can hope for an exponential decay of the approximation error for a larger class of functions.

Acknowledgments The authors gratefully acknowledge support by the German Research Foundation in the framework of the RTG 2088. Further, the authors thank the reviewers for many helpful comments to improve the presentation of the results in this paper.

References

1. Adamjan, V., Arov, D., Krein, M.: Analytic properties of the Schmidt pairs of a Hankel operator and the generalized Schur-Takagi problem. Math. USSR Sb. **86**, 34–75 (1971)
2. Andersson, F., Carlsson, M., de Hoop, M.: Sparse approximation of functions using sums of exponentials and AAK theory. J. Approx. Theory **163**, 213–248 (2011)
3. Baechler, G., Scholefield, A., Baboulaz, L., Vetterli, M.: Sampling and exact reconstruction of pulses with variable width. IEEE Trans. Signal Process. **65**(10), 2629–2644 (2017)
4. Barone, P.: On the distribution of poles of Padé approximants to the Z-transform of complex Gaussian white noise. J. Approx. Theory **132**(2), 224–240 (2005)
5. Beylkin, G., Monzón, L.: On approximation of functions by exponential sums. Appl. Comput. Harmon. Anal. **19**, 17–48 (2005)
6. Braess, D., Hackbusch, W.: Approximation of $1/x$ by exponential sums in $[1, \infty)$. IMA J. Numer. Anal. **25**, 685–697 (2005)
7. Bresler, Y., Macovski, A.: Exact maximum likelihood parameter estimation of superimposed exponential signals in noise. IEEE Trans. Acoust. Speech Signal Process. **34**(5), 1081–1089 (1986)

8. Chunaev, P., Danchenko, V.: Approximation by amplitude and frequency operators. J. Approx. Theory **207**, 1–31 (2016)
9. Cuyt, A., Tsai, M.N., Verhoye, M., Lee, W.S.: Faint and clustered components in exponential analysis. Appl. Math. Comput. **327**, 93–103 (2018)
10. Dragotti, P., Vetterli, M., Blu, T.: Sampling moments and reconstructing signals of finite rate of innovation: Shannon meets Strang–Fix. IEEE Trans. Signal Process. **55**(5), 1741–1757 (2007)
11. Drmač, Z.: SVD of Hankel matrices in Vandermonde-Cauchy product form. Electron. Trans. Numer. Anal. **44**, 593–623 (2015)
12. Hackbusch, W.: Computation of best l^{∞} exponential sums for $1/x$ by Remez' algorithm. Comput. Vis. Sci. **20**(1–2), 1–11 (2019)
13. Hauer, J., Demeure, C., Scharf, L.: Initial results in Prony analysis of power system response signals. IEEE Trans. Power Syst. **5**(1), 80–89 (1990)
14. Hua, Y., Sarkar, T.: On the total least squares linear prediction method for frequency estimation. IEEE Trans. Acoust. Speech Signal Process. **38**(12), 2186–2189 (1990)
15. Lang, M.C.: Least-squares design of IIR filters with prescribed magnitude and phase responses and a pole radius constraint. IEEE Trans. Signal Process. **48**(11), 3109–3121 (2000)
16. Levin, D.: Behavior preserving extension of univariate and bivariate functions. In: Hoggan, P. (ed.) Electronic Structure Methods with Applications to Experimental Chemistry, vol. 68, pp. 19–42. Proceedings of MEST 2012. Academic Press, Chennai (2014)
17. Manolakis, D., Ingle, V., Kogon, S.: Statistical and Adaptive Signal Processing. McGraw-Hill, Boston (2005)
18. Markovsky, I.: Low-Rank Approximation: Algorithms, Implementation, Applications, 2nd edn. Springer, Berlin (2018)
19. Osborne, M., Smyth, G.: A modified Prony algorithm for fitting functions defined by difference equations. SIAM J. Sci. Stat. Comput. **12**, 362–382 (1991)
20. Osborne, M., Smyth, G.: A modified Prony algorithm for exponential function fitting. SIAM J. Sci. Comput. **16**(1), 119–138 (1995)
21. Peter, T., Plonka, G.: A generalized Prony method for reconstruction of sparse sums of eigenfunctions of linear operators. Inverse Prob. **29**(2) (2013)
22. Plonka, G., Pototskaia, V.: Application of the AAK theory for sparse approximation of exponential sums (2016). Preprint. http://arxiv.org/pdf/1609.09603
23. Plonka, G., Pototskaia, V.: Computation of adaptive Fourier series by sparse approximation of exponential sums. J. Fourier Anal. Appl. **25**(4), 1580–1608 (2019)
24. Plonka, G., Tasche, M.: Prony methods for recovery of structured functions. GAMM-Mitteilungen **37**(2), 239–258 (2014)
25. Plonka, G., Stampfer, K., Keller, I.: Reconstruction of stationary and non-stationary signals by the generalized Prony method. Anal. Appl. **17**(2), 179–210 (2019)
26. Poh, K., Marziliano, P.: Compressive sampling of EEG signals with finite rate of innovation. EURASIP J. Adv. Signal Process. **2010**, 183105 (2010)
27. Potts, D., Tasche, M.: Parameter estimation for exponential sums by approximate Prony method. Signal Process. **90**(5), 1631–1642 (2010)
28. Potts, D., Tasche, M.: Parameter estimation for multivariate exponential sums. Electron. Trans. Numer. Anal. (40), 204–224 (2013)
29. Potts, D., Tasche, M.: Parameter estimation for nonincreasing exponential sums by Prony-like methods. Linear Algebra Appl. **439**(4), 1024–1039 (2013)
30. Potts, D., Tasche, M.: Error estimates for the ESPRIT algorithm. In: Large Truncated Toeplitz Matrices, Toeplitz Operators, and Related Topics, pp. 621–648. Birkhäuser, Basel (2017)
31. Roy, R., Kailath, T.: Esprit estimation of signal parameters via rotational invariance techniques. IEEE Trans. Acoust. Speech Signal Process. **37**, 984–995 (1989)
32. Skrzipek, M.R.: Signal recovery by discrete approximation and a Prony-like method. J. Comput. Appl. Math. **326**, 193–203 (2017)
33. Stampfer, K., Plonka, G.: The generalized operator-based Prony method. Constr. Approx. (2020). https://doi.org/10.1007/s00365-020-09501-6

34. Stoica, P., Moses, R.L.: Spectral analysis of signals. Pearson Prentice Hall, Upper Saddle River (2005)
35. Urigen, J., Blu, T., Dragotti, P.: FRI sampling with arbitrary kernels. IEEE Trans. Signal Process. **61**(21), 5310–5323 (2013)
36. Usevich, K., Markovsky, I.: Variable projection for affinely structured low-rank approximation in weighted 2-norms. J. Comput. Appl. Math. **272**, 430–448 (2014)
37. Vetterli, M., Marziliano, P., Blu, T.: Sampling signals with finite rate of innovation. IEEE Trans. Signal Process. **50**(6), 1417–1428 (2002)
38. Zhang, R., Plonka, G.: Optimal approximation with exponential sums by a maximum likelihood modification of Pronys method. Adv. Comput. Math. **45**(3), 1657–1687 (2019)

On Eigenvalue Distribution of Varying Hankel and Toeplitz Matrices with Entries of Power Growth or Decay

Gidon Kowalsky and Doron S. Lubinsky

Abstract We study the distribution of eigenvalues of varying Toeplitz and Hankel matrices such as $\left[a_{n+k-j}\right]_{j,k}$ and $\left[a_{n+k+j}\right]_{j,k}$ where a_n behaves roughly like n^{β} for some non-0 complex number β, and $n \to \infty$. This complements earlier work on these matrices when the coefficients $\{a_n\}$ arise from entire functions.

Keywords Toeplitz matrices · Hankel matrices · Eigenvalue distribution

1 Introduction and Results

The distribution of eigenvalues of Toeplitz matrices $\left[c_{k-j}\right]_{1\leq j,k\leq n}$ is a much studied topic, especially when their entries are trigonometric moments [1, 2, 5, 7, 9, 18, 19, 26, 29, 30]. There is a classic paper of Widom [28] dealing with both finite and infinite Hankel matrices $\left[c_{j+k}\right]$. There is a large literature on random Hankel and Toeplitz matrices, see for example, [3, 10, 12, 13, 21, 22]. Generalizations of Toeplitz matrix sequences are considered and studied in [7].

Our interest arises from classical function theory and Padé approximation. There is a connection to complex function theory: Polya [20] proved that if $f(z) = \sum_{j=0}^{\infty} a_j/z^j$ can be analytically continued to a function analytic in the complex plane outside a set of logarithmic capacity $\tau \geq 0$, then

$$\limsup_{n\to\infty} \left|\det\left[a_{n-j+k}\right]_{1\leq j,k\leq n}\right|^{1/n^2} \leq \tau.$$

There are many extensions of this result [4, 16].

In the recent paper [16], we analyzed distribution of the eigenvalues of such matrices under appropriate hypotheses on

G. Kowalsky (✉) · D. S. Lubinsky
School of Mathematics, Georgia Institute of Technology, Atlanta, GA, USA
e-mail: gkowalsky3@gatech.edu; lubinsky@math.gatech.edu

G. E. Fasshauer et al. (eds.), *Approximation Theory XVI*, Springer Proceedings in Mathematics & Statistics 336, https://doi.org/10.1007/978-3-030-57464-2_8

$$q_j = \frac{a_{j-1}a_{j+1}}{a_j^2}.$$

The motivation comes from Padé approximation for functions such as

$$f(z) = \sum_{j=0}^{\infty} z^j / (j!)^{1/\alpha}, \ \alpha > 0, \tag{1.1}$$

for which (cf. [14, 15])

$$q_j = \exp\left(-\frac{1}{\alpha j} + O\left(\frac{1}{j^2}\right)\right).$$

More generally, we considered series

$$f(z) = \sum_{j=0}^{\infty} a_j z^j,$$

that satisfy

$$q_j = \frac{a_{j-1}a_{j+1}}{a_j^2} = \exp\left(-\frac{1}{\rho_j}\left(1 + o\left(\rho_j^{-1/2}\right)\right)\right),$$

with appropriate smoothly increasing or decreasing sequences $\{\rho_j\}$ of positive numbers. We proved, under mild conditions on $\{\rho_j\}$, the following assertions about the eigenvalues $\{\lambda_{nj}\}_{j=1}^{n}$ of the normalized matrix $\frac{1}{a_n}\left[a_{n+k-j}\right]_{1\leq j,k\leq n}$:

1. The eigenvalue of largest modulus satisfies

$$\max_{1\leq j\leq n} \left|\lambda_{nj}\right| = \sqrt{2\pi\rho_n}\,(1 + o(1)).$$

2. The set of all limit points of $\left\{\lambda_{nj}/\sqrt{2\pi\rho_n}\right\}_{1\leq j\leq n, n\geq 1}$ is [0, 1].
3. The scaled zero counting measures

$$\mu_n = \frac{1}{n}\sum_{j=1}^{n}\left(\operatorname{Re}\lambda_{nj}\right)\delta_{\lambda_{nj}/\sqrt{2\pi\rho_m}}$$

admit the weak convergence

$$d\mu_n \xrightarrow{*} |\pi \log t|^{-1/2}\,dt \tag{1.2}$$

in the sense that for each function f defined and continuous in an open subset of the plane containing [0, 1],

$$\lim_{n\to\infty} \int f \, d\mu_n = \int_0^1 f(t) \, |\pi \log t|^{-1/2} \, dt. \tag{1.3}$$

The hypotheses in [16] treat a broad array of entire functions of zero, finite positive, or infinite order, and also some power series of finite radius of convergence. However the hypotheses exclude the case where the coefficients have power growth or decay. It is the purpose of this paper to study that case. The general sequences of Toeplitz matrices in [7] differ from our situation in that our sequences of varying matrices require a different normalization as $n \to \infty$, and a different formulation for the eigenvalue counting measures. Moreover, in Widom's paper [28], the matrices treated have the form $\left[c_{j+k}\right]_{0\le j,k\le n}$, whereas in this paper the top left-hand corner element is a_m with m growing to ∞, so the results and methods are different. We consider the Hankel matrices

$$H_{mn} = \left[a_{m+k+j}\right]_{0\le j,k\le n-1}$$

and Toeplitz matrices

$$T_{mn} = \left[a_{m+k-j}\right]_{1\le j,k\le n}$$

where a_n behaves roughly like n^{β}.

Our approach is also quite different from that in [16], due to the different growth rates. There we used a similarity transformation on T_{mn} and showed that the eigenvalues of T_{mn}/a_m behaved like those of the matrix $E_{mn} = -\left[e^{-\frac{(j-k)^2}{2\rho_n}}\right]_{1\le j,k\le n}$. There roughly $O\left(\sqrt{n}\right)$ central bands of the matrix dominate and one can compute the asymptotics of the trace of E_{mn}^k for each fixed $k = 0, 1, 2, \ldots$. This approach fails for the sequences we consider here, as all bands contribute, and indeed we get a different weak limit from that above.

2 Hankel Matrices

In this section, we state our results for Hankel matrices $\left[a_{m+j+k}\right]_{0\le j,k\le n-1}$ where the a_j grow or decay like j^{β}. Of course if β is real, these matrices are real and symmetric, so have real eigenvalues. In the special case, where $\beta < 0$ and $a_j = j^{\beta}$, these matrices are actually positive definite, so have positive eigenvalues. Indeed this follows directly from the fact that for $\beta < 0$ and $j \ge 1$.

$$j^{\beta}=\frac{1}{\Gamma(-\beta)}\int_0^1 s^j\left(\log\frac{1}{s}\right)^{-\beta-1}s^{-1}ds.$$

This identity in turn follows from the standard integral for the gamma function

$$\Gamma(-\beta)=\int_0^{\infty}t^{-\beta-1}e^{-t}dt$$

by the substitution $s=e^{-t/j}$. Our first result allows possibly complex β. As above we let

$$H_{mn}=\left[a_{m+j+k}\right]_{0\leq j,k\leq n-1}. \tag{2.1}$$

We also let $\Lambda\left(H_{mn}/a_m\right)$ denote the collection of all eigenvalues of H_{mn}/a_m, and form the weighted counting measure

$$\mu_{mn}=\frac{1}{n^2}\sum_{\lambda\in\Lambda(H_{mn}/a_m)}\lambda^2\delta_{\lambda/n}. \tag{2.2}$$

Thus μ_{mn} places mass $\left(\frac{\lambda}{n}\right)^2$ at $\frac{1}{n}\lambda$ for each eigenvalue λ of H_{mn}/a_m. This is rather different from the usual eigenvalue counting measures, but is needed in our situation. The weighting reflects the fact that eigenvalues of H_{mn}/a_m tend to cluster around 0. For general sequences of Hankel and other matrices, this clustering effect has been extensively explored—see [6, 8, 23, 27].

Theorem 2.1 *Fix $k\geq 1$ and $R>0$. Assume $m=m(n)\to\infty$ in such a way that $m/n\to R$ as $n\to\infty$. Assume that $\beta\in\mathbb{C}$ and given $R>0$, we have as $n\to\infty$, uniformly for $0\leq\ell\leq Rm$,*

$$\frac{a_{m+\ell}}{a_m}=\left(1+\frac{\ell}{m}\right)^{\beta}(1+o(1)). \tag{2.3}$$

Then

(I)

$$\limsup_{n\to\infty}\frac{1}{n}\sup\left\{|\lambda|:\lambda\in\Lambda\left(H_{mn}/a_m\right)\right\}$$
$$\leq\int_0^1\max_{0\leq y\leq 1}\left(1+\frac{x+y}{R}\right)^{\operatorname{Re}\beta}dx. \tag{2.4}$$

In particular, the supports of $\left\{\mu_{mn}\right\}_{n\geq 1}$ are contained in a compact set independent of n.

(II)

$$\limsup_{n\to\infty} \left|\mu_{mn}\right| (\mathbb{C}) \leq \int_0^1 \int_0^1 \left(1 + \frac{x+y}{R}\right)^{2\,\mathrm{Re}\,\beta} dx\, dy. \tag{2.5}$$

(III) For $k \geq 1$,

$$\lim_{n\to\infty} \frac{1}{n^k} Tr\left(\left[\frac{H_{mn}}{a_m}\right]^k\right) = c_k, \tag{2.6}$$

where

$$c_k = R^k \int_0^{1/R} \int_0^{1/R} \dots \int_0^{1/R} (1 + t_1 + t_2)^\beta \dots$$
$$(1 + t_{k-1} + t_k)^\beta (1 + t_k + t_1)^\beta\, dt_1 dt_2 \dots dt_k. \tag{2.7}$$

Consequently for $k \geq 0$,

$$\lim_{n\to\infty} \int \lambda^k d\mu_{mn}(\lambda) = c_{k+2}. \tag{2.8}$$

Corollary 2.2 *Assume that* β *is real and all* $\{a_j\}$ *are real. Then there is a finite positive measure* ω *with compact support on the real line such that for all functions* f *continuous on the real line with compact support,*

$$\lim_{n\to\infty} \int f(t)\, d\mu_{mn}(t) = \int f(t)\, d\omega(t). \tag{2.9}$$

The measure ω *is uniquely determined by the moment conditions*

$$\int t^k d\omega(t) = c_{k+2}, k \geq 0.$$

Remarks

(a) Note that (2.3) is satisfied if $a_n = n^\beta b_n$, where $\frac{b_{n+\ell}}{b_n} = 1 + o(1)$ for $0 \leq \ell \leq Rm$. For example this is true if $a_n = n^\beta (\log n)^\gamma (\log\log n)^\kappa$ for some γ, κ.
(b) If we do not assume that the $\{a_j\}$ are real, then we can only prove convergence for functions f analytic in a ball center 0 of large enough radius, as in Corollary 3.2 below.
(c) It is obviously of interest to find an explicit form for ω. There is a classic technique for simplices that provides an explicit value for similar Dirichlet-Liouville multiple integrals [11, 25], but it does not seem to work for cubes.
(d) Note that our eigenvalue counting measure μ_{mn} has a different normalization and scaling to standard ones, so we cannot apply standard results such as in [7].

We prove Theorem 2.1 and Corollary 2.2 in Sect. 4.

3 Toeplitz Matrices

As above, we let

$$T_{mn} = \left[a_{m+k-j}\right]_{1 \leq j,k \leq n}.$$

Here we set $a_j = 0$ if $j < 0$. We also let

$$\nu_{mn} = \frac{1}{n^2} \sum_{\lambda \in \Lambda(T_{mn}/a_m)} \lambda^2 \delta_{\lambda/n}. \tag{3.1}$$

We prove:

Theorem 3.1 *Let $R \geq 1$. Assume $m = m(n) \to \infty$ in such a way that $m/n \to R$ as $n \to \infty$. Let $\beta \in \mathbb{C}$. Assume that given $\varepsilon \in (0, 1)$, we have as $n \to \infty$, uniformly for $-m(1-\varepsilon) \leq \ell \leq (R-1)m$,*

$$\frac{a_{m+\ell}}{a_m} = \left(1 + \frac{\ell}{m}\right)^{\beta} (1 + o(1)). \tag{3.2}$$

If $R = 1$, we assume in addition that $\operatorname{Re} \beta > -1$*and*

$$\lim_{\varepsilon \to 0+} \left(\limsup_{n \to \infty} \frac{1}{n |a_n|} \sum_{j=1}^{[\varepsilon n]} |a_j| \right) = 0. \tag{3.3}$$

Then

(I)

$$\limsup_{n \to \infty} \frac{1}{n} \sup\{|\lambda| : \lambda \in \Lambda(T_{mn}/a_m)\} \leq \int_0^1 \max_{0 \leq y \leq 1} \left(1 + \frac{x-y}{R}\right)^{\operatorname{Re} \beta} dx.$$

In particular, the supports of $\{\nu_{mn}\}_{n \geq 1}$ are contained in a compact set independent of n.

(II)

$$\limsup_{n \to \infty} |\nu_{mn}|(\mathbb{C}) \leq \int_0^1 \int_0^1 \left(1 + \frac{x-y}{R}\right)^{2 \operatorname{Re} \beta} dx\, dy.$$

(III) For $k \geq 1$,

$$\lim_{n\to\infty}\frac{1}{n^k}Tr\left(\left[\frac{T_{mn}}{a_m}\right]^k\right)=d_k,$$

where

$$d_k=R^k\int_0^{1/R}\int_0^{1/R}\dots\int_0^{1/R}(1+t_1-t_2)^\beta\dots$$
$$(1+t_{k-1}-t_k)^\beta(1+t_k-t_1)^\beta\,dt_1dt_2\dots dt_k.$$

Consequently for $k\geq 0$,

$$\lim_{n\to\infty}\int\lambda^k d\nu_{mn}(\lambda)=d_{k+2}. \tag{3.4}$$

Corollary 3.2 *There is a finite complex measure* ω *with compact support in the plane such that for all functions* f *analytic in the ball center* 0, *radius* $\int_0^1\max_{0\leq y\leq 1}\left(1+\frac{x-y}{R}\right)^{\mathrm{Re}\,\beta}dx$,

$$\lim_{n\to\infty}\int f(t)\,d\nu_{mn}(t)=\int f(t)\,d\omega(t)\,. \tag{3.5}$$

The measure ω *admits the moment conditions*

$$\int t^k d\omega(t)=d_{k+2},\,k\geq 0.$$

Here in the case $R=1$, *we assume* $\mathrm{Re}\,\beta>-1$.

We note that it is not clear if the complex valued measure ω is uniquely determined by the moment conditions, as it is supported in the complex plane. We prove the results of this section in Sect. 5.

4 Proof of Theorem 2.1 and Corollary 2.2

Proof of Theorem 2.1(I) It follows from Gershgorin's Theorem [17, p. 146] that every eigenvalue λ of H_{mn}/a_m satisfies

$$\frac{|\lambda|}{n}\leq\max_{0\leq j\leq n-1}\frac{1}{n}\sum_{k=0}^{n-1}\left|\frac{a_{m+k+j}}{a_m}\right|.$$

Our hypothesis (2.3) gives uniformly for $0\leq j,k\leq n-1$,

$$\left|\frac{a_{m+k+j}}{a_m}\right| = \left|\left(1+\frac{k+j}{m}\right)^{\beta}(1+o(1))\right|$$
$$= \left(1+\frac{k+j}{Rn(1+o(1))}\right)^{\operatorname{Re}\beta}(1+o(1))$$
$$\leq \max_{1\leq \ell\leq n}\left(1+\frac{k+\ell}{Rn}\right)^{\operatorname{Re}\beta}(1+o(1)),$$

so that

$$\frac{|\lambda|}{n} \leq \frac{1}{n}\sum_{k=0}^{n-1}\max_{0\leq y\leq 1}\left(1+\frac{k}{Rn}+\frac{y}{R}\right)^{\operatorname{Re}\beta}+o(1)$$
$$\to \int_0^1 \max_{0\leq y\leq 1}\left(1+\frac{x}{R}+\frac{y}{R}\right)^{\operatorname{Re}\beta}dx$$

as $n\to\infty$. ■

Proof of Theorem 2.1(II) By Schur's Inequality [17, p. 142],

$$|\mu_{mn}|(\mathbb{C}) = \frac{1}{n^2}\sum_{\lambda\in\Lambda(H_{mn}/a_m)}|\lambda|^2 \leq \frac{1}{n^2}\sum_{j,k=0}^{n-1}\left|\frac{a_{m+j+k}}{a_m}\right|^2$$
$$= \frac{1}{n^2}\sum_{j,k=0}^{n-1}\left|\left(1+\frac{j+k}{m}\right)^{\beta}(1+o(1))\right|^2$$
$$= \frac{1}{n^2}\sum_{j,k=0}^{n-1}\left(1+\frac{j+k}{Rn}\right)^{2\operatorname{Re}\beta}(1+o(1))$$
$$\to \int_0^1\int_0^1\left(1+\frac{x+y}{R}\right)^{2\operatorname{Re}\beta}dx\,dy$$

as $n\to\infty$. ■

Proof of Theorem 2.1(III) Now

$$\frac{1}{n^k}Tr\left(\left[\frac{H_{mn}}{a_m}\right]^k\right)$$
$$= \frac{1}{n^k}\sum_{j_1=0}^{n-1}\sum_{j_2=0}^{n-1}\cdots\sum_{j_k=0}^{n-1}\frac{a_{m+j_1+j_2}}{a_m}\frac{a_{m+j_2+j_3}}{a_m}\cdots\frac{a_{m+j_{k-1}+j_k}}{a_m}\frac{a_{m+j_k+j_1}}{a_m}$$

$$= \frac{1}{n^k} \sum_{j_1=0}^{n-1} \sum_{j_2=0}^{n-1} \cdots \sum_{j_k=0}^{n-1} \left(1 + \frac{j_1 + j_2}{m}\right)^\beta \left(1 + \frac{j_2 + j_3}{m}\right)^\beta \cdots$$

$$\left(1 + \frac{j_k + j_1}{m}\right)^\beta (1 + o(1))$$

$$= \frac{1}{n^k} \sum_{j_1=0}^{n-1} \sum_{j_2=0}^{n-1} \cdots \sum_{j_k=0}^{n-1} \left(1 + \frac{j_1 + j_2}{nR(1 + o(1))}\right)^\beta \left(1 + \frac{j_2 + j_3}{nR(1 + o(1))}\right)^\beta \cdots$$

$$\left(1 + \frac{j_k + j_1}{nR(1 + o(1))}\right)^\beta (1 + o(1))$$

$$= \frac{1}{n^k} \sum_{j_1=0}^{n-1} \sum_{j_2=0}^{n-1} \cdots \sum_{j_k=0}^{n-1} \left(1 + \frac{j_1 + j_2}{nR}\right)^\beta \left(1 + \frac{j_2 + j_3}{nR}\right)^\beta \cdots$$

$$\left(1 + \frac{j_k + j_1}{nR}\right)^\beta + o(1),$$

since each of the n^k terms are bounded independently of n and each index j_i, $1 \leq i \leq k$. The sum in the last line is a Riemann sum for the multiple integral

$$\int_0^1 \int_0^1 \cdots \int_0^1 \left(1 + \frac{x_1 + x_2}{R}\right)^\beta \cdots$$

$$\left(1 + \frac{x_{k-1} + x_k}{R}\right)^\beta \left(1 + \frac{x_k + x_1}{R}\right)^\beta dx_1 dx_2 \dots dx_k$$

and so we obtain the result (2.7), after making the substitution $x_j = Rt_j$ for $1 \leq j \leq k$. Finally, from (2.2),

$$\int \lambda^j d\mu_{mn}(\lambda) = \frac{1}{n^{j+2}} Tr\left(\left[\frac{H_{mn}}{a_m}\right]^{j+2}\right).$$

Then (2.8) follows. ■

Proof of Corollary 2.2 Firstly as H_{mn}/a_m is real and symmetric, all its eigenvalues are real. It follows that μ_{mn} is a positive measure supported on the real line. Moreover, Theorem 2.1 shows that the supports of all μ_{mn} are contained in a bounded interval independent of n, while their total mass is bounded independent of n. By Helly's Theorem (or if you prefer the Banach-Alaoglu Theorem) every subsequence of $\{\mu_{mn}\}$ contains another subsequence converging weakly to some positive measure ω with compact support in the real line. It follows from Theorem 2.1(III) that for $j \geq 0$,

$$\int t^j d\omega(t) = c_{j+2}.$$

As the Hausdorff moment problem [24] (or moment problem for a bounded interval) has a unique solution, ω is independent of the subsequence. Then the full sequence $\{\mu_{mn}\}$ converges weakly to ω. ■

For the largest eigenvalue for this positive case, we prove:

Lemma 4.1 *Assume β is real and all $\{a_j\}$are real. Let $\lambda_{\max}$ denote the largest eigenvalue of H_{mn}/a_m. Then*

$$\liminf_{n\to\infty} \frac{1}{n}\lambda_{\max} \geq \int_0^1 \int_0^1 \left(1+\frac{x+y}{R}\right)^{\beta} dx\, dy$$

and

$$\limsup_{n\to\infty} \frac{1}{n}\lambda_{\max} \leq \left(\int_0^1 \int_0^1 \left(1+\frac{x+y}{R}\right)^{2\beta} dx\, dy\right)^{1/2}.$$

Proof As H_{mn}/a_m is real symmetric, its largest eigenvalue $\lambda_{\max}$ satisfies

$$\lambda_{\max} = \sup\left\{\sum_{j,k=0}^{n-1} \frac{a_{m+j+k}}{a_m} x_j x_k : \sum_{j=0}^{n-1} x_j^2 = 1\right\}.$$

Choosing all $x_j = \frac{1}{\sqrt{n}}$, we see much as above that

$$\begin{aligned}\liminf_{n\to\infty} \frac{1}{n}\lambda_{\max} &\geq \lim_{n\to\infty} \frac{1}{n^2}\sum_{j,k=0}^{n-1}\left(1+\frac{j+k}{Rn}\right)^{\beta}(1+o(1))\\ &= \int_0^1 \int_0^1 \left(1+\frac{x+y}{R}\right)^{\beta} dx\, dy.\end{aligned}$$

In the other direction, two applications of the Cauchy-Schwarz inequality give, if $\sum_{j=0}^{n-1} x_j^2 = 1$,

$$\begin{aligned}&\left|\sum_{j=0}^{n-1}\sum_{k=0}^{n-1} \frac{a_{m+j+k}}{a_m} x_j x_k\right|\\ &\leq \sum_{j=0}^{n-1} |x_j| \left(\sum_{k=0}^{n-1}\left(\frac{a_{m+j+k}}{a_m}\right)^2\right)^{1/2} \left(\sum_{k=0}^{n-1} x_k^2\right)^{1/2}\end{aligned}$$

$$\leq \left(\sum_{j=0}^{n-1} \sum_{k=0}^{n-1} \left(\frac{a_{m+j+k}}{a_m} \right)^2 \right)^{1/2} \left(\sum_{j=0}^{n-1} x_j^2 \right)^{1/2},$$

so much as above,

$$\limsup_{n \to \infty} \frac{1}{n} \lambda_{\max}$$

$$\leq \lim_{n \to \infty} \left(\frac{1}{n^2} \sum_{j,k=0}^{n-1} \left(1 + \frac{j+k}{Rn} \right)^{2\beta} (1 + o(1)) \right)^{1/2}$$

$$= \left(\int_0^1 \int_0^1 \left(1 + \frac{x+y}{R} \right)^{2\beta} dx\, dy \right)^{1/2}.$$

■

5 Proof of Theorem 3.1 and Corollary 3.2

Toeplitz matrices are more delicate, as reflected both in the hypotheses and proofs. In the sequel, we let

$$\phi(\varepsilon) = \limsup_{n \to \infty} \frac{1}{n\,|a_n|} \sum_{j-1}^{[\varepsilon n]+1} |a_j|, \; \varepsilon \in [0, 1].$$

If $R = 1$, our hypothesis (3.3) is that $\phi(\varepsilon) \to 0$ as $\varepsilon \to 0+$.

Proof of Theorem 3.1(I) It follows from Gershgorin's Theorem that every eigenvalue λ of T_{mn}/a_m satisfies

$$\frac{|\lambda|}{n} \leq \max_{1 \leq j \leq n} \frac{1}{n} \sum_{k=1}^{n} \left| \frac{a_{m+k-j}}{a_m} \right|. \tag{5.1}$$

Assume first $R > 1$. We can use our asymptotic (3.2) to deduce that

$$\frac{|\lambda|}{n} \leq \max_{1 \leq j \leq n} \frac{1}{n} \sum_{k=1}^{n} \left| \left(1 + \frac{k-j}{m} \right)^{\beta} (1 + o(1)) \right|$$

$$\leq \max_{1 \leq j \leq n} \frac{1}{n} \sum_{k=1}^{n} \left(1 + \frac{k-j}{m} \right)^{\operatorname{Re} \beta} + o(1)$$

$$\leq \max_{1\leq j\leq n} \frac{1}{n}\sum_{k=1}^{n}\left(1+\frac{k-j}{Rn\left(1+o\left(1\right)\right)}\right)^{\operatorname{Re}\beta}+o\left(1\right)$$

$$\leq \frac{1}{n}\sum_{k=1}^{n}\max_{0\leq y\leq 1}\left(1+\frac{k}{Rn}-\frac{y}{R}\right)^{\operatorname{Re}\beta}+o\left(1\right)$$

$$\to \int_{0}^{1}\max_{0\leq y\leq 1}\left(1+\frac{x-y}{R}\right)^{\operatorname{Re}\beta}dx.$$

Now suppose that $R=1$. Choose a subsequence $\mathcal{S}$ of integers n and then for $n\in\mathcal{S}$, choose $j=j\left(n\right)\in\left[1,n\right]$, such that

$$\limsup_{n\to\infty}\left(\max_{1\leq j\leq n}\frac{1}{n}\sum_{k=1}^{n}\left|\frac{a_{m+k-j}}{a_m}\right|\right)=\lim_{n\to\infty,n\in\mathcal{S}}\frac{1}{n}\sum_{k=1}^{n}\left|\frac{a_{m+k-j(n)}}{a_m}\right|. \tag{5.2}$$

By choosing a further subsequence, which we also denote by $\mathcal{S}$, we may assume that for some $\alpha\in\left[0,1\right]$,

$$\lim_{n\to\infty}\frac{j\left(n\right)}{n}=\alpha.$$

Fix $\varepsilon\in\left(0,\frac{1}{2}\right)$. Observe that if $k-j\geq-\left(1-\varepsilon\right)m$, we can apply (3.2). Here as $n\to\infty$, this inequality is asymptotically equivalent to $k\geq\left(\alpha+\varepsilon-1\right)n\left(1+o\left(1\right)\right)$. Then for $n\in\mathcal{S}$ and $j=j\left(n\right)$,

$$\frac{1}{n}\sum_{\substack{k:1\leq k\leq n\\ \text{and } k-j\geq-(1-\varepsilon)m}}\left|\frac{a_{m+k-j}}{a_m}\right|$$

$$\leq \frac{1}{n}\sum_{k\leq n:k\geq\max\{1,(\alpha+\varepsilon-1)n(1+o(1))\}}\left|\left(1+\frac{k-j}{m}\right)^{\beta}\left(1+o\left(1\right)\right)\right|$$

$$\leq \frac{1}{n}\sum_{k\leq n:k\geq\max\{1,(\alpha+\varepsilon-1)n(1+o(1))\}}\left(1+\frac{k-\alpha n\left(1+o\left(1\right)\right)}{n\left(1+o\left(1\right)\right)}\right)^{\operatorname{Re}\beta}+o\left(1\right)$$

$$=\int_{\max\{0,\alpha+\varepsilon-1\}}^{1}\left(1+x-\alpha\right)^{\operatorname{Re}\beta}dx+o\left(1\right).$$

Next, recall that $a_j=0$ for $j<0$. If $k-j\leq-\left(1-\varepsilon\right)m$, then $m+k-j\leq\varepsilon m$. Then as $m/n\to 1$ as $n\to\infty$, we have for large enough n and $j\geq 1$,

$$\frac{1}{n}\sum_{\substack{k:1\leq k\leq n\\ \text{and } k-j\leq -(1-\varepsilon)m}}\left|\frac{a_{m+k-j}}{a_m}\right|$$
$$\leq \frac{1+o(1)}{m|a_m|}\sum_{\ell=1}^{[\varepsilon m]+1}|a_\ell| \leq \phi(\varepsilon)+o(1).$$

Adding the two sums together, we obtain

$$\limsup_{n\to\infty}\left(\max_{1\leq j\leq n}\frac{1}{n}\sum_{k=1}^{n}\left|\frac{a_{m+k-j}}{a_m}\right|\right)$$
$$\leq \int_{\max\{0,\alpha+\varepsilon-1\}}^{1}(1+x-\alpha)^{\mathrm{Re}\,\beta}\,dx+\phi(\varepsilon).$$

Letting $\varepsilon\to 0+$, and using Dominated Convergence, we obtain

$$\limsup_{n\to\infty}\left(\max_{1\leq j\leq n}\frac{1}{n}\sum_{k=1}^{n}\left|\frac{a_{m+k-j}}{a_m}\right|\right)\leq \int_{\max\{0,\alpha-1\}}^{1}(1+x-\alpha)^{\mathrm{Re}\,\beta}\,dx$$
$$\leq \int_0^1\max_{0\leq y\leq 1}(1+x-y)^{\mathrm{Re}\,\beta}\,dx.$$

So we obtain the result for $R=1$. ■

Proof of Theorem 3.1(II) As in the proof of Theorem 2.1(II), Schur's inequality gives

$$|\nu_{mn}|(\mathbb{C})=\frac{1}{n^2}\sum_{\lambda\in\Lambda(T_{mn}/a_m)}|\lambda|^2\leq \frac{1}{n^2}\sum_{j,k=0}^{n-1}\left|\frac{a_{m+k-j}}{a_m}\right|^2.$$

Suppose first $R>1$. Then for large enough n, if $0\leq j,k\leq n-1$,

$$\begin{aligned} m+k-j &\geq Rn(1+o(1))-n+1\\ &\geq (R-1)n+o(n)\\ &\geq \frac{R-1}{R}m+o(m), \end{aligned}$$

so uniformly for such j,k, (3.2) gives

$$\frac{a_{m+k-j}}{a_m}=\left(1+\frac{k-j}{Rn}\right)^{\beta}(1+o(1)). \tag{5.3}$$

Then

$$\begin{aligned} |\nu_{mn}|(\mathbb{C}) &\leq \frac{1}{n^2}\sum_{j,k=0}^{n-1}\left|\frac{a_{m+k-j}}{a_m}\right|^2 \\ &\leq \frac{1}{n^2}\sum_{j,k=0}^{n-1}\left(1+\frac{k-j}{Rn}\right)^{2\operatorname{Re}\beta}(1+o(1)) \\ &\to \int_0^1\int_0^1\left(1+\frac{y-x}{R}\right)^{2\operatorname{Re}\beta}dx\,dy \end{aligned}$$

as $n\to\infty$. Next, let $R=1$. Much as above, we can see that given $\varepsilon\in(0,1)$,

$$\frac{1}{n^2}\sum_{0\leq j,k\leq n-1:k-j\geq-(1-\varepsilon)m}\left|\frac{a_{m+k-j}}{a_m}\right|^2$$

may be bounded above by a Riemann sum for the integral

$$\int\int_{\{(x,y):x,y\in[0,1]\text{ and }y-x\geq-(1-\varepsilon)\}}(1+y-x)^{2\operatorname{Re}\beta}\,dx\,dy$$

multiplied by $1+o(1)$. To deal with the tail sum, first observe that as $m=m(n)=m(1+o(1))$,

$$\begin{aligned} \frac{1}{n}\sum_{j=1}^{3n}\frac{|a_j|}{|a_m|} &\leq (1+o(1))\,\phi\left(\frac{1}{4}\right)+\frac{1}{n}\sum_{j=\left[\frac{1}{4}n\right]}^{3n}\left|\frac{a_j}{a_n}\right| \\ &\leq (1+o(1))\,\phi\left(\frac{1}{4}\right)+\frac{1+o(1)}{n}\sum_{\ell=\left[\frac{1}{4}n\right]-n}^{2n}\left|\frac{a_{n+\ell}}{a_n}\right| \\ &\leq (1+o(1))\,\phi\left(\frac{1}{4}\right)+\frac{1+o(1)}{n}\sum_{\ell=\left[\frac{1}{4}n\right]-n}^{2n}\left|1+\frac{\ell}{n}\right|^{\operatorname{Re}\beta}(1+o(1)) \\ &\leq (1+o(1))\,\phi\left(\frac{1}{4}\right)+(1+o(1))\int_{-3/4}^{2}|1+x|^{\operatorname{Re}\beta}\,dx. \end{aligned}$$

It follows that for some C independent of m,n,

$$\frac{1}{n}\sum_{j=1}^{3n}\frac{|a_j|}{|a_m|}\leq C. \tag{5.4}$$

Then

$$\begin{aligned}
&\frac{1}{n^2} \sum_{0 \le j,k \le n-1 : k-j \le -(1-\varepsilon)m} \left| \frac{a_{m+k-j}}{a_m} \right|^2 \\
&\le \left(\frac{1}{n} \sup_{1 \le \ell \le 2m} \left| \frac{a_\ell}{a_m} \right| \right) \left(\frac{1}{n} \sum_{\ell=1}^{[\varepsilon m]} \left| \frac{a_\ell}{a_m} \right| \right) \\
&\le C\phi(\varepsilon),
\end{aligned}$$

in view of (5.4). This and the estimate above give

$$\begin{aligned}
&\limsup_{n \to \infty} |\nu_{mn}| (\mathbb{C}) \\
&= \limsup_{n \to \infty} \frac{1}{n^2} \sum_{\lambda \in \Lambda(T_{mn}/a_m)} |\lambda|^2 \\
&\le \int \int_{\{(x,y):x,y \in [0,1] \text{ and } y-x \ge -(1-\varepsilon)\}} (1+y-x)^{2\,\mathrm{Re}\,\beta}\, dx\, dy + C\phi(\varepsilon).
\end{aligned}$$

Letting $\varepsilon \to 0+$ and using our hypothesis (3.3) gives the result. ■

Proof of Theorem 3.1(III)

Step 1 Suppose first $R > 1$. Then for large enough n, we have (5.3) and also

$$\sup_{1 \le j, \ell \le n} \left| \frac{a_{m+j-\ell}}{a_m} \right| = O(1). \tag{5.5}$$

Then

$$\begin{aligned}
&\frac{1}{n^k} Tr \left(\left[\frac{T_{mn}}{a_m} \right]^k \right) \\
&= \frac{1}{n^k} \sum_{j_1=1}^{n} \sum_{j_2=1}^{n} \cdots \sum_{j_k=1}^{n} \frac{a_{m+j_2-j_1}}{a_m} \frac{a_{m+j_3-j_2}}{a_m} \cdots \frac{a_{m+j_k-j_{k-1}}}{a_m} \frac{a_{m+j_1-j_k}}{a_m} \\
&= \frac{1}{n^k} \sum_{j_1=1}^{n} \sum_{j_2=1}^{n} \cdots \sum_{j_k=1}^{n} \left(1 + \frac{j_2 - j_1}{m} \right)^\beta \left(1 + \frac{j_3 - j_2}{m} \right)^\beta \cdots \\
&\quad \left(1 + \frac{j_1 - j_k}{m} \right)^\beta (1 + o(1)) \\
&= \frac{1}{n^k} \sum_{j_1=1}^{n} \sum_{j_2=1}^{n} \cdots \sum_{j_k=1}^{n} \left(1 + \frac{j_2 - j_1}{Rn(1+o(1))} \right)^\beta \left(1 + \frac{j_3 - j_2}{Rn(1+o(1))} \right)^\beta \cdots
\end{aligned}$$

$$\left(1+\frac{j_1-j_k}{Rn\,(1+o\,(1))}\right)^{\beta}(1+o\,(1))$$

$$=\frac{1}{n^k}\sum_{j_1=1}^{n}\sum_{j_2=1}^{n}\cdots\sum_{j_k=1}^{n}\left(1+\frac{j_2-j_1}{Rn}\right)^{\beta}\left(1+\frac{j_3-j_2}{Rn}\right)^{\beta}\cdots$$

$$\left(1+\frac{j_1-j_k}{Rn}\right)^{\beta}+o\,(1)\,.$$

The sum in the last line is a Riemann sum for the multiple integral

$$\int_0^1\int_0^1\cdots\int_0^1\left(1+\frac{x_2-x_1}{R}\right)^{\beta}\left(1+\frac{x_3-x_2}{R}\right)^{\beta}\cdots$$

$$\left(1+\frac{x_1-x_k}{R}\right)^{\beta}dx_1dx_2\ldots dx_k$$

and so we obtain the result, after making the substitution $x_j=Rt_j$ for $1\leq j\leq k$.

Step 2 Now we turn to the more delicate case where $R=1$ and $\mathrm{Re}\,\beta>-1$. Fix $\varepsilon>0$. We observe that if $k-j\geq -m\,(1-\varepsilon)$, then we have (5.3). Then identifying $j_{k+1}=j_1$,

$$\frac{1}{n^k}\sum_{\substack{1\leq j_1,j_2,\ldots,j_k\leq n\\ \text{all } j_{i+1}-j_i\geq -m(1-\varepsilon)}}^{n}\frac{a_{m+j_2-j_1}}{a_m}\frac{a_{m+j_3-j_2}}{a_m}\cdots\frac{a_{m+j_k-j_{k-1}}}{a_m}\frac{a_{m+j_1-j_k}}{a_m}$$

$$=\frac{1}{n^k}\sum_{\substack{1\leq j_1,j_2,\ldots,j_k\leq n\\ \text{all } j_{i+1}-j_i\geq -m(1-\varepsilon)}}^{n}\left(1+\frac{j_2-j_1}{m}\right)^{\beta}\left(1+\frac{j_3-j_2}{m}\right)^{\beta}\cdots \tag{5.6}$$

$$\left(1+\frac{j_1-j_k}{m}\right)^{\beta}(1+o\,(1))$$

$$=\frac{1}{n^k}\sum_{\substack{1\leq j_1,j_2,\ldots,j_k\leq n\\ \text{all } j_{i+1}-j_i\geq -m(1-\varepsilon)}}^{n}\left(1+\frac{j_2-j_1}{n}\right)^{\beta}\left(1+\frac{j_3-j_2}{n}\right)^{\beta}\cdots$$

$$\left(1+\frac{j_1-j_k}{n}\right)^{\beta}+o\,(1)$$

$$=\int\cdots\int_{\mathcal{S}}(1+x_2-x_1)^{\beta}(1+x_3-x_2)^{\beta}\cdots$$

$$(1+x_1-x_k)^{\beta}\,dx_1dx_2\ldots dx_k+o\,(1)$$

where $\mathcal{S}=\left\{(x_1,x_2,\ldots,x_k)\in[0,1]^k : x_{j+1}-x_j\geq -(1-\varepsilon)\text{ for each } j\right\}$. Here we identify $x_{k+1}=x_1$. To treat the remaining terms in the sum where at least one $j_{i+1} - j_i \leq -m(1-\varepsilon)$, we proceed as follows: necessarily $j_{i+1} \leq n - m + \varepsilon m \leq 2\varepsilon m$, for large enough n, while $1 \leq m + j_{i+1} - j_i \leq \varepsilon m$, so

$$\frac{1}{n} \sum_{j_i : j_{i+1} - j_i \leq -m(1-\varepsilon)} \left| \frac{a_{m+j_{i+1}-j_i}}{a_m} \right| \leq \frac{1+o(1)}{n} \frac{1}{|a_m|} \sum_{\ell=1}^{[\varepsilon m]} |a_\ell| \leq (1+o(1))\,\phi(\varepsilon).$$

Then

$$\frac{1}{n^k} \sum_{\substack{1 \leq j_1, j_2, \ldots, j_k \leq n \\ \text{for some } i,\ j_{i+1} - j_i \geq -m(1-\varepsilon)}}^{n} \left| \frac{a_{m+j_2-j_1}}{a_m} \frac{a_{m+j_3-j_2}}{a_m} \cdots \frac{a_{m+j_k-j_{k-1}}}{a_m} \frac{a_{m+j_1-j_k}}{a_m} \right|$$
$$\leq C^{k-1}(1+o(1))\,\phi(\varepsilon),$$

recall (5.4). We now combine this with (5.6) and then let $\varepsilon \to 0+$ to get the result. Also (3.4) follows from (3.1). ■

Proof of Corollary 3.2 Since $\{\nu_{mn}\}$ have support in a compact set independent of n and total mass bounded independent of n, we can choose weakly convergent subsequences with limit ω. (One can think of applying Helly's Theorem to the decomposition of μ_{mn} into first real and imaginary parts and then positive and negative parts of each of those.) All weak limits of subsequences have the same moments $\{d_{j+2}\}_{j\geq 0}$. We have that if f is a polynomial,

$$\lim_{n\to\infty} \int P(t)\,d\nu_{mn}(t) = \int P(t)\,d\omega(t).$$

Note that the same limit holds for the full sequence of integers because all weak limits ω have the same power moments. As such polynomials are dense in the class of functions analytic in any ball, the result follows. ■

Acknowledgments Research supported by NSF grant DMS1800251 and Georgia Tech Mathematics REU Program NSF Grant DMS181843.

References

1. Basor, E., Morrison, K.: The Fisher-Hartwig conjecture and Toeplitz eigenvalues. Linear Algebra Appl. **202**, 129–142 (1994)
2. Böttcher, A., Silbermann, B.: Introduction to Large Truncated Toeplitz Matrices, Springer, New York (1999)
3. Bryc, W., Dembo, A., Jiang, Spectral measure of large random Hankel, Markov and Topelitz matrices. Ann. Probab. **34**, 1–38 (2006)

4. Edrei, A.: Sur les determinants récurrants et les singularites d'une fonction donnée par son developpement de Taylor. Compos. Math. **7**, 20–88 (1939)
5. Erhardt, T., Shao, B.: Asymptotic behavior of variable-coefficient Toeplitz determinants. J. Fourier Anal. Appl. **7**, 71–92 (2001)
6. Fasino, D., Tilli, P.: Spectral clustering of block multilevel Hankel matrices. Linear Algebra Appl. **306**, 155–163 (2000)
7. Garoni, C., Serra-Capizzano, S.: Generalized Locally Toeplitz Sequences: Theory and Applications, vol. 1. Springer, Berlin (2017)
8. Golinskii, L., Serra-Capizzano, S.: The asymptotic properties of the spectrum of nonsymetrically perturbed Jacobi matrices. J. Approx. Theory **144**, 84–102 (2007)
9. Grenander, U., Szegő, G.: Toeplitz Forms. Chelsea, New York (1958)
10. Hammond, C., Miller, S.J.: Distribution of eigenvalues for the ensemble of real symmetric Toeplitz matrices. J. Theor. Probab. **18**, 537–566 (2005)
11. Jose Caro, F., Nagar, D.: Evaluation of matrix Liouville-Dirichlet integrals using Laplace transform. Integral Transforms Spec. Funct. **17**, 245–255 (2006)
12. Krasovsky, I.: Aspects of Toeplitz determinants. In: Lenz, D., Sobieczky, F., Woss, W. (eds.) Boundaries and Spectra of Random Walks. Progress in Probability. Birkhauser, Basel (2011)
13. Liu, D.-Z., Sun, X., Wang, Z.-D.: Fluctuation of eigenvalues for random Toeplitz and related matrices. Electron. J. Probability **17**, paper no. 95, 22 pp. (2012)
14. Lubinsky, D.S.: Uniform convergence of rows of the Padé table for functions with smooth Maclaurin series coefficients. Constr. Approx. **3**, 307–330 (1987)
15. Lubinsky, D.S.: Padé tables of entire functions of very slow and smooth growth II. Constr. Approx. **4**, 321–339 (1988)
16. Lubinsky, D.S.: Universality of distribution of eigenvalues of Toeplitz matrices with smooth entries. Manuscript
17. Marcus, M., Minc, H.: A Survey of Matrix Theory and Matrix Inequalities. Dover, New York (1965)
18. Mascarenhas, H., Silbermann, B.: Sequences of variable-coefficient Toeplitz matrices and their singular values. J. Funct. Anal. **270**, 1479–1500 (2016)
19. Mejlbo, L., Schmidt, P.: On the eigenvalues of generalized Topelitz matrices. Math. Scand. **10**, 5–16 (1962)
20. Polya, G.: Über gewisse notwendige Determinantenkriterein fur die Forsetzbarkeit einer Potenzreihe. Math. Ann. **99**, 687–706 (1928)
21. Sen, A., Virag, B.: Absolute continuity of the limiting eigenvalue distribution of the random Toeplitz matrix. Elect. Commun. Probab. **16**, 606–711 (2011)
22. Sen, A., Virag, B.: The top eigenvalue of the random Toeplitz matrix and the sine kernel., Ann. Probab. **41**, 4050–4079 (2013)
23. Serra-Capizzano, S., Bertaccini, D., Golub, G.: How to deduce a proper eigenvalue cluster from a proper singular value cluster in the nonnormal case. SIAM J. Matrix Anal. Appl. **27**, 82–86 (2005)
24. Shohat, J., Tamarkin, J.: The problem of moments. American Mathematical Society, Rhode Island (1943)
25. Sivazlian, B.D.: On a multivariate extension of the beta and gamma distributions. SIAM J. Appl. Math. **41**, 205–209 (1981)
26. Tilli, P.: Some results on complex Toeplitz eigenvalues. J. Math. Anal. Appl. **239**, 390–401 (1999)
27. Tyrtyshnikov, E.: How bad are Hankel matrices?. Numer. Math. **67**, 261–269 (1994)
28. H. Widom, Hankel matrices. Trans. Amer. Math. Soc. **121**, 1–35 (1966)
29. Zabroda, O., Simonenko, I.: Asymptotic invertibility and the collective asymptotic spectral behavior of generalized one-dimensional discrete convolutions. Funct. Anal. Appl. **38**, 65–66 (2004)
30. Zamarashkin, N., Tyrtyshnikov, E.: Distribution of eigenvalues and singular values of Toeplitz matrices under weakened conditions on the generating function. Math. Sbornik **188**, 1191–1201 (1997)

On the Gradient Conjecture for Quadratic Polynomials

Tom McKinley and Boris Shekhtman

Abstract The gradient conjecture asserts that for homogeneous polynomials f and p the equality $p(\nabla f) = 0$ implies $p(\nabla) f = 0$. We verify this conjecture for quadratic polynomials and present a few applications to density problems and characterization of derivation operator.

Keywords Homogeneous polynomial · Gradient · Algebraic dependency

1 Introduction

Gradient conjecture was formulated and verified in a few specific cases in [9].

Conjecture 1 (Gradient Conjecture) Let p and f be homogeneous polynomials in d variables such that

$$p(\nabla f) = 0 \tag{1}$$

then

$$p(\nabla) f = 0 \tag{2}$$

Here ∇ stands for the gradient, i.e.,

$$\nabla = \left(\frac{\partial}{\partial x_1}, \ldots, \frac{\partial}{\partial x_d} \right).$$

T. McKinley · B. Shekhtman (✉)
Department of Mathematics and Statistics, University of South Florida, Tampa, FL, USA
e-mail: shekhtma@usf.edu

G. E. Fasshauer et al. (eds.), *Approximation Theory XVI*, Springer Proceedings in Mathematics & Statistics 336, https://doi.org/10.1007/978-3-030-57464-2_9

Hence the assumption of the conjecture (1) says that f satisfies a certain nonlinear first order PDE while the conclusion asserts that f satisfies a higher order linear PDE with constant coefficients.

It was shown in [9] that the conjecture holds in 4 variables and also holds when p depends on 3 variables while f depends on d variables.

The main result of this paper is to verify the conjecture in case when the polynomial p is quadratic. In particular we show that if

$$\sum_{i=1}^{d} A_i \left(\frac{\partial f}{\partial x_i}\right)^2 = 0$$

for some constants A_i then

$$\sum_{i=1}^{d} A_i \left(\frac{\partial^2 f}{\partial x_i^2}\right) = 0.$$

Setting the constants $A_i = 1$ we conclude that if

$$\sum_{i=1}^{d} \left(\frac{\partial f}{\partial x_i}\right)^2 = 0$$

then the polynomial f is harmonic.

This conjecture has strong correlation with a question posed by **Allan Pinkus and Bronislav Wajnryb.** In their survey paper [11] they introduced and studied the density of the space

$$\mathcal{P}(f) := \operatorname{span}\left\{(f(\mathbf{x}+\mathbf{b}))^k : \mathbf{b} \in \mathbb{R}^d, k \in \mathbb{Z}_+\right\} \tag{3}$$

for a fixed polynomial f.

They showed that if $\mathcal{P}(f)$ is not dense in the space $C(\mathbb{R}^n)$ then the Hessian determinant of f is identically zero and asked if the converse is true for homogeneous polynomials f.

Observe that the Hessian determinant is the determinant of the Jacobian map of $\nabla f : \mathbb{k}^n \to \mathbb{k}^n$. Hence the condition is equivalent to the vanishing of the Jacobian which, in turn, holds if and only if the first partial derivative of f are polynomially dependent (cf. [3]). Thus (1) holds for some polynomial p. The homogeneity of f implies that p can be chosen to be homogeneous.

By the Chain rule it follows from (1) that $p(\nabla)\left(f^k\right) = 0$ for all $k \in \mathbb{Z}_+$ and the gradient conjecture would imply that $p(\nabla) f^k = 0$ for all $k \in \mathbb{Z}_+$ and thus $p(\nabla) g = 0$ for all $g \in \mathcal{P}(f)$. And this (cf. [8]) implies that $\mathcal{P}(f)$ is not dense in $C(\mathbb{R}^n)$.

2 Notations and Historic Preliminaries

Since the results of this paper are valid for complex polynomials as well as real polynomials, we will now switch to polynomials over complex field.

We use $\mathbb{C}[x_1, \ldots, x_d] = \mathbb{C}[\mathbf{x}]$ to denote the set of all polynomials in d variables $x_1, \ldots, x_d$ with complex coefficients. For a polynomial $f \in \mathbb{C}[\mathbf{x}]$ we use f_i to denote the partial derivative $\frac{\partial f}{\partial x_i}$ and $f_{i,j}$ to denote $\frac{\partial^2 f}{\partial x_i \partial x_j}$ The Hessian matrix $H_f(\mathbf{x})$, is the matrix of second partial derivatives

$$H_f(\mathbf{x}) = \left(f_{i,j}(\mathbf{x})\right)_{i,j=1}^{d}$$

while $\mathcal{H}_f(\mathbf{x})$ denotes the determinant of $H_f(\mathbf{x})$.

As was noted earlier $\mathcal{H}_f(\mathbf{x})$ vanishes identically if and only if there exists a polynomial $p \in \mathbb{C}[\mathbf{x}]$ such that

$$p(\nabla f) = p(f_1, \ldots, f_d) = 0.$$

If f is a homogeneous then p could also be chosen to be homogeneous.

The study of such polynomials was initiated by Hesse (cf. [6]) and finalized by Gordan and Nöther [5] as well as [10]. Gordan and Nöther the result holds for in four variables and disproved the claim in 5 variables:

Theorem 1 (Gordan and Nöther [5]) *If $d \leq 4$ and the Hessian determinant of f vanishes then $f_1, \ldots, f_d$ are linearly dependent.*

An easy counterexample is due to Perazzo [10]

$$f(x_1, x_2, x_3, x_4, x_5) := x_1 x_4^2 + x_2 x_4 x_5 + x_3 x_5^2$$

By direct calculation we see that the first partial are linearly independent and the equation

$$f_1 f_3 - (f_2)^2 = 0$$

holds.

In the last 20 years the subject became popular again (cf. [1, 2, 4, 7, 12]) and many of the results in [5] had been verified and rewritten in a modern language. In particular we will need the following (cf. [2]):

Theorem 2 *Suppose that p and f satisfy* (1), *Then for every* $\mathbf{x} \in \mathbb{R}^d$

$$f(\mathbf{x}) = f(\mathbf{x} + t \nabla p(\nabla f(\mathbf{x}))) \textit{ for all } t \in \mathbb{R}$$

and for every $i = 1, \ldots, d$

$$f_i(\mathbf{x}) = f_i(\mathbf{x} + t \bigtriangledown p(\nabla f(\mathbf{x}))) \text{ for all } t \in \mathbb{R}. \tag{4}$$

It simply asserts that for every fixed $\mathbf{x} \in \mathbb{R}^d$ the line

$$\mathbf{x} + t \bigtriangledown p(\nabla f)(\mathbf{x}) = \mathbf{x} + t(p_1(\nabla f(\mathbf{x})), \ldots, p_d(\nabla f(\mathbf{x})))$$

is a characteristic of the partial differential equation (1).

3 Main Results

Theorem 3 *Let* p *and* f *satisfy* (1). *Then the matrix* $H_p(\nabla f(\mathbf{x}))H_f(\mathbf{x})$ *is nilpotent.*

Proof The proof involves somewhat tedious computation using the chain rule. From (4) we obtain

$$\begin{aligned}
f_{i,j}(\mathbf{x}) &= \frac{\partial f_i(\mathbf{x})}{\partial x_j} = \frac{\partial}{\partial x_j} f_i\left((x_1, \ldots, x_d) + t\left(p_1(\nabla f(\mathbf{x})), \ldots, p_d(\nabla f(\mathbf{x}))\right)\right) \\
&= \sum_{k=1}^{d} f_{i,k}(\mathbf{x} + t \bigtriangledown p(\nabla f(\mathbf{x}))) \frac{\partial}{\partial x_j}(x_k + t p_k(\nabla f(\mathbf{x}))) \\
&= \sum_{k=1}^{d} \left[f_{i,k}(\mathbf{x} + t \bigtriangledown p(\nabla f(\mathbf{x})))\right] \left[\delta_{jk} + t \frac{\partial}{\partial x_j}(p_k(f_1, \ldots, f_d))\right] \\
&= \sum_{k=1}^{d} \left[f_{i,k}(\mathbf{x} + t \bigtriangledown p(\nabla f(\mathbf{x})))\right] \left[\delta_{jk} + t \sum_{m=1}^{d} p_{k,m}(\nabla f(\mathbf{x})) f_{m,j}\right]
\end{aligned}$$

where

$$\delta_{jk} = \begin{cases} 1 \text{ if } j = k \\ 0 \text{ if } j \neq k \end{cases}$$

Interpreting the terms $f_{i,j}$ as entries in the Hessian H_f and $p_{k,m}$ as entries in the matrix H_p we conclude that

$$H_f(\mathbf{x}) = H_f(\mathbf{x} + t \bigtriangledown p(\nabla f(\mathbf{x}))) \left(I + t H_p(\nabla f(\mathbf{x})\right) H_f(\mathbf{x}).$$

Notice that $H_f(\mathbf{x} + t \bigtriangledown p(\nabla f(\mathbf{x})))$ is a matrix with polynomial entries in the variable t. Hence

$$H_f(\mathbf{x} + t \bigtriangledown p(\nabla f(\mathbf{x}))) = H_f(\mathbf{x}) + t M_1(\mathbf{x}) + t^2 M_2(\mathbf{x}) + \cdots + t^N M_N(\mathbf{x})$$

for some matrices $M_1(\mathbf{x}), \ldots, M_N(\mathbf{x})$ with $N = \deg f - 2$. Hence

$$\begin{aligned} H_f(\mathbf{x}) &= \left(H_f(\mathbf{x}) + tM_1(\mathbf{x}) + t^2 M_2(\mathbf{x}) + \cdots + t^N M_N(\mathbf{x})\right)\left(I + tH_p(\nabla f(\mathbf{x}))H_f(\mathbf{x})\right) \\ &= H_f(\mathbf{x}) + t\left(M_1(\mathbf{x}) + H_p(\nabla f(\mathbf{x}))H_f(\mathbf{x})\right) \\ &\quad + \sum_{s=1}^{N+1} t^{s+1}\left(M_{s+1}(\mathbf{x}) + M_s(\mathbf{x})H_p(\nabla f(\mathbf{x}))H_f(\mathbf{x})\right). \end{aligned}$$

With an understanding that matrices H_f and M_s depend on $\mathbf{x}$ and the matrices H_p depend on $\nabla f(\mathbf{x})$ we can rewrite the last identity as

$$\begin{aligned} H_f &= H_f + t\left(M_1 + H_f H_p H_f\right) + t^2(M_2 + M_1 H_p H_f) \\ &\quad + \cdots + t^s(M_s + M_{s-1}H_p H_f) + \cdots + t^{N+1} M_N H_p H_f. \end{aligned}$$

And since the matrix H_f on the left of this identity does not depend on t, the matrix coefficients of the positive powers of t are equal to zero.

Thus $M_1 = -H_f H_p H_f$, $M_2 = H_f H_p H_f H_p H_f = H_f \left(H_p H_f\right)^2, \ldots, M_s = (-1)^s H_f \left(H_p H_f\right)^s$ and, since $M_N H_p H_f = 0$ we have $(-1)^N H_f \left(H_p H_f\right)^N = 0$. Therefore

$$H_p\left(H_f\left(H_p H_f\right)^N\right) = \left(H_p H_f\right)^{N+1} = 0$$

which proves the theorem.

As a corollary we obtain the main result:

Theorem 4 *Let* $p(x_1, \ldots x_d) = \sum_{k,j=1, i \geq j}^{d} A_{kj} x_k x_j$ *be a quadratic polynomial for some constant coefficient* $A_{i,j}$. *Then the gradient conjecture holds.*

Proof In this case the Hessian $H_p(\nabla f(\mathbf{x}))$ is just a constant symmetric matrix

$$H_p(\nabla f(\mathbf{x})) = \begin{pmatrix} 2A_{1,1} & A_{1,2} & \ldots & A_{1,d} \\ A_{2,1} & 2A_{2,2} & \ldots & A_{2,d} \\ \vdots & \vdots & \ddots & \vdots \\ A_{d,1} & A_{d,2} & \ldots & 2A_{d,d} \end{pmatrix}$$

while $H_f(\mathbf{x})$ is the symmetric matrix

$$H_f(\mathbf{x}) = \begin{pmatrix} f_{1,1}(\mathbf{x}) & f_{1,2}(\mathbf{x}) & \ldots & f_{1,d}(\mathbf{x}) \\ f_{2,1}(\mathbf{x}) & f_{2,2}(\mathbf{x}) & \ldots & f_{2,d}(\mathbf{x}) \\ \vdots & \vdots & \ddots & \vdots \\ f_{d,1}(\mathbf{x}) & f_{d,2}(\mathbf{x}) & \ldots & f_{d,d}(\mathbf{x}) \end{pmatrix}.$$

By the previous theorem

$$0 = \operatorname{tr}\left(H_p(\nabla f(\mathbf{x}))H_f(\mathbf{x})\right) = \sum_{k=1}^{d}\sum_{j=1}^{d}\left(1+\delta_{j,k}\right)A_{k,j}f_{j,k}$$

$$= \sum_{k=1}^{d} 2A_{k,k}f_{k,k} + \sum_{k,j=1,k\neq j}^{d} A_{kj}f_{k,j}$$

where, as usual,

$$\delta_{j,k} := \begin{cases} 1 \text{ if } j = k \\ 0 \text{ if } j \neq k \end{cases}.$$

Observe that the matrices H_p and H_f are symmetric, hence

$$\sum_{k,j=1,k\neq j}^{d} A_{kj}f_{k,j} = 2\sum_{k,j=1,k>j}^{d} A_{k,j}f_{k,j}$$

and

$$0 = 2\left(\sum_{k,j=1,i\geq j}^{d} A_{kj}f_{k,j}\right)$$

which is the desired conclusion.

The following corollary to this theorem is a partial resolution of the question posed by Pinkus and Wajnryb:

Theorem 5 *Let f be a homogeneous polynomial in d variable and p be a homogeneous quadratic polynomial in the same variables. If $p(\nabla f) = 0$ then $\mathcal{P}(f)$ defined by* (3) *is not dense in $C(\mathbb{R}^d)$.*

4 Application to Derivations Operators

Next we will present an interesting interpretation of the Theorem 4 to the derivation operators.

Let A be a subalgebra of $\mathbb{C}[x_1, \ldots, x_d]$.

Definition 1 A linear operator $L : A \to \mathbb{C}[x_1, \ldots, x_d]$ is called a derivation if $L(fg) = fLg + gLf$ for all $f, g \in A$.

Theorem 6 *Let $A(f)$ be a subalgebra of $\mathbb{C}[x_1, \ldots, x_d]$ generated by a single homogeneous polynomial $f \in \mathbb{C}[x_1, \ldots, x_d]$ and let $L = p(\nabla)$ for some homogeneous polynomial of order 2. Then L is a derivation on $A(f)$ if and only if $L = 0$ on $A(f)$.*

Proof Observe that

$$\frac{\partial^2}{\partial x_k \partial x_j} f^2 = \frac{\partial}{\partial x_k}\left(2f\frac{\partial}{\partial x_j}f\right) = 2\left(\frac{\partial}{\partial x_k}f\right)\left(\frac{\partial}{\partial x_j}f\right) + 2f\frac{\partial^2}{\partial x_k \partial x_j}f$$

Hence

$$p(\nabla)f^2 = \left(\sum A_{k,j}\frac{\partial^2}{\partial x_k \partial x_j}\right)f^2 = p(\nabla f) + 2f(p(\nabla)f). \tag{5}$$

If $p(\nabla)$ is a derivation on $A(f)$ then

$$p(\nabla)f^2 = 2fp(\nabla)f$$

which together with (5) implies $p(\nabla)f = 0$ and thus $p(\nabla)f^2 = 0$. Inductively we obtain

$$p(\nabla)f^m = mf^{m-1}(p(\nabla)f) = 0$$

which proves the theorem.

From this it follows that the gradient conjecture is equivalent to the following:

Conjecture 2 Let p be a homogeneous polynomial. The $p(\nabla)$ is a non-trivial derivation on $A(f)$ if and only if p is of order one.

5 One More Case of the Gradient Conjecture

We finish this paper by verifying the gradient conjecture in one more case:

Theorem 7 *Let p and f be homogeneous polynomials on $\mathbb{C}^d$ such that $p(\nabla f) = 0$ and*

$$f = \sum_{j=1}^{d} l_j^m$$

where l_j's are linear forms on $\mathbb{C}^d$ that are linearly independent. Then $p(\nabla)f = 0$.

Proof Suppose that

$$l_j(\mathbf{x}) = \sum_{k=1}^{d} a_{k,j} x_k.$$

Then

$$\nabla l_j^m = m l_j^{m-1} \left(a_{1,j}, \dots, a_{d,j}\right)$$

and

$$p(\nabla f) = p\left(m \sum_{j=1}^{d} l_j^{m-1}(a_{1,j}, \dots, a_{d,j})\right) = 0$$

for all $(x_1, \dots, x_d) \in \mathbb{C}^d$. Since the linear forms are linearly independent we can find $\mathbf{x} = (x_1, \dots, x_d) \in \mathbb{C}^d$ such that $l_t(\mathbf{x}) = \delta_{t,j}$.

Then

$$0 = p(\nabla f)(\mathbf{x}) = p\left(a_{1,t}, \dots, a_{d,t}\right).$$

Notice that

$$\frac{\partial^n}{\partial x_1^{\alpha_1} \dots \partial x_d^{\alpha_d}} l_t^m = \frac{m!}{(m-n)!} l_t^{m-n} \cdot a_{1,t}^{\alpha_1} \cdot \dots \cdot a_{d,t}^{\alpha_d} \tag{6}$$

and since p is homogeneous

$$p(\nabla) l_t^m = \frac{m!}{(m-n)!} l_t^{m-n} p\left(a_{1,t}, \dots, a_{d,t}\right) = 0$$

by (6). In other words $l_t^m \in \ker p(\nabla)$ for every $t = 1, \dots, d$ and thus $f \in \ker p(\nabla)$.

Remark 1 Notice that, while every homogeneous polynomial can be written as the sum of powers of linear forms, those linear forms, unfortunately, are not necessarily linearly independent.

Acknowledgments We would like to express our gratitude to Dima Khavinson and Razvan Teodorescu for their interest in the conjecture and contribution to our understanding of non-linear PDEs.

We would also like to thank the referees for many suggestions that improved the paper.

References

1. Ciliberto, C., Russo, F., Simis, A.: Homaloidal hypersurfaces and hypersurfaces with vanishing hessian. Adv. Math. **218**(6), 1759–1805 (2008). http://dx.doi.org/10.1016/j.aim.2008.03.025
2. de Bondt, M., Watanabe, J.: On the theory of Gordan–Nöther on homogeneous forms with zero Hessian (improved version). In: PRAAG-2018: Polynomial Rings and Affine Algebraic Geometry (Tokyo, February 12–16, 2018), Proceedings in Mathematics and Statistics. Springer, Switzerland (2020). arXiv:1703.07624
3. Ehrenborg, R., Rota, G.C.: Apolarity and canonical forms for homogeneous polynomials. Eur. J. Comb. **14**(3), 157–181 (1993). http://dx.doi.org/10.1006/eujc.1993.1022
4. Gondim, R., Russo, F.: On cubic hypersurfaces with vanishing hessian. J. Pure Appl. Algebra **219**(4), 779–806 (2015). http://dx.doi.org/10.1016/j.jpaa.2014.04.030
5. Gordan, P., Nöther, M.: Ueber die algebraischen Formen, deren Hesse'sche determinante identisch verschwindet. Math. Ann. **10**(4), 547–568 (1876). http://dx.doi.org/10.1007/BF01442264
6. Hesse, O.: Zur Theorie der ganzen homogenen Functionen. J. Reine Angew. Math. **56**, 263–269 (1859). http://dx.doi.org/10.1515/crll.1859.56.263
7. Lossen, C.: When does the hessian determinant vanish identically? (On Gordan and Noether's proof of Hesse's claim). Bull. Braz. Math. Soc. (N.S.) **35**(1), 71–82 (2004). http://dx.doi.org/10.1007/s00574-004-0004-0
8. McKinley, T., Shekhtman, B.: On a problem of Pinkus and Wajnryb regarding density of multivariate polynomials. Proc. Am. Math. Soc. **145**, 185–190 (2017). http://dx.doi.org/10.1090/proc/13196
9. McKinley, T., Shekhtman, B.: On polynomials with vanishing hessians and some density problems. In: Fasshauer, G., Schumaker, L. (eds.) Approximation Theory XV: San Antonio 2016. Springer Proceedings in Mathematics and Statistics, vol. 201, pp. 269–277. Springer, Cham (2017)
10. Perazzo, U.: Sulle variet'a cubiche la cui hessiana svanisce identicamente. Giornale di Matematiche (Battaglini) **38**, 337–354 (1900)
11. Pinkus, A., Wajnryb, B.: On a problem of approximation by means of multidimensional polynomials. Uspekhi Mat. Nauk **50**, 89–110 (1995). (Russian Math. Surveys 50 319–340 (1995, in English))
12. Russo, F.: Hypersurfaces with Vanishing Hessian. In: Lecture Notes of the Unione Matematica Italiana, pp. 177–220. Springer, Switzerland (2016)

Balian-Low Theorems in Several Variables

Michael Northington V and Josiah Park

Abstract Recently, Nitzan and Olsen showed that Balian-Low theorems (BLTs) hold for discrete Gabor systems defined on $\mathbb{Z}_d$. Here we extend these results to a multivariable setting. Additionally, we show a variety of applications of the Quantitative BLT, proving in particular nonsymmetric BLTs in both the discrete and continuous setting for functions with more than one argument. Finally, in direct analogy of the continuous setting, we show the Quantitative Finite BLT implies the Finite BLT.

Keywords Frames · Gabor systems · Riesz bases · Time-frequency analysis · Uncertainty principles · Balian-Low theorems

1 Introduction

Gabor systems are fundamental objects in time-frequency analysis. Given a set $\Lambda \subset \mathbb{R}^{2l}$ and a function $g \in L^2(\mathbb{R}^l)$, the Gabor system $G(g, \Lambda)$ is defined as

$$G(g, \Lambda) = \{g(x - m)e^{2\pi i n \cdot x}\}_{(m,n)\in\Lambda}.$$

When Λ is taken to be $\mathbb{Z}^{2l}$, $G(g) = G(g, \mathbb{Z}^{2l})$ is referred to as the *integer lattice Gabor system* generated by g. The Balian-Low theorem (BLT) and its generalizations are uncertainty principles concerning the generator g of such a system in the case that $G(g, \Lambda)$ forms a Riesz basis.

Theorem 1.1 (BLTs) *Let $g \in L^2(\mathbb{R})$ and suppose that the Gabor system $G(g) = G(g, \mathbb{Z}^2)$ is a Riesz basis for $L^2(\mathbb{R})$.*

(i) If $1 < p < \infty$ and $\frac{1}{p} + \frac{1}{q} = 1$, then either,

---

M. Northington V (✉) · J. Park
Georgia Institute of Technology, Atlanta, GA, USA
e-mail: mcnv3@gatech.edu; j.park@gatech.edu

G. E. Fasshauer et al. (eds.), *Approximation Theory XVI*, Springer Proceedings in Mathematics & Statistics 336, https://doi.org/10.1007/978-3-030-57464-2_10

$$\int_{\mathbb{R}} |x|^p |g(x)|^2 dx = \infty \ or \ \int_{\mathbb{R}} |\xi|^q |\widehat{g}(\xi)|^2 d\xi = \infty.$$

(ii) If g is compactly supported, then

$$\int_{\mathbb{R}} |\xi| |\widehat{g}(\xi)|^2 d\xi = \infty.$$

This part also holds with g and $\widehat{g}$ interchanged.

The first theorem, stated independently by Balian [2] and Low [12], was the *symmetric* (i.e., $p = q = 2$) case of the theorem above and originally was stated only for orthonormal bases. The first proofs contained a common error, and a new, correct proof came later from Battle [3]. Soon afterwards, Coifman, Daubechies, and Semmes [8] completed the argument in the original proofs and extended the result to all Riesz bases. The second part of Theorem 1.1 was originally given by Benedetto, Czaja, Powell, and Sterbenz [6], while Gautam [9] extended the BLT to the full range of *nonsymmetric* (i.e., $p \neq q$) cases above.

The Balian-Low Theorem has been generalized in many ways. For example, Gröchenig, Han, Heil, and Kutyniok [10] extended the symmetric Balian-Low theorem to multiple variables.

Theorem 1.2 (Theorem 9, [10]) *Let $g \in L^2(\mathbb{R}^l)$ and consider the Gabor system $G(g, \mathbb{Z}^{2l}) = \{g(x - m)e^{2\pi i n \cdot x}\}_{(m,n) \in \mathbb{Z}^{2l}}$. If $G(g, \mathbb{Z}^{2l})$ is a Riesz basis for $L^2(\mathbb{R}^l)$, then for any $1 \leq k \leq l$, either*

$$\int_{\mathbb{R}^l} |x_k|^2 |g(x)|^2 dx = \infty \ or \ \int_{\mathbb{R}^l} |\xi_k|^2 |\widehat{g}(\xi)|^2 d\xi = \infty.$$

Another important generalization is the following *Quantitative BLT* of Nitzan and Olsen, which quantitatively bounds the time-frequency localization of a square-integrable function.

Theorem 1.3 (Theorem 1, [13]) *Let $g \in L^2(\mathbb{R})$ be such that $G(g)$ is a Riesz basis for $L^2(\mathbb{R})$. Then, for any $R, L \geq 1$, we have*

$$\int_{|x| \geq R} |g(x)|^2 dx + \int_{|\xi| \geq L} |\widehat{g}(\xi)|^2 d\xi \geq \frac{C}{RL}, \tag{1}$$

where the constant C only depends on the Riesz basis bounds for $G(g)$.

This result has also been extended to Gabor systems in $L^2(\mathbb{R}^l)$ in [15]. (See Theorem 5.1 below.) The Quantitative BLT is a strong result. In particular, a function satisfying (1) automatically satisfies the conclusions of both parts of Theorem 1.1. Later, we will use the Quantitative BLT and its higher dimensional analog to show that nonsymmetric versions of Theorem 1.2 hold for $\mathbb{R}^l$, $l \geq 2$.

In applications Gabor systems are used in signal analysis to give alternate representations of data with desirable properties. Often it is useful to have window functions which measure locally in time for efficiency while capturing local frequency information simultaneously. Uncertainty theorems like the BLT limit how well localized a window can be in the time and frequency domains. This led Lammers and Stampe [11] to conjecture that finite versions of the BLT should hold for discrete Gabor systems. The essence of this question was answered in one dimension by Nitzan and Olsen [14] who showed that versions of both the BLT and the quantitative BLT exist for discrete Gabor systems.

In the finite setting, instead of functions in $L^2(\mathbb{R})$, complex–valued sequences defined on $\mathbb{Z}_d = \mathbb{Z}/d\mathbb{Z}$ act as the object of study, where $d = N^2$ for some $N \in \mathbb{N}$. It is sometimes useful to fix representatives of $\mathbb{Z}_d$ in connection with the view of $\mathbb{Z}_d$ as a discretization of $\mathbb{R}$, and a convenient choice in what follows is $I_d = [-d/2, d/2) \bigcap \mathbb{Z} = \{-\lfloor \frac{d}{2} \rfloor, \ldots, d - \lfloor \frac{d}{2} \rfloor - 1\}$. Such sequences b may be thought of as samples of a continuous function g defined on $[-\frac{N}{2}, \frac{N}{2}]$ at integer multiples of $1/N$ so that $b(j) = g(j/N)$ for $j \in I_d$.

Let ℓ_2^d denote the set of complex-valued sequences on $\mathbb{Z}_d = \mathbb{Z}/d\mathbb{Z}$ with the norm

$$\|b\|^2_{\ell\_2^d} = \frac{1}{N} \sum_{j \in \mathbb{Z}_d} |b(j)|^2. \tag{2}$$

With this normalization and the sampling view noted above, $\|b\|^2_{\ell_2^d}$ approximates $\|g\|^2_{L^2[-\frac{N}{2}, \frac{N}{2}]}$. We define the *discrete Gabor system generated by* b, denoted $G_d(b)$, to be,

$$G_d(b) = \{b(j - nN)e^{2\pi i \frac{mj}{N}}\}_{(n,m) \subset [0,\ldots,N-1]^2} = \{b(j - n)e^{2\pi i \frac{mj}{d}}\}_{(n,m) \in (N\mathbb{Z}_d)^2}.$$

Here $N\mathbb{Z}_d = \{Nj : j \in \mathbb{Z}\} \mod d$ so that $\#(N\mathbb{Z}_d) = N$. This definition lines up with the definition of $G(g)$ above, as shifting g by n corresponds to shifting b by nN, and modulation of g, $g(x)e^{2\pi i m x}$, corresponds to a new sequence $b(j)e^{2\pi i m j/N}$.

To formulate the (symmetric) BLT in a finite setting, it is useful to consider an equivalent condition to the conclusion of the BLT which is in terms of the distributional derivatives of g and $\widehat{g}$, Dg and $D\widehat{g}$. In particular, the condition

$$\int_{\mathbb{R}} |x|^2 |g(x)|^2 dx = \infty \text{ or } \int_{\mathbb{R}} |\xi|^2 |\widehat{g}(\xi)|^2 d\xi = \infty \tag{3}$$

is equivalent to

$$Dg \notin L^2(\mathbb{R}) \text{ or } D\widehat{g} \notin L^2(\mathbb{R}). \tag{4}$$

For finite generators, $b \in \ell_2^d$, we instead work with differences,

$$\Delta b = \{b(j+1) - b(j)\}_{j \in \mathbb{Z}_d},$$

and note that $N\Delta b$ approximates the derivative of g. We normalize the discrete Fourier transform of b by

$$\mathcal{F}_d(b)(k) = \frac{1}{N} \sum_{j \in \mathbb{Z}_d} b(j) e^{-2\pi i \frac{jk}{d}},$$

so that $\mathcal{F}_d$ is an isometry on ℓ_2^d. Then the quantity

$$\|N\Delta b\|_{\ell_2^d}^2 + \|N\Delta \mathcal{F}_d(b)\|_{\ell_2^d}^2$$

acts as a discrete counterpart to the expressions in Eq. (4). Recall that a sequence $\{h_n\}$ is a Riesz basis for a separable Hilbert space, $\mathcal{H}$, if and only if it is complete in $\mathcal{H}$ and there exists constants $0 < A \leq B < \infty$ such that

$$A\left(\sum_n |c_n|^2\right) \leq \left\|\sum_n c_n h_n\right\|_{\mathcal{H}} \leq B\left(\sum_n |c_n|^2\right), \tag{5}$$

for any sequence (equivalently, a Riesz basis is the image of an orthonormal basis under a bounded invertible operator on $\mathcal{H}$). Here A and B are referred to as the lower and upper Riesz basis bounds, respectively. We say that b *generates an A,B-Gabor Riesz basis* if $G_d(b)$ is a basis for ℓ_2^d with Riesz basis bounds A and B.

The following *Finite BLT* of Nitzan and Olsen shows optimal bounds on the growth of this quantity for the class of sequences which generate Gabor Riesz bases with fixed Riesz basis bounds.

Theorem 1.4 (Theorem 4.2, [14]) *For $0 < A \leq B < \infty$, there exists a constant $c_{AB} > 0$, depending only on A and B, such that for any $N \geq 2$ and for any $b \in \ell_2^d$ which generates an A, B-Gabor Riesz basis for ℓ_2^d,*

$$c_{AB} \log(N) \leq \|N\Delta b\|_{\ell_2^d}^2 + \|N\Delta \mathcal{F}_d(b)\|_{\ell_2^d}^2.$$

Conversely, there exists a constant C_{AB} such that for any $N \geq 2$, there exists $b \in \ell_2^d$ which generates and A, B-Gabor Riesz basis for ℓ_2^d such that

$$\|N\Delta b\|_{\ell_2^d}^2 + \|N\Delta \mathcal{F}_d(b)\|_{\ell_2^d}^2 \leq C_{AB} \log(N).$$

Nitzan and Olsen also show that the continuous BLT, Theorem 1.1, follows from this discrete version and that the following *Finite Quantitative BLT* also holds.

Theorem 1.5 (Theorem 5.3, [14]) *Let $A, B > 0$. There exists a constant $C_{AB} > 0$ such that the following holds. Let $N \geq 200\sqrt{B/A}$ and let $b \in \ell_2^d$ generate an A, B-*

Gabor Riesz basis. Then, for all positive integers $1 \leq Q, R \leq (N/16)\sqrt{A/B}$, *we have*

$$\frac{1}{N}\sum_{j=NQ}^{d-1}|b(j)|^2 + \frac{1}{N}\sum_{k=NR}^{d-1}|\mathcal{F}_d b(k)|^2 \geq \frac{C_{AB}}{QR}.$$

1.1 Extension to Several Variables

The first goal of this paper is to extend Theorems 1.4 and 1.5 to several variables, which we state below in Theorems 1.6 and 1.7.

We consider complex-valued sequences on $\mathbb{Z}_d^l = \mathbb{Z}_d \times \cdots \times \mathbb{Z}_d$ for $l \geq 1$, and we denote the set of all such sequences as $\ell_2^{d,l}$. The view of these sequences as samples of a continuous $g \in L^2([-\frac{N}{2}, \frac{N}{2}]^l)$, where $b(\mathbf{j}) = g(\mathbf{j}/N)$ for $\mathbf{j} = (j_1, \ldots, j_l) \in I_d^l$ leads to the normalization

$$\|b\|^2_{\ell_2^{d,l}} = \frac{1}{N^l}\sum_{\mathbf{j}\in\mathbb{Z}_d^l}|b(\mathbf{j})|^2 = \frac{1}{N^l}\sum_{\mathbf{j}\in I_d^l}|b(\mathbf{j})|^2.$$

The discrete Fourier transform, $\mathcal{F}_{d,l}$, on $\ell_2^{d,l}$, is given by

$$\mathcal{F}_{d,l}(b)(\mathbf{k}) = \frac{1}{N^l}\sum_{\mathbf{j}\in\mathbb{Z}_d^l} b(\mathbf{j})e^{-2\pi i\frac{\mathbf{j}\cdot\mathbf{k}}{d}}.$$

Under this normalization, $\mathcal{F}_{d,l}$ is an isometry on $\ell_2^{d,l}$. The Gabor system generated by b, $G_{d,l}(b)$ is given by

$$G_{d,l}(b) = \{b(\mathbf{j} - N\mathbf{n})e^{2\pi i\frac{\mathbf{j}\cdot\mathbf{m}}{N}}\}_{(\mathbf{n},\mathbf{m})\in\{0,\ldots,N-1\}^{2l}} = \{b(\mathbf{j} - \mathbf{n})e^{2\pi i\frac{\mathbf{j}\cdot\mathbf{m}}{d}}\}_{(\mathbf{n},\mathbf{m})\in(N\mathbb{Z}_d)^{2l}}.$$

For any $k \in \{1, \ldots, l\}$, let $\Delta_k : \ell_2^{d,l} \to \ell_2^{d,l}$ be defined by

$$\Delta_k b(\mathbf{j}) = b(\mathbf{j} + \mathbf{e}_k) - b(\mathbf{j}),$$

where $\{\mathbf{e}_k\}_{k\in\{1,\ldots,l\}}$ is the standard orthonormal basis for $\mathbb{R}^l$. Then $N\Delta_k b$ approximates the partial derivative $\frac{\partial g}{\partial x_k}$.

We have the following generalization of Theorem 1.4.

Theorem 1.6 *Fix constants* $0 < A \leq B < \infty$. *With the same constants* c_{AB} *and* C_{AB} *from Theorem 1.4, for* $N \geq 2$, $1 \leq k \leq l$, *and for any* $b \in \ell_2^{d,l}$ *which generates an* A, B*-Gabor Riesz basis for* $\ell_2^{d,l}$, *we have*

$$c_{AB}\log(N) \le \|N\Delta_k b\|^2_{\ell_2^{d,l}} + \|N\Delta_k \mathcal{F}_{d,l}(b)\|^2_{\ell_2^{d,l}}.$$

Conversely, for $N \ge 2$ and $1 \le k \le l$, there exists $b \in \ell_2^{d,l}$ which generates an A, B-Gabor Riesz basis such that

$$\|N\Delta_k b\|^2_{\ell_2^{d,l}} + \|N\Delta_k \mathcal{F}_{d,l}(b)\|^2_{\ell_2^{d,l}} \le C_{AB}\log(N).$$

We provide a direct proof of Theorem 1.6 in Sect. 3. In Sect. 4, we extend Theorem 1.5 in the following way. For simplicity of notation, for $t > 0$, we let $\{|j_k| \ge t\}$ denote the set $\{\mathbf{j} \in I_d^l : |j_k| \ge t\}$.

Theorem 1.7 *Let $A, B > 0$ and $l \in \mathbb{N}$. There exists a constant $C > 0$ depending only on A, B, and l, such that the following holds. Let $N \ge 200\sqrt{B/A}$ and let $b \in \ell_2^{d,l}$ generate an A, B-Gabor Riesz basis for $\ell_2^{d,l}$. Then, for any $1 \le k \le l$ and all integers $1 \le Q, R \le (N/16)\sqrt{A/B}$, we have*

$$\frac{1}{N^l}\sum_{|j_k| \ge \frac{NR}{2}} |b(\mathbf{j})|^2 + \frac{1}{N^l}\sum_{|j_k| \ge \frac{NQ}{2}} |\mathcal{F}_{d,l}b(\mathbf{j})|^2 \ge \frac{C}{QR}.$$

1.2 Finite Nonsymmetric BLTs

In Sect. 5, we prove nonsymmetric versions of the finite BLT. In the process, we show that symmetric and nonsymmetric versions of the finite BLT follow as corollaries of the finite quantitative BLT (Theorem 1.7), as long as N is sufficiently large.

Theorem 1.8 (Nonsymmetric Finite BLT) *Let $A, B > 0$ and $1 < p, q < \infty$ be such that $\frac{1}{p} + \frac{1}{q} = 1$. There exists a constant $C > 0$, depending only on A, B, p and q such that the following holds. Let $N \ge 200\sqrt{B/A}$. Then, for any $b \in \ell_2^{d,l}$ which generates an A, B-Gabor Riesz basis for $\ell_2^{d,l}$,*

$$C\log(N) \le \frac{1}{N^l}\sum_{\mathbf{j}\in I_d^l}\left|\frac{j_k}{N}\right|^p |b(\mathbf{j})|^2 + \frac{1}{N^l}\sum_{\mathbf{j}\in I_d^l}\left|\frac{j_k}{N}\right|^q |b(\mathbf{j})|^2.$$

Remark 1 Theorem 1.8 gives a finite dimensional version of the nonsymmetric BLT for parameters satisfying $1 < p, q < \infty$. Thus, it is a finite dimensional analog of part (i) of Theorem 1.1 in all dimensions. In Sect. 5 we extend this result to the case where either p or q is ∞, thus giving a finite dimensional analog of part (ii) of Theorem 1.1 for all dimensions. In the same section, a generalization of this result is demonstrated for pairs (p, q) such that $\frac{1}{p} + \frac{1}{q} \neq 1$. (See Theorem 5.3.)

Remark 2 It is readily checked that the $p = q = 2$ version of Theorem 1.8 is equivalent to Theorem 1.6, so the proof of Theorem 1.8 gives an alternative proof of Theorem 1.6 for $N \geq 200\sqrt{B/A}$. In particular, the proof shows that the Finite Quantitative BLT implies the finite symmetric (and nonsymmetric) BLT.

1.3 Applications of the Continuous Quantitative BLT

In Sect. 5, we also prove several results related to functions of continuous arguments. We first state the simplest of these results, a generalization of Theorem 1.2 to nonsymmetric weights.

Theorem 1.9 *Let $g \in L^2(\mathbb{R}^l)$ and suppose that $G(g) = G(g, \mathbb{Z}^{2l})$ is a Riesz basis for $L^2(\mathbb{R}^l)$. For any $1 \leq k \leq \infty$, the following must hold.*

(i) If $1 < p < \infty$ and $\frac{1}{p} + \frac{1}{q} = 1$, then either

$$\int_{\mathbb{R}^l} |x_k|^p |g(x)|^2 dx = \infty \quad or \quad \int_{\mathbb{R}^l} |\xi_k|^q |\widehat{g}(\xi)|^2 d\xi = \infty.$$

(ii) If g is compactly supported, then

$$\int_{\mathbb{R}^l} |\xi_k| |\widehat{g}(\xi)|^2 d\xi = \infty.$$

This part also holds with g and $\widehat{g}$ interchanged.

In addition we are able to show more concrete estimates on the growth of related quantities, and we also may remove the assumption that $\frac{1}{p} + \frac{1}{q} = 1$.

Theorem 1.10 *Suppose $1 \leq p, q < \infty$ and let $g \in L^2(\mathbb{R}^l)$ be such that $G(g) = G(g, \mathbb{Z}^{2l}) = \{e^{2\pi i n \cdot x} g(x - m)\}_{(m,n) \in \mathbb{Z}^{2l}}$ is a Riesz basis for $L^2(\mathbb{R}^l)$. Let $\tau = \frac{1}{p} + \frac{1}{q}$. Then, there is a constant C depending only on the Riesz basis bounds of $G(g)$ such that for any $1 \leq k \leq l$ and any $2 \leq T < \infty$, the following inequalities hold.*

(i) If $\tau = \frac{1}{p} + \frac{1}{q} < 1$, then

$$\frac{C(1 - 2^{\tau - 1})}{(1 - \tau)} T^{1-\tau} \leq \int_{\mathbb{R}^l} \min(|x_k|^p, T) |g(x)|^2 dx + \int_{\mathbb{R}^l} \min(|\xi_k|^q, T) |\widehat{g}(\xi)|^2 d\xi.$$

(ii) If $\tau = \frac{1}{p} + \frac{1}{q} = 1$, then

$$C \log(T) \leq \int_{\mathbb{R}^l} \min(|x_k|^p, T) |g(x)|^2 dx + \int_{\mathbb{R}^l} \min(|\xi_k|^q, T) |\widehat{g}(\xi)|^2 d\xi.$$

(iii) If $\tau = \frac{1}{p} + \frac{1}{q} > 1$, *then*

$$\frac{C}{\tau - 1} \leq \int_{\mathbb{R}^l} |x_k|^p |g(x)|^2 dx + \int_{\mathbb{R}^l} |\xi_k|^q |\widehat{g}(\xi)|^2 d\xi.$$

When the bound $2 \leq T < \infty$ is replaced by $1 < \gamma \leq T < \infty$, the bound $\frac{C(1-2^{\tau-1})}{(1-\tau)} T^{1-\tau}$ in part *(i)* can be replaced by $\frac{C(1-\gamma^{\tau-1})}{(1-\tau)} T^{1-\tau}$. In Sect. 5 we extend this theorem to the case where either $p = \infty$ or $q = \infty$.

The first and second inequalities in Theorem 1.10 quantify the growth of 'localization' quantities in terms of cutoff weights of the form $\min(|x_k|^p, T)$. The log term in the second inequality shows a connection between the continuous BLT and its finite dimensional versions. The last inequality, on the other hand, shows that generators of Gabor Riesz bases must satisfy a Heisenberg type uncertainty principle for every $0 < p \leq 2$. A similar inequality is known to hold for arbitrary $L^2(\mathbb{R})$ functions by a result of Cowling and Price [7]. However, for generators of Gabor Riesz bases, we have explicit estimates on the dependence of the constant on τ and the result here is stated for higher dimensions.

2 Preliminaries: The Zak Transform and Quasiperiodic Functions

The Zak transform is an essential tool for studying lattice Gabor systems. The discrete Zak transform $Z_{d,l}$ of $b \in \ell_2^{d,l}$ for $(\mathbf{m}, \mathbf{n}) \in \mathbb{Z}_d^{2l}$ is given by

$$Z_{d,l}(b)(\mathbf{m}, \mathbf{n}) = \sum_{\mathbf{j} \in \{0,\ldots,N-1\}^l} b(\mathbf{m} - N\mathbf{j}) e^{2\pi i \frac{\mathbf{n}\cdot\mathbf{j}}{N}} = \sum_{\mathbf{j} \in N\mathbb{Z}_d^l} b(\mathbf{m} - \mathbf{j}) e^{2\pi i \frac{\mathbf{n}\cdot\mathbf{j}}{d}}.$$

The following properties show that $Z_{d,l}(b)$ encodes basis properties of $G_{d,l}(b)$, while retaining information about 'smoothness' (see the remark following Proposition 2.1) of b and $\mathcal{F}_{d,l}(b)$. Note that $Z_{d,l}(b)(\mathbf{m}, \mathbf{n})$ is defined for $(\mathbf{m}, \mathbf{n}) \in \mathbb{Z}_d^{2l}$ and is d-periodic in each of its $2l$ variables. However, the Zak transform satisfies even stronger periodicity conditions. In fact, $Z_{d,l}(b)$ is *N-quasiperiodic on* $\mathbb{Z}_d^{2l}$, that is

$$Z_{d,l}(b)(\mathbf{m} + N\mathbf{e}_k, \mathbf{n}) = e^{2\pi i \frac{n_k}{N}} Z_{d,l}(b)(\mathbf{m}, \mathbf{n}), \tag{6}$$

$$Z_{d,l}(b)(\mathbf{m}, \mathbf{n} + N\mathbf{e}_k) = Z_{d,l}(b)(\mathbf{m}, \mathbf{n}).$$

Let $S_N = \{0, \ldots, N-1\}$. Then, the quasi-periodicity conditions above show that $Z_{d,l}(b)$ is completely determined by its values on S_N^{2l}.

We will use the notation $\ell_2(S_N^{2l})$ to denote the set of sequences $W(\mathbf{m}, \mathbf{n})$ defined on S_N^{2l} with norm given by

$$\|W\|_{\ell_2(S_N^{2l})}^2 = \frac{1}{N^{2l}} \sum_{(\mathbf{m},\mathbf{n})\in S_N^{2l}} |W(\mathbf{m},\mathbf{n})|^2,$$

where here we keep the variables **m** and **n** separate due to the connection with the Zak transform. The normalization is chosen so that if W is a sampling of a function $h(\mathbf{x}, \mathbf{y})$ on $[0, 1]^{2l}$, then $\|W\|_{\ell_2(S_N^{2l})}$ approximates the $L^2([0, 1]^{2l})$ norm of h.

The Zak transform has many other important properties, some of which we collect in the next proposition. Arguments for these facts are standard and presented in [1] and [14], for instance.

Proposition 2.1 *Let* $b \in \ell_2^{d,l}$.

(i) $Z_{d,l}$ *is a unitary mapping from* $\ell_2^{d,l}$ *onto* $\ell_2(S_N^{2l})$.
(ii) A sequence $b \in \ell_2^{d,l}$ *generates an* A, B*-Gabor Riesz basis for* $\ell_2^{d,l}$ *if and only if* $Z_{d,l}(b)$ *satisfies*

$$A \le |Z_{d,l}(b)(\mathbf{m},\mathbf{n})|^2 \le B, \text{ for } (\mathbf{m},\mathbf{n}) \in \mathbb{Z}_d^{2l}.$$

(iii) Let $\widehat{b} = \mathcal{F}_{d,l}(b)$. *Then,*

$$Z_{d,l}(\widehat{b})(\mathbf{m},\mathbf{n}) = e^{2\pi i \frac{\mathbf{m}\cdot\mathbf{n}}{d}} Z_{d,l}(b)(-\mathbf{n},\mathbf{m}).$$

(iv) For $a, b \in \ell_2^{d,l}$ *define* $(a * b)(\mathbf{k}) = \frac{1}{N^l}\sum_{j\in\mathbb{Z}_d^l} a(\mathbf{k}-\mathbf{j})b(\mathbf{j})$. *Then,*

$$Z_{d,l}(a * b)(\mathbf{m},\mathbf{n}) = \frac{1}{N^l}\sum_{\mathbf{j}\in\mathbb{Z}_d^l} b(\mathbf{j}) Z_{d,l}(a)(\mathbf{m}-\mathbf{j},\mathbf{n}) = (Z_{d,l}(a) *_1 b)(\mathbf{m},\mathbf{n}),$$

where $*_1$ *denotes convolution of* b *with respect to the first set of variables of* $Z_{d,l}(a)$, **m**, *keeping the second set,* **n**, *fixed.*

Remark 3 We will be interested in the 'smoothness' of b and $Z_{d,l}(b)$ for $b \in \ell_2^{d,l}$. Since these are functions on discrete sets, smoothness is not well defined, but we use the term in relation to the size of norms of certain difference operators defined on $\ell_2^{d,l}$ and $\ell_2(S_N^{2l})$, which mimic norms of partial derivatives of differentiable functions.

For $1 \le k \le l$ and any N-quasiperiodic function on $\mathbb{Z}_d^l$, let Δ_k, Γ_k be defined as follows:

$$\Delta_k W(\mathbf{m},\mathbf{n}) = W(\mathbf{m}+\mathbf{e_k},\mathbf{n}) - W(\mathbf{m},\mathbf{n}),$$

$$\Gamma_k W(\mathbf{m},\mathbf{n}) = W(\mathbf{m},\mathbf{n}+\mathbf{e_k}) - W(\mathbf{m},\mathbf{n}).$$

For $b \in \ell_2^{d,l}$ define $\alpha_k(b)$ and $\beta_k(b)$ by

$$\alpha_k(b) = \|N\Delta_k b\|^2_{\ell_2^{d,l}} + \|N\Delta_k \mathcal{F}_{d,l}(b)\|^2_{\ell_2^{d,l}},$$

$$\beta_k(b) = \frac{1}{N^{2l}} \sum_{(\mathbf{m},\mathbf{n})\in S_N^{2l}} |N\Delta_k Z_{d,l}(b)(\mathbf{m},\mathbf{n})|^2 + \frac{1}{N^{2l}} \sum_{(\mathbf{m},\mathbf{n})\in S_N^{2l}} |N\Gamma_k Z_{d,l}(b)(\mathbf{m},\mathbf{n})|^2.$$

The following proposition shows that $\alpha_k(b)$ and $\beta_k(b)$ are essentially equivalently sized. Proposition 4.1 in [14] proves this for the case $l = k = 1$, and it is readily checked that the proof carries over directly to the $l > 1$ setting.

Proposition 2.2 *Let $B > 0$ and let $b \in \ell_2^{d,l}$ be such that $|Z_{d,l}(b)(\mathbf{m},\mathbf{n})|^2 \leq B$ for all $(\mathbf{m},\mathbf{n}) \in \mathbb{Z}_d^{2l}$. Then, for all integers $N \geq 2$ and any $1 \leq k \leq l$, we have*

$$\frac{1}{2}\beta_k(b) - 8\pi^2 B \;\leq\; \alpha_k(b) \;\leq\; 2\beta_k(b) + 8\pi^2 B.$$

We thus see that in order to bound $\alpha_k(b)$ as in Theorem 1.6, it is sufficient to bound $\beta_k(b)$. For $b \in \ell_2^d = \ell_2^{d,1}$, let $\beta(b) = \beta_1(b)$, and let

$$\beta_{A,B}(N) = \inf\{\beta(b)\},$$

where the infimum is taken over all $b \in \ell_2^d$ such that b generates an A, B-Gabor Riesz basis.

Theorem 2.1 (Theorem 4.2, [14]) *There exist constants $0 < c_{AB} \leq C_{AB} < \infty$ such that for all $N \geq 2$, we have*

$$c_{AB}\log(N) \;\leq\; \beta_{A,B}(N) \;\leq\; C_{AB}\log(N).$$

To prove the lower bound in this theorem (as is done in [14]), one may examine the argument of the Zak transform of a sequence $b \in \ell_2^d$ which generates a basis with Riesz basis bounds A and B over finite dimensional lattice-type structures in the square $\{0, \ldots, N\}^2$. Due to the N-quasiperiodicity conditions satisfied by $Z_{d,l}(b)$ this argument is forced to 'jump' at some step between neighboring points along these lattice-type sets (See Lemma 3.1 and 3.4 in [14]). Due to the Riesz basis assumption and part (iv) of Proposition 2.1, jumps in the argument of $Z_{d,l}(b)$ correspond directly to jumps in $Z_{d,l}(b)$ (see Corollary 3.6 in [14]). By counting the number of lattice-type sets which are disjoint, a logarithmic lower bound is given for the number of jumps in $Z_{d,l}(b)$ corresponding to jumps in the argument, which gives the lower bound in Theorem 2.1.

The proof of the upper bound involves an explicit construction of the argument of a unimodular function, W, on S_N^2. Since the Zak transform is a unitary, invertible mapping between ℓ_2^d and $\ell_2(S_N^2)$, there is a corresponding $\tilde{b} \in \ell_2^d$ so that $G(\tilde{b})$ is an orthonormal basis (which can be scaled to form a Riesz basis with bounds A and B for any A and B) and such that $\tilde{b}$ satisfies $Z_d(\tilde{b}) = W$. For this construction, $\beta(\tilde{b})$ can be bounded directly to show the upper bound in the theorem.

3 Proof of Theorem 1.6

Based on Proposition 2.2, to prove Theorem 1.6 it is sufficient to show that Theorem 2.1 extends from ℓ_2^d to $\ell_2^{d,l}$. We show this below, and in particular that by restricting the Zak transform of a multi-variable sequence to the kth variable in each component, we can directly use Theorem 2.1 to prove the multi-variable version of the lower bound. Similarly, we show that by taking suitable products of the constructed $\tilde{b}$ function mentioned above, we can also extend the logarithmic upper bound to higher dimensions.

Let

$$\beta_{A,B,k}(N,l) = \inf\{\beta_k(b)\},$$

where the infimum is over all $b \in \ell_2^{d,l}$ which generate an A, B-Gabor Riesz basis for $\ell_2^{d,l}$.

Theorem 3.1 *For the same constants* $0 < c_{AB} \leq C_{AB} < \infty$ *as Theorem 2.1, for all* $N \geq 2$, *and for any* $1 \leq k \leq l$, *we have*

$$c_{AB} \log(N) \;\leq\; \beta_{A,B,k}(N,l) \;\leq\; C_{AB} \log(N).$$

Proof For notational convenience, we show both the lower and upper bound with $k = 1$, but a similar argument applies for any $1 \leq k \leq l$.

Lower Bound Let $b \in \ell_2^{d,l}$ generate an A, B-Gabor Riesz basis for $\ell_2^{d,l}$.

Let $\mathbf{m} = (m_1, \mathbf{m}')$ and $\mathbf{n} = (n_1, \mathbf{n}')$ for fixed $(\mathbf{m}', \mathbf{n}') \in S_N^{2(l-1)}$ and define

$$T(m_1, n_1) \;-\; T_{\mathbf{m}',\mathbf{n}'}(m_1, n_1) \;-\; Z_{d,l}(b)((m_1, \mathbf{m}'), (n_1, \mathbf{n}')).$$

Then, T satisfies

$$T(m_1 + N, n_1) = Z_{d,l}(b)(\mathbf{m} + N\mathbf{e}_1, \mathbf{n}) \;=\; e^{2\pi i \frac{n_1}{N}} T(m_1, n_1),$$
$$T(m_1, n_1 + N) = Z_{d,l}(b)(\mathbf{m}, \mathbf{n} + N\mathbf{e}_1) \;=\; T(m_1, n_1),$$

so T is N-quasiperiodic on $\mathbb{Z}_d^2$ (see Eq. 6). By the unitary property of the Zak transform (Proposition 2.1), there exists a $b_1 \in \ell_2^d$ so that $T = Z_{d,1}(b_1)$, and since $A \leq |T(m_1, n_1)|^2 \leq B$ for any $(m_1, n_1) \in \mathbb{Z}_d^2$, the same property shows that $G_d(b_1)$ is a Riesz basis for ℓ_2^d with bounds A and B. Thus, Theorem 2.1 shows that

$$C_{AB} \log(N) \leq \sum_{(m_1,n_1)\in S_N^2} |\Delta_1 T_{\mathbf{m}',\mathbf{n}'}(m_1, n_1)|^2 + \sum_{(m_1,n_1)\in S_N^2} |\Gamma_1 T_{\mathbf{m}',\mathbf{n}'}(m_1, n_1)|^2.$$

Since the choice of $(\mathbf{m}', \mathbf{n}') \in S_N^{2(l-1)}$ was arbitrary, this bound holds for any such choice.

Thus, computing $\beta_1(b)$, we find

$$\frac{1}{N^{2(l-1)}} \sum_{(\mathbf{m}',\mathbf{n}')\in S_N^{2(l-1)}} \left[\sum_{(m_1,n_1)\in S_N^2} |\Delta T_{\mathbf{m}',\mathbf{n}'}(m_1, n_1)|^2 + \sum_{(m_1,n_1)\in S_N^2} |\Gamma T_{\mathbf{m}',\mathbf{n}'}(m_1, n_1)|^2 \right]$$
$$\geq C_{AB} \log(N),$$

since the bound holds for each term inside the brackets, and $\beta_1(b)$ is simply an average of these terms. Taking an infimum over all acceptable $b \in \ell_2^{d,l}$ proves the lower bound.

Upper Bound To prove the upper bound, we adapt the construction used to prove the one-dimensional upper bound in [14] to higher dimensions. The sequence used in this construction builds on a continuous construction first given in [5]. Note that it suffices to prove the result for orthonormal bases, as the result for Riesz bases follows by scaling the constructed generator by the Riesz basis bounds.

In Section 4.3 of [14], it is shown that there is a constant $C > 0$ such that for any $N \geq 2$, there exists a $b \in \ell_2^d$ such that $G_d(b)$ is an orthonormal basis for ℓ_2^d and

$$\beta(b) = \sum_{(m,n)\in S_N^2} \left|\Delta Z_{d,1}(b)(m,n)\right|^2 + \sum_{(m,n)\in S_N^2} \left|\Gamma Z_{d,1}(b)(m,n)\right|^2 \leq C \log(N).$$

For $\mathbf{j} \in \mathbb{Z}_d^l$, let $b_l(\mathbf{j}) = b(j_1)b(j_2)\cdots b(j_l)$. Then,

$$Z_{d,l}(b_l)(\mathbf{m},\mathbf{n}) = Z_{d,1}(b)(m_1,n_1)\cdots Z_{d,1}(b)(m_l,n_l).$$

Since $G_d(b)$ is an orthonormal basis for ℓ_2^d, $Z_{d,l}(b_l)$ is unimodular, and therefore, $G_{d,l}(b_l)$ is an orthonormal basis for $\ell_2^{d,l}$ by Proposition 2.1. We have, $\beta_1(b_l)$ is equal to

$$\frac{1}{N^{2(l-1)}} \sum_{(\mathbf{m}',\mathbf{n}')\in \mathbb{Z}_N^{2(l-1)}} \left[\sum_{(m_1,n_1)\in S_N^2} \left|\Delta Z_{d,1}(b)(m_1, n_1)\right|^2 + \sum_{(m_1,n_1)\in S_N^2} \left|\Gamma Z_{d,1}(b)(m_1, n_1)\right|^2 \right]$$
$$\leq C \log(N).$$

Theorem 1.6 follows by combining Theorem 3.1 with Proposition 2.2.

4 Proof of Theorem 1.7

In establishing a Finite Quantitative BLT for several variables, we follow a similar argument used to prove the one variable version (from [14]), but there are some necessary updates to certain parts of the proof. We include the details here for completeness.

We start with a straightforward bound on the 'smoothness' of $Z_{d,l}(b * \phi)$. This observation is analogous to Lemma 2.6 of [14]. Let $\|\phi\|_{\ell_1^{d,l}} = \frac{1}{N^l}\sum_{\mathbf{j}\in\mathbb{Z}_d^l}|\phi(\mathbf{j})|$, and for $a, b \in \ell_2^{d,l}$, recall that $(a * b)(\mathbf{k}) = \frac{1}{N^l}\sum_{\mathbf{j}\in\mathbb{Z}_d^l} a(\mathbf{k}-\mathbf{j})b(\mathbf{j})$.

Lemma 4.1 *Suppose $b, \phi \in \ell_2^{d,l}$ are such that $|Z_{d,l}(b)|^2 \le B$ everywhere. Then, for any integer t,*

$$|Z_{d,l}(b*\phi)(\mathbf{m}+t\mathbf{e}_k,\mathbf{n}) - Z_{d,l}(b*\phi)(\mathbf{m},\mathbf{n})| \le \frac{\sqrt{B}|t|}{N}\|N\Delta_k\phi\|_{\ell_1^{d,l}}.$$

Proof From Proposition 2.1, we have

$$Z_{d,l}(b*\phi)(\mathbf{m},\mathbf{n}) = \frac{1}{N^l}\sum_{\mathbf{j}\in\mathbb{Z}_d^l}\phi(\mathbf{j})Z_{d,l}(b)(\mathbf{m}-\mathbf{j},\mathbf{n}) = Z_{d,l}(b) *_1 \phi(\mathbf{m},\mathbf{n}).$$

Therefore, we have

$$\begin{aligned}
&|Z_{d,l}(b*\phi)(\mathbf{m}+t\mathbf{e}_k,\mathbf{n}) - Z_{d,l}(b*\phi)(\mathbf{m},\mathbf{n})| \\
&\le \sum_{s=0}^{t-1}|Z_{d,l}(b*\phi)(\mathbf{m}+(s+1)\mathbf{e}_k,\mathbf{n}) - Z_{d,l}(b*\phi)(\mathbf{m}+s\mathbf{e}_k,\mathbf{n})| \\
&= \sum_{s=0}^{t-1}\left|\frac{1}{N^l}\sum_{\mathbf{j}\in\mathbb{Z}_d^l} Z_{d,l}(b)(\mathbf{j},\mathbf{n})[\phi(\mathbf{m}+(s+1)\mathbf{e}_k-\mathbf{j}) - \phi(\mathbf{m}+s\mathbf{e}_k-\mathbf{j})]\right| \\
&\le \sum_{s=0}^{t-1}\frac{\sqrt{B}}{N^l}\sum_{\mathbf{j}\in\mathbb{Z}_d^l}|\Delta_k\phi(\mathbf{m}+s\mathbf{e}_k-\mathbf{j})| = \frac{\sqrt{B}}{N}t\|N\Delta_k\phi\|_{\ell_1^{d,l}}.
\end{aligned}$$

□

Next we extend the following Lemma 5.2 of [14] to higher dimensions. The adjustments to this lemma for the higher dimensional setting are minimal, however we state the one-dimensional and multi-variable versions separately for comparison.

Lemma 4.2 (Lemma 5.2, [14]) *Let $A, B > 0$ and $N \ge 200\sqrt{B/A}$. There exist positive constants $\delta = \delta(A)$ and $C = C(A, B)$ such that the following holds (with $d = N^2$). Let*

(i) $Q, R \in \mathbb{Z}$ *such that* $1 \le Q, R \le (N/16) \cdot \sqrt{A/B}$,
(ii) $\phi, \psi \in \ell_2^d$ *such that* $\sum_n |\Delta\phi(n)| \le 10R$ *and* $\sum_n |\Delta\psi(n)| \le 10Q$,
(iii) $b \in \ell_2^d$ *such that* $A \le |Z_d(b)|^2 \le B$.

Then, there exists a set $S \subset ([0, N-1] \cap \mathbb{Z})^2$ *of size* $|S| \ge CN^2/QR$ *such that all* $(u, v) \in S$ *satisfy either*

$$|Z_d(b)(u, v) - Z_d(b * \phi)(u, v)| \ge \delta, \qquad or \tag{7}$$

$$|Z_d(\mathcal{F}_d b)(u, v) - Z_d((\mathcal{F}_d b) * \psi)(u, v)| \ge \delta. \tag{8}$$

Lemma 4.3 *Let* $A, B > 0$, $1 \le k \le l$, *and* $N \ge 200\sqrt{B/A}$. *There exist positive constants* $\delta = \delta(A)$ *and* $C = C(A, B)$, *such that the following holds. Let*

(i) $Q, R \in \mathbb{Z}$ *be such that* $1 \le Q, R \le \frac{N}{16}\sqrt{\frac{A}{B}}$
(ii) $\phi, \psi \in \ell_2^{d,l}$ *be such that* $\|N\Delta_k\phi\|_{\ell_1^{d,l}} \le 10R$ *and* $\|N\Delta_k\psi\|_{\ell_1^{d,l}} \le 10Q$
(iii) $b \in \ell_2^{d,l}$ *be such that* $A \le |Z_{d,l}(b)|^2 \le B$.

Then, there exists a set $S \subset ([0, N-1] \cap \mathbb{Z})^{2l}$ *of size* $|S| \ge CN^{2l}/QR$ *such that all* $(\mathbf{u}, \mathbf{v}) \in S$ *satisfy either*

$$|Z_{d,l}(b)(\mathbf{u}, \mathbf{v}) - Z_{d,l}(b * \phi)(\mathbf{u}, \mathbf{v})| \ge \delta, \ or \tag{9}$$

$$|Z_{d,l}(\mathcal{F}_{d,l} b)(\mathbf{u}, \mathbf{v}) - Z_{d,l}((\mathcal{F}_{d,l} b) * \psi)(\mathbf{u}, \mathbf{v})| \ge \delta. \tag{10}$$

Proof Without loss of generality, we prove this for $k = 1$.

As in Lemma 5.2 of [14], let $\delta_1 = 2\sqrt{A}\sin(\pi(\frac{1}{4} - \frac{1}{200}))$. Also, choose K and L to be the smallest integers satisfying

$$\frac{200\sqrt{B}R}{9\delta_1} \le K \le N \quad \text{and} \quad \frac{\sqrt{B}}{\delta_1}\max\left\{\frac{200Q}{9}, 80\pi\right\} \le L \le N.$$

For $s, t \in \mathbb{Z}$, let

$$\sigma_s = \left\lceil \frac{sN}{K} \right\rceil, \quad \text{and} \quad \omega_t = \left\lceil \frac{tN}{L} \right\rceil,$$

and let $\Sigma = \inf_s\{\sigma_{s+1} - \sigma_s\} \ge \left\lfloor \frac{N}{K} \right\rfloor \ge \frac{N}{K}$, $\Omega = \inf_t\{\omega_{t+1} - \omega_t\} \ge \frac{N}{L}$. Then, we have

$$\Sigma\Omega \ge C_1\frac{N^2}{QR},$$

where C_1 can be chosen to be

$$C_1 = \left[(\frac{200\sqrt{B}}{9\delta_1} + 1)(\frac{\sqrt{B}}{\delta_1} \max(\frac{200}{9}, 80\pi) + 1) \right]^{-1}.$$

We recall the following definition from [14]. For $(u, v) \in ([0, \Sigma - 1] \cap \mathbb{Z}) \times ([0, \Omega - 1] \cap \mathbb{Z})$, let

$$\mathrm{Lat}(u, v) = \{(u + \sigma_s, v + \omega_t) : (s, t) \in ([0, K - 1] \cap \mathbb{Z}) \times ([0, L - 1] \cap \mathbb{Z})\},$$

and

$$\mathrm{Lat}^*(u, v) = \{(N - v - \omega_t, u + \sigma_s) : (s, t) \in ([0, K - 1] \cap \mathbb{Z}) \times ([0, L - 1] \cap \mathbb{Z})\}.$$

Note that $\mathrm{Lat}(u, v)$ and $\mathrm{Lat}(u', v')$ are disjoint for distinct (u, v) and (u', v'), and similarly for $\mathrm{Lat}^*(u, v)$. However, it is possible that $\mathrm{Lat}(u, v) \cap \mathrm{Lat}^*(u', v') \neq \emptyset$ for some (u, v) and (u', v').

Now similarly, for any $(\mathbf{m}', \mathbf{n}') \in ([0, N - 1] \cap \mathbb{Z})^{2(l-1)}$, let

$$\mathrm{Lat}_{(\mathbf{m}', \mathbf{n}')}(u, v) = \{((m_1, \mathbf{m}'), (n_1, \mathbf{n}')) : (m_1, n_1) \in \mathrm{Lat}(u, v)\},$$

and

$$\mathrm{Lat}^*_{(\mathbf{m}', \mathbf{n}')}(u, v) = \{((n_1, N - \mathbf{n}'), (m_1, \mathbf{m}')) : (n_1, m_1) \in \mathrm{Lat}^*(u, v)\}.$$

Here, by $N - \mathbf{n}'$ we mean $(N - n'_1, N - n'_2, \ldots, N - n'_{l-1})$. We have that $\mathrm{Lat}_{(\mathbf{m}', \mathbf{n}')}(u, v) \cap \mathrm{Lat}_{(\mathbf{m}'', \mathbf{n}'')}(u', v') = \emptyset$ unless it holds that $((u, \mathbf{m}'), (v, \mathbf{n}')) = ((u', \mathbf{m}''), (v', \mathbf{n}''))$, and similar properties for $\mathrm{Lat}^*_{(\mathbf{m}', \mathbf{n}')}(u, v)$.

Now, fix $(\mathbf{m}', \mathbf{n}') \in ([0, N - 1] \cap \mathbb{Z})^{2(l-1)}$, and consider

$$T(m_1, n_1) = T_{\mathbf{m}', \mathbf{n}'}(m_1, n_1) = Z_{d,l}(b)((m_1, \mathbf{m}'), (n_1, \mathbf{n}')),$$

for $(m_1, n_1) \in \mathbb{Z}_d^2$. Note that T is N-quasiperiodic on $\mathbb{Z}_d^2$, and satisfies $A \leq |T|^2 \leq B$.

For each $(u, v) \in ([0, \Sigma - 1] \cap \mathbb{Z}) \times ([0, \Omega - 1] \cap \mathbb{Z})$, Corollary 3.6 of [14] guarantees at least one point $(s, t) \in ([0, K - 1] \cap \mathbb{Z}) \times ([0, L - 1] \cap \mathbb{Z})$ so that either

$$|T(u + \sigma_{s+1}, v + \omega_t) - T(u + \sigma_s, v + \omega_t)| \geq \delta_1, \text{ or} \tag{11}$$

$$|T(u + \sigma_s, v + \omega_{t+1}) - T(u + \sigma_s, v + \omega_t)| \geq \delta_1. \tag{12}$$

We now make a claim which will furnish the last part of the proof of the lemma. □

Claim For u, v, σ_s, ω_t, $\mathbf{m}'$ and $\mathbf{n}'$ as above,

(i) If (11) is satisfied, then there exists $(\mathbf{a}, \mathbf{b}) \in \mathrm{Lat}_{(\mathbf{m}',\mathbf{n}')}(u, v)$ so that (9) is satisfied for $\delta = \frac{\delta_1}{20}$.
(ii) If (12) is satisfied, then there exists $(\mathbf{a}, \mathbf{b}) \in \mathrm{Lat}^*_{(\mathbf{m}',\mathbf{n}')}(u, v)$ so that (10) is satisfied for $\delta = \frac{\delta_1}{40}$

Before proving this claim, we show how to complete the proof of the lemma. For a fixed $(\mathbf{m}', \mathbf{n}')$ there are $\Sigma\Omega \geq C_1 \frac{N^2}{QR}$ distinct choices of (u, v) to consider and each of them either falls in part (*i*) or (*ii*) of the claim. Let $S^1_{(\mathbf{m}',\mathbf{n}')}$ be the set of (u, v) points which fall into category (*i*), and similarly let $S^2_{(\mathbf{m}',\mathbf{n}')}$ be the set of (u, v) points which fall into category (*ii*). Then, for either $i = 1, 2$, we must have

$$|S^i_{(\mathbf{m}',\mathbf{n}')}| \geq \frac{C_1 N^2}{2QR}. \tag{13}$$

Now, there are $N^{2(l-2)}$ possible choices of $(\mathbf{m}', \mathbf{n}')$. Let S_1 be the set of all $(\mathbf{m}', \mathbf{n}')$ such that (13) is satisfies with $i = 1$, and let S_2 be the set of all $(\mathbf{m}', \mathbf{n}')$ such that (13) is satisfied with $i = 2$. So at least one of S_1 or S_2 must contain $N^{2(l-2)}/2$ elements.

In the case that S_1 contains this many elements (the S_2 case is nearly identical and left to the reader), since $\mathrm{Lat}_{(\mathbf{m}',\mathbf{n}')}(u, v)$ are disjoint for distinct $((u, \mathbf{m}'), (v, \mathbf{n}'))$, we find at least $\frac{C_1 N^{2l}}{4QR} = C\frac{N^{2l}}{QR}$ distinct points all satisfy (9) if $i = 1$. The lemma is then proved conditioning on the claim above. We then establish finally the two part claim.

Proof of Claim For both parts we use properties of the Zak transform detailed in Proposition 2.1. First we show part (*i*). Let $H(u, v) = Z_{d,l}(b * \phi)((u, \mathbf{m}'), (v, \mathbf{n}'))$. Note that Lemma 4.1 and the assumptions on R and $\|N\Delta_1\phi\|_{\ell_1^{d,l}}$ imply that for any integer t satisfying $t \leq \frac{2N}{K}$,

$$|H(u + t, v) - H(u, v)| \leq \tfrac{2\sqrt{B}}{K}\|N\Delta_1\phi\|_{\ell_1^{d,l}} \leq \tfrac{20\sqrt{B}R}{K} \leq \tfrac{9\delta_1}{10}. \tag{14}$$

So, if (11) is satisfied, using (14), we have

$$\begin{aligned}\delta_1 &\leq |T(u + \sigma_{s+1}, v + \omega_t) - T(u + \sigma_s, v + \omega_t)| \\ &\leq |T(u + \sigma_{s+1}, v + \omega_t) - H(u + \sigma_{s+1}, v + \omega_t)| \\ &\quad + \frac{9\delta_1}{10} + |T(u + \sigma_s, v + \omega_t) - H(u + \sigma_s, v + \omega_t)|.\end{aligned}$$

Upon rearranging terms, we find

$$\begin{aligned}\tfrac{\delta_1}{10} &\leq |T(u + \sigma_{s+1}, v + \omega_t) \\ &- H(u + \sigma_{s+1}, v + \omega_t)| + |T(u + \sigma_s, v + \omega_t) - H(u + \sigma_s, v + \omega_t)|,\end{aligned}$$

which shows that (9) is satisfied for $\delta' = \frac{\delta_1}{20}$, and for either $((u+\sigma_{s+1}, \mathbf{m}'), (v+\omega_t, \mathbf{n}'))$ or $((u+\sigma_s, \mathbf{m}'), (v+\omega_t, \mathbf{n}'))$. If $(u+\sigma_{s+1}, v+\omega_t)$ is not in $\mathrm{Lat}(u, v)$, by the N-quasiperiodicity of T, we may find another point in $\mathrm{Lat}(u, v)$ which satisfies the same bound.

Now we prove part *(ii)*. Letting $\hat{b} = \mathcal{F}_{d,l}(b)$, we have,

$$\begin{aligned}
\delta_1 &\le |T(u+\sigma_s, v+\omega_{t+1}) - T(u+\sigma_s, v+\omega_t)| \\
&= |Z_{d,l}(b)((u+\sigma_s, \mathbf{m}'), (v+\omega_{t+1}, \mathbf{n}')) - Z_{d,l}(b)((u+\sigma_s, \mathbf{m}'), (v+\omega_t, \mathbf{n}'))| \\
&= |Z_{d,l}(\hat{b})((-v-\omega_{t+1}, -\mathbf{n}'), (u+\sigma_s, \mathbf{m}')) \\
&\qquad - e^{-2\pi i(\omega_{t+1}-\omega_t)(u+\sigma_s)/d} Z_{d,l}(\hat{b})((-v-\omega_t, -\mathbf{n}'), (u+\sigma_s, \mathbf{m}'))| \\
&= |Z_{d,l}(\hat{b})((N-v-\omega_{t+1}, N-\mathbf{n}'), (u+\sigma_s, \mathbf{m}')) \\
&\qquad - e^{-2\pi i(\omega_{t+1}-\omega_t)(u+\sigma_s)/d} Z_{d,l}(\hat{b})((N-v-\omega_t, N-\mathbf{n}'), (u+\sigma_s, \mathbf{m}'))|,
\end{aligned}$$

where we have used that $Z_{d,l}(b)(\mathbf{m}, \mathbf{n}) = e^{2\pi i \mathbf{m}\cdot\mathbf{n}/d} Z_{d,l}(\widehat{b})(-\mathbf{n}, \mathbf{m})$ in the second step, and for the last step we have used N-quasiperiodicity.

Let $\widetilde{T}(v, u) = Z_{d,l}(\hat{b})((v, N-\mathbf{n}'), (u, \mathbf{m}'))$, and $\widetilde{H}(v, u) = Z_{d,l}(\hat{b}*\psi)((v, N-\mathbf{n}'), (u, \mathbf{m}'))$. Then,

$$\begin{aligned}
\delta_1 &\le |\widetilde{T}(N-v-\omega_{t+1}, u+\sigma_s) - e^{-2\pi i(\omega_{t+1}-\omega_t)(u+\sigma_s)/d} \widetilde{T}(N-v-\omega_t, u+\sigma_s)| \\
&\le |\widetilde{T}(N-v-\omega_{t+1}, u+\sigma_s) - \widetilde{T}(N-v-\omega_t, u+\sigma_s)| + \frac{\delta_1}{20}.
\end{aligned}$$

Combining these, we see that

$$\frac{19}{20}\delta_1 \le |\widetilde{T}(N-v-\omega_{t+1}, u+\sigma_s) - \widetilde{T}(N-v-\omega_t, u+\sigma_s)|.$$

Arguing as in the first case above, and replacing H by $\widetilde{H}$ and T by $\widetilde{T}$, we find that either $((N-v-\omega_{t+1}, N-\mathbf{n}'), (u, \mathbf{m}'))$, or $((N-v-\omega_t, N-\mathbf{n}'), (u, \mathbf{m}'))$ satisfy (10), with $\delta = \frac{\delta_1}{40}$. Again, using quasi-periodicity, we can guarantee that there is a point in $\mathrm{Lat}^*_{(\mathbf{m}',\mathbf{n}')}(u, v)$ satisfying (10). □

Finally, we follow the construction of [14] to create the functions ϕ and ψ appearing in the previous lemma (Lemma 4.3) which in turn are used to prove Theorem 1.7. Let $\rho : \mathbb{R} \to \mathbb{R}$ be the inverse Fourier transform of

$$\hat{\rho}(\xi) = \begin{cases} 1, & |\xi| \le 1/2 \\ 2(1-\xi sgn(\xi)), & 1/2 \le |\xi| \le 1\,. \\ 0, & |\xi| \ge 1 \end{cases}$$

For $f \in L^2(\mathbb{R})$ satisfying $\sup_{t\in\mathbb{R}} |t^2 f(t)| < \infty$ and $\sup_{\xi\in\mathbb{R}} |\xi^2 \widehat{f}(t)| < \infty$, let

$$P_N f(t) = \sum_{k=-\infty}^{\infty} f(t + kN)$$

and for an N-periodic continuous function h, let

$$S_N h = \{h(j/N)\}_{j=0}^{d-1}.$$

Let $\rho_R(t) = R\rho(Rt)$. Fix $1 \le k \le l$, and for $\mathbf{j} \in I_d^l$ define the vector $\mathbf{j}' = (j_1, \dots, j_{k-1}, j_{k+1}, \dots, j_l) \in I_d^{l-1}$, and let

$$\phi_{R,k}(\mathbf{j}) = N^{l-1} \delta_{\mathbf{j}',\mathbf{0}} \left(S_N P_N \rho_R(j_k)\right).$$

Now $\phi_{R,k}(\mathbf{j})$ is equal to $(S_N P_N \rho_R(j_k))$ when $j_i = 0$ for each $i \ne k$, and is zero otherwise.

Lemma 4.4 *Let $\phi_{R,k}$ be as above for a positive integer R. Then,*

$$\|N\Delta_k \phi_{R,k}\|_{\ell_1^{d,l}} \le 10R.$$

Proof We have

$$\|N\Delta_k \phi_{R,k}\|_{\ell_1^{d,l}} = \frac{1}{N^l} \sum_{\mathbf{j} \in I_d^l} N|\Delta_k \phi_{R,k}(\mathbf{j})| = \sum_{j_k \in I_d} |\Delta S_N P_N \rho_R(j_k)|.$$

Lemma 2.10 and Lemma 5.1 of [14] show that the right hand side is bounded by $10R$. □

We now have sufficient tools to prove the Finite Quantitative BLT, Theorem 1.7.

Proof (Theorem 1.7) For simplicity we show the result for $k = 1$. Let R and Q be integers such that $1 \le R, Q \le (N/16)\sqrt{A/B}$. Let $\phi = \phi_{R,1}$ and $\psi = \phi_{Q,1}$, and note that Lemma 4.1 shows that

$$\|N\Delta_1 \phi\|_{\ell_1^{d,l}} \le 10R, \text{ and } \|N\Delta_1 \psi\|_{\ell_1^{d,l}} \le 10Q.$$

Proposition 2.8 of [14], and the fact that $\mathcal{F}_d(N\delta_{j,0})(k) = 1$ for all $k \in I_d$, shows that

$$\begin{aligned} \mathcal{F}_{d,l}(\phi)(\mathbf{k}) &= \mathcal{F}_d(S_N P_N \rho_R)(k_1) \\ &= (S_N P_N \mathcal{F}(\rho_R))(k_1) = (S_N P_N \widehat{\rho}(\cdot/R))(k_1), \end{aligned} \tag{15}$$

and since $R < N/2$, then $0 \le \widehat{\phi} = \mathcal{F}_{d,l}(\phi) \le 1$. Also, $\widehat{\phi}(\mathbf{k}) = 1$ for any $\mathbf{k}$ which satisfies $k_1 \in [-RN/2, RN/2]$, independent of the values of $k_2, \dots, k_l$, that is, for any $\mathbf{k} \in S_{NR,1}$. The same holds for $\widehat{\psi} = \mathcal{F}_{d,l}(\psi)$ with Q replacing R. Applying Lemma 4.3, we find a constant C such that

$$\frac{CN^{2l}}{QR} \leq \sum_{(\mathbf{m},\mathbf{n})\in\ell_2(S_N^{2l})} |Z_{d,l}(b)(\mathbf{m},\mathbf{n}) - Z_{d,l}(b*\phi)(\mathbf{m},\mathbf{n})|^2$$

$$+ \sum_{(\mathbf{m},\mathbf{n})\in\ell_2(S_N^{2l})} |Z_{d,l}(\widehat{b})(\mathbf{m},\mathbf{n}) - Z_{d,l}(\widehat{b}*\psi)(\mathbf{m},\mathbf{n})|^2,$$

where here we have let $\widehat{b} = \mathcal{F}_{d,l}(b)$. Using that $Z_{d,l}$ and $\mathcal{F}_{d,l}$ are both isometries and the properties of ϕ and ψ listed above, we have

$$\frac{C}{QR} \leq \|Z_{d,l}(b) - Z_{d,l}(b*\phi)\|^2_{\ell_2(S_N^{2l})} + \|Z_{d,l}(\widehat{b}) - Z_{d,l}(\widehat{b}*\psi)\|^2_{\ell_2(S_N^{2l})}$$

$$= \|b - b*\phi\|^2_{\ell_2^{d,l}} + \|\widehat{b} - \widehat{b}*\psi\|^2_{\ell_2^{d,l}}$$

$$= \|\widehat{b}(1-\widehat{\phi})\|^2_{\ell_2^{d,l}} + \|b(1-\widehat{\psi})\|^2_{\ell_2^{d,l}}$$

$$\leq \frac{1}{N^l} \sum_{|j_1|\geq \frac{NR}{2}} |\mathcal{F}_d b(\mathbf{j})|^2 + \frac{1}{N^l} \sum_{|j_1|\geq \frac{NQ}{2}} |b(\mathbf{j})|^2.$$

□

5 Nonsymmetric Finite BLT and Applications of the Quantitative BLTs

In this penultimate section, we prove the nonsymmetric finite BLT, Theorem 1.8, and the uncertainty principles of Theorem 1.10. We show each of these follows from a version of the Quantitative BLT, however, the details of the proof of Theorem 1.8 are more difficult due to subtleties from discreteness. For this reason, we first prove Theorem 1.10 which shows the central idea of both proofs without the added technical difficulty.

First, we state the higher dimensional quantitative BLT of [15]. For notational simplicity, we write $\{|x_k| \geq s\}$ to mean $\{x \in \mathbb{R}^l : |x_k| \geq s\}$ in situations where the dependence on l is clear.

Theorem 5.1 (Theorem 1, [15]) *Let $g \in L^2(\mathbb{R}^l)$ be such that the Gabor system generated by g*

$$G(g) = \{e^{2\pi i n\cdot x} g(x-m)\}_{(m,n)\in\mathbb{Z}^{2l}}$$

is a Riesz basis for $L^2(\mathbb{R}^l)$. Let $R, Q \geq 1$ be real numbers. Then, there is a constant C which only depends on the Riesz basis bounds of $G(g)$ such that for any $1 \leq k \leq l$

$$\int_{|x_k|\geq R} |g(x)|^2 dx + \int_{|\xi_k|\geq Q} |\widehat{g}(\xi)|^2 d\xi \geq \frac{C}{RQ}. \tag{16}$$

Remark 4 In [15], the conclusion of this theorem is stated where the integrals in (16) are taken over $\mathbb{R}^l \setminus \mathcal{R}$ and $\mathbb{R}^l \setminus \mathcal{Q}$, respectively, where $\mathcal{Q}$ and $\mathcal{R}$ are finite volume rectangles in $\mathbb{R}^d$. However, a straightforward limiting argument shows that the result holds after removing 'infinite volume' rectangles, as in the statement above.

Proof (Theorem 1.10) We will prove this for $k = 1$ without loss of generality.

Let $1 \leq S < \infty$, and choose $R = S^{1/p}$ and $Q = S^{1/q}$. Note $1 \leq R, Q < \infty$ for any value of S. Theorem 5.1 then shows that for C only depending on the Riesz basis bounds of $G(f)$,

$$\frac{C}{S^\tau} = \frac{C}{S^{\frac{1}{p}+\frac{1}{q}}} \leq \int_{|x_1|\geq S^{1/p}} |g(x)|^2 dx + \int_{|\xi_1|\geq S^{1/q}} |\widehat{g}(\xi)|^2 d\xi. \tag{17}$$

In each case, the result follows by integrating both sides of (17) over a particular set of S values, and then using Tonelli's Theorem to interchange the order of integration.

Case 1: $\tau = \frac{1}{p} + \frac{1}{q} < 1$. We have,

$$\begin{aligned} C\frac{(1-2^{\tau-1})}{1-\tau}T^{1-\tau} &= C\int_1^T S^{-\tau} dS \\ &\leq \int_{\mathbb{R}^{l-1}} \int_0^T \int_{|x_1|\geq S^{1/p}} |g(x_1, x')|^2 dx_1 dS dx' \\ &+ \int_{\mathbb{R}^{l-1}} \int_0^T \int_{|\xi_1|\geq S^{1/q}} |\widehat{g}(\xi_1, \xi')|^2 d\xi_1 dS d\xi' \\ &\leq \int_{\mathbb{R}^l} \int_0^{\min(|x_1|^p, T)} |g(x)|^2 dS dx \\ &+ \int_{\mathbb{R}^l} \int_0^{\min(|\xi_1|^q, T)} |\widehat{g}(\xi)|^2 dS d\xi \\ &= \int_{\mathbb{R}^l} \min(|x_1|^p, T)|g(x)|^2 dx + \int_{\mathbb{R}^l} \min(|\xi_1|^q, T)|\widehat{g}(\xi)|^2 d\xi. \end{aligned}$$

Case 2: $\tau = \frac{1}{p} + \frac{1}{q} = 1$. Similarly, we have

$$C \log T = C\int_1^T S^{-1} dS$$

$$\leq \int_{\mathbb{R}^{l-1}} \int_0^T \int_{|x_1| \geq S^{1/p}} |g(x_1, x')|^2 dx_1 dS dx'$$

$$+ \int_{\mathbb{R}^{l-1}} \int_0^T \int_{|\xi_1| \geq S^{1/q}} |\widehat{g}(\xi_1, \xi')|^2 d\xi_1 dS d\xi'$$

$$\leq \int_{\mathbb{R}^l} \min(|x_1|^p, T) |g(x)|^2 dx + \int_{\mathbb{R}^l} \min(|\xi_1|^q, T) |\widehat{g}(\xi)|^2 d\xi.$$

Case 3: $\tau = \frac{1}{p} + \frac{1}{q} > 1$. Finally, in this case

$$\frac{C}{\tau - 1} = C \int_1^\infty S^{-\tau} dS$$

$$\leq \int_{\mathbb{R}^{l-1}} \int_0^\infty \int_{|x_1| \geq S^{1/p}} |g(x_1, x')|^2 dx_1 dS dx'$$

$$+ \int_{\mathbb{R}^{l-1}} \int_0^\infty \int_{|\xi_1| \geq S^{1/q}} |\widehat{g}(\xi_1, \xi')|^2 d\xi_1 dS d\xi'$$

$$= \int_{\mathbb{R}^l} |x_1|^p |g(x)|^2 dx + \int_{\mathbb{R}^l} |\xi_1|^q |\widehat{g}(\xi)|^2 d\xi.$$

□

The following result generalizes part (ii) of Theorem 1.1.

Theorem 5.2 *Suppose* $1 \leq p < \infty$, *and* $g \in L^2(\mathbb{R}^l)$ *is such that* $G(g) = \{e^{2\pi i n \cdot x} g(x - m)\}_{(m,n) \in \mathbb{Z}^{2l}}$ *is a Riesz basis for* $L^2(\mathbb{R}^l)$ *and* g *is supported in* $(-M, M)^l$. *Then, there exists a constant* C *depending only on the Riesz basis bounds of* $G(g)$ *such that for any* $1 \leq k \leq 1$ *and any* $2 \leq T \leq \infty$ *each of the below hold.*

(i) If $p > 1$, *then*

$$\frac{C(1 - 2^{1/p-1})}{M(1 - 1/p)} \leq \int_{\mathbb{R}^l} \min(|\xi_k|^p, T) |\widehat{g}(\xi)|^2 d\xi.$$

(ii) If $p = 1$, *then*

$$\frac{C \log(T)}{M} \leq \int_{\mathbb{R}^l} \min(|\xi_k|, T) |\widehat{g}(\xi)|^2 d\xi.$$

(iii) If $p < 1$, *then*

$$\frac{C}{M(1/p - 1)} \leq \int_{\mathbb{R}^l} |\xi_k|^p, |\widehat{g}(\xi)|^2 d\xi.$$

This result also holds when g and $\widehat{g}$ are interchanged.

The proof is nearly identical to that of Theorem 1.10, after noticing that by applying the quantitative BLT with $R = M$, the integral related to $|g(x)|^2$ is zero due to the support assumption. Note that letting $T \to \infty$ in part (ii) gives part (ii) of Theorem 1.9.

Finally, we focus on the finite nonsymmetric BLT. For $1 \le p, q < \infty$ and $b \in \ell_2^{d,l}$, let

$$\alpha_k^{p,q}(b) = \frac{1}{N^l} \sum_{\mathbf{j} \in \mathbb{Z}_d^l} \left| \frac{j_k}{N} \right|^p |b(\mathbf{j})|^2 + \frac{1}{N^l} \sum_{\mathbf{j} \in \mathbb{Z}_d^l} \left| \frac{j_k}{N} \right|^q |\mathcal{F}_{d,l} b(\mathbf{j})|^2.$$

To give a finite dimensional analog of part (ii) of Theorem 1.1, it will be convenient to define $\alpha_k^{p,\infty}(b)$ and $\alpha_k^{\infty,q}(b)$ as

$$\alpha_k^{p,\infty}(b) = \frac{1}{N^l} \sum_{\mathbf{j} \in \mathbb{Z}_d^l} \left| \frac{j_k}{N} \right|^p |b(\mathbf{j})|^2, \quad \alpha_k^{\infty,q}(b) = \frac{1}{N^l} \sum_{\mathbf{j} \in \mathbb{Z}_d^l} \left| \frac{j_k}{N} \right|^q |\mathcal{F}_{d,l} b(\mathbf{j})|^2.$$

Theorem 5.3 *Let $A, B > 0$ and $1 \le p, q < \infty$ and let $\tau = \frac{1}{p} + \frac{1}{q}$. Assume $b \in \ell_2^{d,l}$ generates an A, B-Gabor Riesz basis for $\ell_2^{d,l}$. There exists a constant $C > 0$, depending only on A, B, p and q such that the following holds. Let $N \ge 200\sqrt{B/A}$.*

(i) If $\tau = \frac{1}{p} + \frac{1}{q} < 1$,

$$C \frac{N^{1-\tau}}{1-\tau} \le \alpha_k^{p,q}(b).$$

(ii) If $\tau = \frac{1}{p} + \frac{1}{q} = 1$,

$$C \log(N) \le \alpha_k^{p,q}(b.)$$

(iii) If $\tau = \frac{1}{p} + \frac{1}{q} > 1$,

$$C \frac{1 - (200/16)^{1-\tau}}{\tau - 1} \le \alpha_k^{p,q}(b).$$

Also, if $\mathcal{F}_{d,l}(b)$ is supported in the set $(-\gamma_N N/2, \gamma_N N/2) \cap \mathbb{Z}$ where $\gamma_N = \lfloor (N/16)\sqrt{A/B} \rfloor$, then parts (i), (ii), and (iii) hold with $\tau = \frac{1}{p}$ and $\alpha^{p,q}(b)$ replaced by $\alpha^{p,\infty}(b)$. Similarly, if b is supported in the set $(-\gamma_N N/2, \gamma_N N/2) \cap \mathbb{Z}$ then parts (i), (ii), and (iii) hold with $\tau = \frac{1}{q}$ and $\alpha^{p,q}(b)$ replaced by $\alpha^{\infty,q}(b)$.

We start with a lemma giving a bound on a typical sum arising in the proof which follows. Similar to above, $\{b > |j_k| \geq a\}$ will be used to denote $\{\mathbf{j} \in I_d^l : b > |j_k| \geq a\}$.

Lemma 5.1 *Let* $1 \leq \nu < \infty$, $N > 200\nu$, $c = 1/(16\nu)$, *and* $\gamma_N = \lfloor cN \rfloor$. *If* $0 < \alpha \leq 1$, *then for any* $b \in \ell_2^{d,l}$, *we have*

$$\sum_{S=1}^{\gamma_N} \sum_{|j_k| \geq NS^\alpha/2} |b(\mathbf{j})|^2 \leq 2^{1/\alpha} \sum_{\mathbf{j} \in \mathbb{Z}^d} \left| \frac{j_k}{N} \right|^{1/\alpha} |b(\mathbf{j})|^2,$$

where C_α *only depends on* α.

Note, we will apply this lemma with $\nu = \sqrt{B/A}$ where A and B are Riesz basis bounds of $G_{d,l}(b)$ for some $b \in \ell_2^{d,l}$. However, this lemma holds regardless of whether $G_{d,l}(b)$ is basis for $\ell_2^{d,l}$.

Proof Rearranging terms, we have

$$\sum_{S=1}^{\gamma_N} \sum_{|j_k| \geq NS^\alpha/2} |b(\mathbf{j})|^2 = \sum_{m=1}^{\gamma_N - 1} m \sum_{\frac{N(m+1)^\alpha}{2} > |j_k| \geq \frac{Nm^\alpha}{2}} |b(\mathbf{j})|^2 + \gamma_N \sum_{|j_k| \geq \frac{N \cdot \gamma_N^\alpha}{2}} |b(\mathbf{j})|^2 \tag{18}$$

Note that for some m, if j_k satisfies $|j_k| \geq \frac{Nm^\alpha}{2}$, then $m \leq 2^{1/\alpha} \left| \frac{j_k}{N} \right|^{1/\alpha}$. Then, from (18), we find

$$\begin{aligned}
\sum_{S=1}^{\gamma_N} \sum_{|j_k| \geq NS^\alpha/2} |b(\mathbf{j})|^2 &\leq 2^{1/\alpha} \sum_{m=1}^{\gamma_N - 1} \sum_{\frac{N(m+1)^\alpha}{2} > |j_k| \geq \frac{Nm^\alpha}{2}} \left| \frac{j_k}{N} \right|^{1/\alpha} |b(\mathbf{j})|^2 \\
&\quad + 2^{1/\alpha} \sum_{|j_k| \geq \frac{N\gamma_N^\alpha}{2}} \left| \frac{j_k}{N} \right|^{1/\alpha} |b(\mathbf{j})|^2 \\
&\leq 2^{1/\alpha} \sum_{\mathbf{j} \in I_d^l} \left| \frac{j_k}{N} \right|^{1/\alpha} |b(\mathbf{j})|^2.
\end{aligned}$$

□

Proof (Theorem 5.3) We prove the result for $k = 1$. We treat the case where p and q are both finite and the case where one of these is infinite separately. Below, we take $\tau = \frac{1}{p} + \frac{1}{q}$.

Case 1: $1 \leq p, q < \infty$. Let S be an integer satisfying $1 \leq S \leq \gamma_N$ where $\gamma_N = \lfloor (N/16)\sqrt{A/B} \rfloor$, and $R = \lceil S^{1/p} \rceil$, $Q = \lceil S^{1/q} \rceil$ if $1 < p, q < \infty$, $R = S$ if $p = 1$, and $Q = S$ if $q = 1$. Note that these choices force $1 \leq R, Q \leq \gamma_N$.

Then, for a constant C only depending on A and B, Theorem 1.7 gives

$$\frac{C/4}{S^\tau} \leq \frac{C}{RQ} \leq \frac{1}{N^l} \sum_{|j_k| \geq \frac{NS^{1/p}}{2}} |b(\mathbf{j})|^2 + \frac{1}{N^l} \sum_{|j_k| \geq \frac{NS^{1/q}}{2}} |\mathcal{F}_{d,l} b(\mathbf{j})|^2.$$

Summing over the values of S in $\{1, \ldots, \gamma_N\}$ and applying Lemma 5.1 with $\tau = \sqrt{B/A}$, to find

$$\frac{C}{4} \sum_{S=1}^{\gamma_N} S^{-\tau} \leq \frac{1}{N^l} \sum_{S=1}^{\gamma_N} \sum_{|j_k| \geq \frac{NS^{1/p}}{2}} |b(\mathbf{j})|^2 + \frac{1}{N^l} \sum_{S=1}^{\gamma_N} \sum_{|j_k| \geq \frac{NS^{1/q}}{2}} |\mathcal{F}_{d,l} b(\mathbf{j})|^2 \leq C' \alpha_k^{p,q}(b)$$

where C' is a constant only depending on p and q. Updating the constant C, (it now depends on A, B, p, q)

$$C \sum_{S=1}^{\gamma_N} S^{-\tau} \leq \alpha_k^{p,q}(b, N, l).$$

The proof of Case 1 follows by noting that

$$\sum_{S=1}^{\gamma_N} S^{-\tau} \geq \begin{cases} C_{\tau,A,B} \frac{N^{1-\tau}}{1-\tau} & 0 < \tau < 1 \\ C_{\tau,A,B} \log(N) & \tau = 1 \\ \frac{(1-(200/16)^{1-\tau})}{1-\tau} & \tau > 1 \end{cases}, \tag{19}$$

where the constants $C_{\tau,A,B}$ depend only on τ, A, and B.

Case 2: One of p or q is ∞. We can assume without loss of generality that $q = \infty$ and $1 \leq p < \infty$. With this in mind, assume b generates an A, B-Gabor Riesz basis for $\ell_2^{d,l}$, and further suppose $\mathcal{F}_{d,l}(b)$ is supported in the set $(-\gamma_N N/2, \gamma_N N/2) \cap \mathbb{Z}$. Then, Theorem 1.7 applied with $Q = \gamma_N$, gives

$$\frac{C}{R\gamma_N} \leq \frac{1}{N^l} \sum_{|j_k| \geq \frac{NR}{2}} |b(\mathbf{j})|^2,$$

where the second sum does not appear due to the support condition on $\mathcal{F}_{d,l}(b)$. As in part (i), let $1 \leq S \leq \gamma_N$ and $R = \lceil S^{1/\alpha} \rceil$ if $1 < p < \infty$ and $R = S$ if $p = 1$. Summing over values of S, and applying Lemma 5.1 we find

$$\frac{C}{2\gamma_N} \sum_{S=1}^{\gamma_N} S^{-\tau} \leq \frac{1}{N^l} \sum_{S=1}^{\gamma_N} \sum_{|j_k| \geq \frac{NS^{1/p}}{2}} |b(\mathbf{j})|^2 \leq 2^p \alpha_k^{p,\infty}(b),$$

and the result follows by combining the constants and another application of Eq. (19). □

6 Further Questions

Upon investigation, similar arguments applied in the one-dimensional Finite BLT apply for several variable analogs. It is interesting to consider the question of whether there are sequences which have the 'best' localization properties, those for which the α_k norm is minimized over the set of all A, B-Gabor Riesz bases. There is a conjecture [11] of Lammers and Stampe which addresses this question and is still open to the authors' knowledge. Also of interest is whether uncertainty principles for different continuous basis systems (e.g. [4]) may be discretized to give similar finite dimensional results.

Another remaining question is related to Theorem 1.9. In [10], a more general version of Theorem 1.2 was shown to hold when $G(g, \mathbb{Z}^{2l})$ is replaced by $G(g, S)$ for any symplectic lattice $S \subset \mathbb{R}^{2l}$. It is not clear to the authors whether Theorem 1.9 also holds in this setting.

References

1. L. Auslander, I. Gertner, R. Tolimieri, The finite Zak transform and the finite Fourier transform. In: The IMA Volumes in Mathematics and its Applications, vol. 39, pp. 21–35. Springer, New York (1992). 1223150
2. R. Balian, Un principe d'incertitude fort en théorie du signal ou en mécanique quantique. In: Comptes Rendus de l'Académie des Sciences, Serie II: Mecanique, Physique, Chimie, Sciences de la Terre et de l'Univers Science Terre, vol. 292 (1981), 1357–1362. 0644367
3. G. Battle, Heisenberg proof of the Balian–Low theorem. Lett. Math. Phys. **15**, 175–177 (1988). 0943990
4. J.J. Benedetto, C. Heil, D. Walnut, Uncertainty principles for time-frequency operators. In: Continuous and Discrete Fourier Transforms, Extension Problems, and Wiener-Hopf Equations. Operation Theory Advantage Application, vol. 58 (1992), pp. 1–25. 1183741
5. J.J. Benedetto, W. Czaja, P. Gadziński, A.M. Powell, The Balian-Low theorem and regularity of Gabor systems. J. Geom. Anal. **13**, 239–254 (2003). 1967026
6. J.J. Benedetto, W. Czaja, A.M. Powell, J. Sterbenz, An endpoint $(1, \infty)$ Balian-Low theorem. Math. Res. Lett. **13**, 467–474 (2006). 2231132
7. M.G. Cowling, J.F. Price, Bandwidth versus time concentration: the Heisenberg-Pauli-Weyl inequality. SIAM J. Math. Anal. **15**, 151–165 (1984). 0728691
8. I. Daubechies, The wavelet transform, time-frequency localization and signal analysis. IEEE Trans. Inform. Theory **39**, 961–1005 (1990). 1066587
9. S.Z. Gautam, A critical exponent Balian-Low Theorem. Math. Res. Lett. **15**(3), 471–483 (2008). arXiv:math/0703905. 2407224
10. K. Gröchenig, D. Han, C. Heil, G. Kutyniok, The Balian-Low theorem for symplectic lattices in higher dimensions. Appl. Comput. Harmon. Anal. **13**, 169–176 (2002). 1942751

11. M. Lammers, S. Stampe, The finite Balian-Low conjecture. In: Proceedings of the 2015 International Conference on Sampling Theory and Applications (SampTA). IEEE, New York (2015), pp. 139–143
12. F. Low, Complete sets of wave packets. In: DeTar, C. et al. (eds.) A Passion for Physics—Essays in Honor of Geoffrey Chew, pp. 17–22. World Scientific, Singapore, (1985)
13. S. Nitzan, J.-F. Olsen, A quantitative Balian-Low theorem. J. Fourier Anal. Appl. **19**, 1078–1092 (2013). arXiv:1205.0163. 3110593
14. S. Nitzan, J.-F. Olsen, Balian-Low type theorems in finite dimensions. Math. Ann. **373**, 643–677 (2019). arXiv:1707.06449. 3968884
15. F. Temur, A quantitative Balian-Low theorem for higher dimensions. Georgian Math. J., to appear (2018). arXiv:1604.05067

Quasi-Interpolant Operators and the Solution of Fractional Differential Problems

Enza Pellegrino, Laura Pezza, and Francesca Pitolli

Abstract Nowadays, fractional differential equations are a well established tool to model phenomena from the real world. Since the analytical solution is rarely available, there is a great effort in constructing efficient numerical methods for their solution. In this paper we are interested in solving boundary value problems having space derivative of fractional order. To this end, we present a collocation method in which the solution of the fractional problem is approximated by a spline quasi-interpolant operator. This allows us to construct the numerical solution in an easy way. We show through some numerical tests that the proposed method is efficient and accurate.

Keywords Fractional differential problem · B-spline · Quasi-interpolant · Collocation method

1 Introduction

In recent years fractional differential equations are becoming a powerful tool to describe real-world phenomena where nonlocality is a key ingredient. Starting from the fundamental paper [3] by Caputo, where the fractional derivative was used for the first time to describe dissipation phenomena in Earth free modes, the literature on fractional models has exploded and now fractional differential equations are used in several fields, like continuum mechanics, signal processing, biophysics (see, [11, 17, 29, 32] and references therein). The Caputo derivative is especially suitable to describe real phenomena since in many ways it behaves like the usual derivative of integer order. In particular, the Caputo derivative of constant functions is zero,

E. Pellegrino
Università di L'Aquila, L'Aquila, Italy
e-mail: enza.pellegrino@univaq.it

L. Pezza · F. Pitolli (✉)
Università di Roma "La Sapienza", Roma, Italy
e-mail: laura.pezza@uniroma1.it; francesca.pitolli@uniroma1.it

G. E. Fasshauer et al. (eds.), *Approximation Theory XVI*, Springer Proceedings in Mathematics & Statistics 336, https://doi.org/10.1007/978-3-030-57464-2_11

which is not true for the Riemann-Liouville derivative [26]. Moreover, initial or boundary conditions can be easily applied [7]. For details on fractional calculus see, for instance [7, 18, 26, 29].

Since the analytical solution of fractional differential equations can be rarely obtained explicitly, to solve these kinds of problems numerical methods are mandatory. There is a huge literature on numerical methods for fractional differential problems (see, [2, 14, 15, 24] and references therein). A crucial point to construct efficient methods is their ability to approximate the nonlocal behavior of the fractional derivative. In this respect, collocation methods that use information of the approximating function in all the discretization interval have received great attention in recent years [12, 19, 21, 22, 25].

In this paper, we present a collocation method based on spline quasi-interpolant operators suitable to solve boundary value differential problems having fractional derivative in space. In particular, we are interested in solving linear boundary value problems of type

$$\begin{cases} D_x^\gamma y(x) + f(x)\, y(x) = g(x)\,, & 0 < x < L\,, \\ \rho_{r0}\, y(0) + \rho_{r1}\, y'(0) + \zeta_{r0}\, y(L) + \zeta_{r1}\, y'(L) = c_r\,, & 1 \le r \le \lceil\gamma\rceil\,, \end{cases} \tag{1}$$

where $\gamma \in \big(\lfloor\gamma\rfloor, \lceil\gamma\rceil\big)$ is a given real number, f and g are continuous given functions, and $\rho_{r0}, \rho_{r1}, \zeta_{r0}, \zeta_{r1}, c_r$ are given parameters. Here, we assume L to be a positive integer. Moreover, we assume the boundary conditions are linearly independent so that the differential problem has a unique solution [7].

The derivative appearing in the differential problem (1) should be intended in the Caputo sense. For a sufficiently smooth function the Caputo fractional derivative is defined as

$$D_x^\gamma\, y(x) := \frac{1}{\Gamma(\lceil\gamma\rceil - \gamma)} \int_0^x \frac{y^{(\lceil\gamma\rceil)}(\xi)}{(x-\xi)^{\gamma-\lceil\gamma\rceil+1}}\, d\xi\,, \tag{2}$$

where Γ is the Euler gamma function

$$\Gamma(\gamma) := \int_0^\infty \xi^{\gamma-1}\, \mathrm{e}^{-\xi}\, d\xi\,.$$

In the method we present in this paper we approximate the solution to the differential problem (1) by a spline quasi-interpolant. Polynomial spline quasi-interpolants are linear operators that are represented as a linear combination of spline basis functions whose coefficients are chosen in order to achieve some special properties, like shape preserving properties or good approximation order [5, 13, 16, 28]. Thus, quasi-interpolants have a greater flexibility with respect to interpolation with the further advantage that they are easy to construct. We show that the proposed method is accurate and efficient since the fractional derivative of the approximating function can be evaluated explicitly. As a consequence, the nonlocal behavior of the fractional derivative can be easily taken into account.

The paper is organized as follows. The main properties of the B-spline basis we use to construct the approximating function are described in Sect. 2 while Sect. 3 is devoted to its fractional derivative. In Sect. 4 we collect the main properties of the Schoenberg–Bernstein operator we use to approximate the solution of Eq. (1). The numerical method we propose is described in Sect. 5 while some numerical results are shown in Sect. 6. Finally, some conclusions are drawn in Sect. 7.

2 The Cardinal B-splines

The cardinal B-splines are piecewise polynomials with breakpoints at the integers [4, 31]. They can be defined as

$$B_n(x) := \frac{1}{n!}\,\Delta^{n+1} x_+^n, \qquad n \geq 0\,, \tag{3}$$

where

$$\Delta^n f(x) := \sum_{\ell=0}^{n} (-1)^\ell \binom{n}{\ell} f(x-\ell)\,, \quad n \in \mathbb{N}\,, \tag{4}$$

is the backward finite difference operator and $x_+^n := \big(\max(0,x)\big)^n$ is the truncated power function.

The integer translates $\{B_n(x-\ell), \ell \in \mathbb{Z}\}$ form a basis for the spline space of degree n on the whole line. A basis for the finite interval $[0, L]$, $L \geq n+1$, can be obtained by restriction, i.e.,

$$\mathcal{B}_n(x) = \{B_n(x-\ell), -n \leq \ell \leq L-1\}\,, \qquad x \in [0, L]\,. \tag{5}$$

We recall that the basis $\mathcal{B}_n(x)$ is totally positive, reproduces polynomials up to degree n and is a partition of unity.

The B-spline bases can be generalized to any sequence of equidistant knots on the interval $[0, L]$ by mapping $x \to h^{-1}x$, where h is the space step:

$$\mathcal{B}_{h,n}(x) = \{B_{h,\ell,n}(x) = B_n(h^{-1}x-\ell), -n \leq \ell \leq h^{-1}L-1\}\,, \qquad x \in [0, L]\,.$$

Thus, $\mathcal{B}_{h,n}$ is a basis for the spline space of degree n having breakpoints at the knots $h\ell$, $0 \leq \ell \leq h^{-1}L$. We observe that for $-n \leq \ell \leq -1$ the functions $B_{h,\ell,n}(x)$ and $B_{h,h^{-1}L+\ell,n}(x)$ are left and right edge functions, respectively. Their support is $[0, h\ell]$ and $[L-h\ell, L]$, respectively. The functions $B_{h,\ell,n}(x)$ with $0 \leq \ell \leq h^{-1}L-n-1$ are interior functions having support $[h\ell, h(\ell+n+1)]$.

3 The Fractional Derivative of the Cardinal B-splines

The fractional derivatives of the B-spline functions are fractional B-splines, i.e., piecewise polynomials of noninteger degree [33]. Their explicit expression can be obtained by applying the Caputo differential operator (2) to the basis functions $B_{h,\ell,n}(x)$ (see [20, 21] for details).

The Caputo derivative of the interior functions and of the right edge functions can be evaluated by the differentiation rule

$$D_x^\gamma\, B_n(x) = \frac{\Delta^{n+1}\, x_+^{n-\gamma}}{\Gamma(n+1-\gamma)}\,, \qquad x \geq 0\,, \qquad 0 < \gamma < n\,, \tag{6}$$

where $x_+^\gamma = (\max(0,x))^\gamma$ is the fractional truncated power function (cf. [21, 33]). This formula generalizes to the noninteger case the well-known differentiation rule for the ordinary derivative of the B-spline

$$B_n^{(m)}(x) = \frac{\Delta^{n+1}\, x_+^{n-m}}{(n-m)!}\,, \qquad 0 \leq m \leq n-1\,. \tag{7}$$

For $-n \leq \ell \leq -1$, the Caputo derivative of the left edge functions is given by Pellegrino et al. [20]

$$D_x^\gamma\, B_{n,\ell}(x) = \frac{\Delta^{n+1}(x-\ell)_+^{n-\gamma}}{\Gamma(n+1-\gamma)} - \sum_{r=0}^{-\ell-1} (-1)^r \binom{n+1}{r} \Bigg(\frac{(x-\ell-r)^{n-\gamma}}{\Gamma(n+1-\gamma)} - \sum_{p=0}^{n-\lceil\gamma\rceil} \frac{(-\ell-r)^{n-\lceil\gamma\rceil-p}\, x^{\lceil\gamma\rceil-\gamma+p}}{(n-\lceil\gamma\rceil-p)!\,\Gamma(\lceil\gamma\rceil-\gamma+p+1)} \Bigg). \tag{8}$$

The fractional derivative of the refined basis functions $B_{h,\ell,n}$ can be evaluated recalling that, for any function f sufficiently smooth, it holds

$$D_x^\gamma\, f(h^{-1}x-\ell) = h^{-\gamma}\, D_{h^{-1}x}^\gamma\, f(h^{-1}x-\ell)$$

(cf. [20]).

4 Quasi-Interpolant Operators

A quasi-interpolant operator is an approximation of a given function that reproduces polynomials up to a given degree. In particular, a spline quasi-interpolant operator is a linear operator of type

$$\mathcal{Q}_n\, y(x) = \sum_{\ell \in \mathbb{Z}} \mu_\ell(y)\, B_n(x - \ell)\,, \tag{9}$$

where $\mu_\ell(y)$, $\ell \in \mathbb{Z}$, are continuous linear functionals that are determined by imposing that $\mathcal{Q}_n\, y$ is exact on polynomials up to degree $m \leq n$. Usually, the functionals $\mu_\ell(y)$ are assumed to be local, i.e., only values of y in some neighborhood of $\sigma_{\ell,n} = supp\, B_n(x - \ell)$ are used to construct $\mu_\ell(y)$. We notice that since $\mu_\ell(y)$ is local and $B_n(x - \ell)$ has compact support, for any $x \in \mathbb{R}$ the sum in (9) is actually a finite sum.

There are several kinds of quasi-interpolant spline operators (see, for instance, [5, 9, 13, 16, 28]). In this paper we consider Bernstein type operators [27] in which the functionals $\mu_\ell(y)$ are suitable values of y evaluated on points belonging to $\sigma_{\ell,n}$. The simplest choice is

$$\mu_\ell(y) = y(\theta_\ell)\,, \tag{10}$$

where

$$\theta_\ell = \ell + \frac{n+1}{2}\,, \qquad \ell \in \mathbb{Z}\,, \tag{11}$$

are the Schoenberg nodes. This choice leads to the Schoenberg–Bernstein operator that reproduces linear functions and has approximation order 1 [30]. Even if the approximation order is poor, the Schoenberg–Bernstein operator has many properties useful in applications. In particular, it is a positive operator that has shape preserving properties. In fact, it enjoys the variation diminishing property, i.e., for any linear function Λ and any function y it holds

$$S^-(\mathcal{Q}_n\, (y - \Lambda)) \leq S^-(y - \Lambda)\,,$$

where $S^-(y)$ denotes the number of strict sign changes of the function y. This property reveals particular attractive in geometric modeling where the approximation of a given set of data is required to reproduce their shape [8].

The operator $\mathcal{Q}_n\, y$ is refinable, i.e., in the spline spaces generated by the B-spline basis $\mathcal{B}_{h,n}$, we can construct the refined operator

$$\mathcal{Q}_{h,n}\, y(x) = \sum_{\ell \in \mathbb{Z}} \mu_{h,\ell}(y)\, B_{h,\ell,n}(x)\,, \tag{12}$$

where $\mu_{h,\ell}(y)$ uses values of y in $supp\, B_{h,\ell,n}$. The functionals $\mu_{h,\ell}(y)$ have expression

$$\mu_{h,\ell}(y) = y(\theta_{h,\ell})\,,$$

where $\theta_{h,\ell} = h\, \theta_\ell$ are the refined Schoenberg nodes.

5 The Quasi-Interpolant Collocation Method

To solve the fractional differential problem (1) we approximate its solution by the refinable Schoenberg–Bernstein operator (12) restricted to the interval $[0, L]$, i.e.,

$$y(x) \approx y_{h,n}(x) = \sum_{\ell=-n}^{N_h} y_{h,n}(\tilde{\theta}_{h,\ell})\, B_{h,\ell,n}(x)\,, \quad N_h = h^{-1}L - 1\,, \qquad x \in [0, L]\,, \tag{13}$$

where $\tilde{\theta}_{h,\ell}$ are the Schoenberg nodes for the interval $[0, L]$. To determine the unknown coefficients $\{y_{h,n}(\tilde{\theta}_{h,\ell}), -n \le \ell \le N_h\}$ we solve the differential problem on a set of *collocation points*. For the sake of simplicity, here we assume the collocation points are a set of equidistant nodes on the interval $[0, L]$ having distance $\delta = 2^{-s}$,

$$X_\delta = \{x_r = \delta\, r, 0 \le r \le N_\delta\}\,, \qquad N_\delta = \delta^{-1} L\,. \tag{14}$$

Thus, collocating Eq. (1) on the nodes X_δ and using (13) we get the linear system

$$\begin{cases} D_x^\gamma\, y_{h,n}(x_r) + f(x_r)\, y_{h,n}(x_r) = g(x_r)\,, \qquad 1 \le r \le N_\delta - 1\,, \\ \rho_{r0}\, y_{h,n}(x_0) + \rho_{r1}\, y'_{h,n}(x_0) + \zeta_{r0}\, y_{h,n}(x_{N_\delta}) + \zeta_{r1}\, y'_{h,n}(x_{N_\delta}) = c_r\,,\; 1 \le r \le \lceil \gamma \rceil\,. \end{cases} \tag{15}$$

Now, let

$$\mathsf{Y}_{h,\delta} = \left[y_{h,n}(\tilde{\theta}_{h,\ell}), -n \le \ell \le N_h\right]^T,$$

be the unknown vector,

$$\mathsf{A}_{h,\delta} = \left[B_{h,\ell,n}(x_r), 1 \le r \le N_\delta - 1, -n \le \ell \le N_h\right]$$

and

$$\mathsf{D}_{h,\delta} = \left[D_x^\gamma\, B_{h,\ell,n}(x_r), 1 \le r \le N_\delta - 1, -n \le \ell \le N_h\right]$$

be the collocation matrices of the refinable basis $\mathcal{B}_{h,n}$ and of its fractional derivative. Then, let

$$\mathsf{F}_\delta = \left[f(x_r), 1 \le r \le N_\delta - 1\right]^T, \qquad \mathsf{G}_\delta = \left[g(x_r), 1 \le r \le N_\delta - 1\right]^T,$$

be the know terms. Finally, we define the parameter vectors

$$\mathsf{R}_k = \left[\rho_{rk}, 1 \le r \le \lceil\gamma\rceil\right]^T, \; k = 0, 1,$$

$$\mathsf{Z}_k = \left[\zeta_{rk}, 1 \le r \le \lceil\gamma\rceil\right]^T, \; k = 0, 1,$$

$$\mathsf{C} = \left[c_r, 1 \le r \le \lceil\gamma\rceil\right]^T,$$

and the vectors containing the boundary values of the basis functions and of their first derivative

$$\mathsf{B}_{h,\delta}(x) = \left[B_{h,\ell,n}(x), -n \le \ell \le N_h\right], \qquad x = 0, L,$$

$$\mathsf{B}'_{h,\delta}(x) = \left[B'_{h,\ell,n}(x), -n \le \ell \le N_h\right], \qquad x = 0, L.$$

Thus, Eq. (15) can be written in matrix form as

$$\begin{cases} (\mathsf{D}_{h,\delta} + \mathsf{F}_\delta \circ \mathsf{A}_{h,\delta})\, \mathsf{Y}_{h,\delta} = \mathsf{G}_\delta, \\ \left(\mathsf{R}_0\, \mathsf{B}_{h,\delta}(0) + \mathsf{R}_1\, \mathsf{B}'_{h,\delta}(0) + \mathsf{Z}_0\, \mathsf{B}_{h,\delta}(L) + \mathsf{Z}_1\, \mathsf{B}'_{h,\delta}(L)\right) \mathsf{Y}_{h,\delta} = \mathsf{C}, \end{cases} \tag{16}$$

Here, $\mathsf{V} \circ \mathsf{A}$ denotes the entrywise product between a vector V and a matrix A meaning that V has to be intended as a matrix having as many columns as A, each column being a replica of the vector V itself. The entries of the matrices $\mathsf{A}_{h,\delta}$ and $\mathsf{D}_{h,\delta}$ can be easily evaluated using formulas given in Sects. 2–3.

The linear system (16) has $N_\delta - 1 + \lceil\gamma\rceil$ equations and $N_h + n + 1$ unknowns. To guarantee the existence of a unique solution the refinement step h, the distance of the collocation points δ and the degree of the B-spline n have to be chosen such that $N_\delta - 1 + \lceil\gamma\rceil \ge N_h + n + 1$ [20]. We notice that the choice $N_\delta - 1 + \lceil\gamma\rceil > N_h + n + 1$ is preferable since in this case there is a greater flexibility in the choice of the degree of the B-spline. In this case we get an overdetermined linear system that can be solved by the least squares method.

Finally, following the same reasoning line as in [20] (cf. also [1]) it can be proved that the collocation method described above is convergent.

Theorem 1 *The collocation method is convergent, i.e.*

$$\lim_{h \to 0} \|y(x) - y_{h,n}(x)\|_\infty = 0,$$

where $\|y(x)\|_\infty = \max_{x \in [0,L]} |y(x)|$.

We notice that since the convergence order of spline collocation methods is related to the approximation properties of the spline spaces, we expect the infinity norm of the error to decrease at least as h^ν, where ν is the smoothness of the known terms, providing that the approximating function and the differential operator are sufficiently smooth.

6 Numerical Tests

6.1 Example 1

In the first test we solve the fractional differential problem

$$\begin{cases} D_x^\gamma y(x) + f(x)\, y(x) = g(x)\,, & 0 < x < 1\,, \\ y(0) + y(1) = 2\,, \end{cases} \tag{17}$$

where $\gamma \in (0, 1)$ is a given real number and

$$f(x) = x^{\frac{1}{2}}\,, \qquad g(x) = \frac{2}{\Gamma(2-\gamma)}\, x^{1-\gamma} + 2x^{\frac{3}{2}}\,.$$

The exact solution is $y(x) = 2x$ so that the collocation method is exact for $n \geq 1$. To compare the exact and the numerical solutions, we evaluate the infinity norm of the error $e_{h,n}(x) = y(x) - y_{h,n}(x)$ as

$$\|e_{h,n}\|_\infty = \max_{0 \leq r \leq \eta N_\delta} |e_{h,n}(x_r)|\,,$$

where $x_r = \delta r/\eta$, $0 \leq r \leq \eta N_\delta$ with $\eta \in \mathbb{N}^+$. In the tests we choose $\eta = 4$. In the table below we list the infinity norm of the error we obtain using the Schoenberg–Bernstein operator with the B-splines of degree $n = 3$ for $h = 1/8$ and $\delta = 1/16$. To give an idea of the conditioning of the final linear system, the condition number $\kappa_{h,n}$ is also shown.

γ	0.25	0.5	0.75
$\|e_{h,3}\|_\infty$	7.33e−15	1.09e−14	2.44e−15
$\kappa_{h,3}$	2.58e+01	1.67e+01	1.64e+01

As expected, the error is in the order of the machine precision.

6.2 Example 2

In the second test we solve the fractional differential problem

$$\begin{cases} D_x^\gamma\, y(x) + y(x) = g(x)\,, & 0 < x < 1\,, \\ y(0) = 0\,, \quad y(1) = 1\,, \end{cases} \tag{18}$$

where $\gamma \in (1, 2)$ is a given real number and

$$g(x) = \frac{\Gamma(\nu + 1)}{\Gamma(\nu + 1 - \gamma)} x^{\nu - \gamma} + x^{\nu} .$$

The exact solution is $y(x) = x^{\nu}$. We approximate the solution by the Schoenberg–Bernstein operator with $n = 4, 5, 6$ when $\nu = 2.5$ and $\gamma = 1.25, 1.5, 1.75$. The infinity norm of the error for different values of h and $\delta = h/2$ is listed in the table below:

$\Vert e_{h,n} \Vert_\infty$ for $\gamma = 1.25$				
h	2^{-3}	2^{-4}	2^{-5}	2^{-6}
$n = 4$	5.91e−05	1.25e−05	2.67e−06	5.74e−07
$n = 5$	3.84e−05	8.19e−06	1.73e−06	3.65e−07
$n = 6$	2.51e−05	5.35e−06	1.13e−06	2.39e−0

$\Vert e_{h,n} \Vert_\infty$ for $\gamma = 1.5$				
h	2^{-3}	2^{-4}	2^{-5}	2^{-6}
$n = 4$	1.05e−04	2.68e−05	6.72e−06	1.67e−06
$n = 5$	6.68e−05	1.71e−05	4.33e−06	1.09e−06
$n = 6$	4.10e−05	1.05e−05	2.66e−06	6.68e−07

$\Vert e_{h,n} \Vert_\infty$ for $\gamma = 1.75$				
h	2^{-3}	2^{-4}	2^{-5}	2^{-6}
$n = 4$	1.37e−04	4.20e−05	1.26e−05	3.72e−06
$n = 5$	8.02e−05	2.50e−05	7.62e−06	2.29e−06
$n = 6$	4.34e−05	1.34e−05	4.06e−06	1.22e−06

As expected, the norm of the error decreases when h decreases. We notice that the error decreases also when n increases.

6.3 Example 3

In the last test we solve the fractional differential problem

$$\begin{cases} D_x^{\gamma} \, y(x) + y(x) = 0 \,, & 0 < x < 1 \,, \\ y(0) = 1 \,, \quad y(1) = E_{\gamma}(-1^{\gamma}) \,, & \end{cases} \tag{19}$$

where $\gamma \in (1, 2)$ is a given real number and

$$E_\gamma(x) = \sum_{\ell \geq 0} \frac{x^\ell}{\Gamma(\gamma\ell + 1)},$$

is the one-parameter Mittag-Leffler function [10]. The exact solution is $y(x) = E_\gamma(-x^\gamma)$. We approximate the solution by the Schoenberg–Bernstein operator with $n = 4, 5, 6$. The infinity norm of the error for different values of h and $\delta = h/2$ when $\gamma = 1.25, 1.5, 1.75$ is listed in the tables below:

$\|e_{h,n}\|_\infty$ for $\gamma = 1.25$				
h	2^{-3}	2^{-4}	2^{-5}	2^{-6}
$n = 4$	9.01e−03	4.55e−03	2.27e−03	1.13e−03
$n = 5$	8.01e−03	4.05e−03	2.03e−03	1.02e−03
$n = 6$	7.06e−03	3.57e−03	1.79e−03	8.98e−04

$\|e_{h,n}\|_\infty$ for $\gamma = 1.5$				
h	2^{-3}	2^{-4}	2^{-5}	2^{-6}
$n = 4$	3.66e−03	1.87e−03	9.41e−04	4.72e−04
$n = 5$	3.11e−03	1.59e−03	8.02e−04	4.03e−04
$n = 6$	2.63e−03	1.34e−03	6.76e−04	3.40e−04

$\|e_{h,n}\|_\infty$ for $\gamma = 1.75$				
h	2^{-3}	2^{-4}	2^{-5}	2^{-6}
$n = 4$	7.91e−04	4.10e−04	2.08e−04	1.05e−04
$n = 5$	6.32e−04	3.28e−04	1.67e−04	8.44e−05
$n = 6$	4.98e−04	2.57e−04	1.30e−04	6.57e−05

Also in this case the norm of the error decreases when h decreases and n increases.

7 Conclusion

We have presented a collocation method based on spline quasi-interpolant operators. The method is easy to implement and has proved to be convergent. The numerical tests show that it is efficient and accurate. The method can be improved in several ways. First of all, we used the truncated B-spline bases to construct the Schoenberg–Bernstein operator. It is well known that truncated bases have a low accuracy in approximating the boundary conditions which can result in a poor approximation of the solution. This problem can be overcome by using B-spline bases with multiple nodes at the endpoints of the interval. The use of this kind of B-splines for the solution of fractional problems has already been considered in [23] where the analytical expression of their fractional derivative is also given. As for the quasi-interpolants, even if the Schoenberg–Bernstein operator produces good results, it is just second order accurate. To increase the approximation order, different quasi-interpolant operators can be used, like the projector quasi-interpolants introduced in [13] or integral or discrete operators [16, 28]. Finally, we notice that the accuracy of the method could also be improved by using Gaussian points instead of equidistant points (cf. [1, 6]). These issues are at present under study.

References

1. Ascher, U.: Discrete least squares approximations for ordinary differential equations. SIAM J. Numer. Anal. **15**, 478–496 (1978)
2. Baleanu, D., Diethelm, K., Scalas, E, Trujillo, J.J.: Fractional Calculus: Models and Numerical Methods. World Scientific, Singapore (2016)
3. Caputo, M.: Linear models of dissipation whose Q is almost frequency independent II. Geophys. J. Int. **13**, 529–539 (1967)
4. de Boor, C.: A Practical Guide to Spline. Springer, Berlin (1978)
5. de Boor, C., Fix, G.: Spline approximation by quasi-interpolants. J. Approx. Theory **8**, 19–45 (1973)
6. de Boor, C., Swartz, B.: Collocation at Gaussian points. SIAM J. Numer. Anal. **10**, 582–606 (1973)
7. Diethelm, K.: The Analysis of Fractional Differential Equations: An Application-Oriented Exposition using Differential Operators of Caputo Type. Springer, Berlin (2010)
8. Goodman, T.N.T.: Total positivity and the shape of curves. In: Gasca M., Micchelli C.A. (eds.) Total Positivity and its Applications, pp. 157–186. Kluwer Academic, Dordecht (1996)
9. Goodman, T.N.T., Sharma, A.: A modified Bernstein–Schoenberg operator. In: Sendov, B.l. (ed.) Constructive Theory of Functions, vol. 87, pp. 166–173. Bulgarian Academy Sciences, Sofia (1988)
10. Gorenflo, R., Kilbas, A.A., Mainardi, F., Rogosin, S.V.: Mittag-Leffler functions, related topics and applications. In: Springer Monographs in Mathematics. Springer, Berlin/Heidelberg (2014)
11. Kilbas, A.A., Srivastava, H.M., Trujillo, J.J.: Theory and applications of fractional differential equations. In: North-Holland Mathematics Studies, vol. 204. Elsevier, Amsterdam (2006)
12. Kolk, M., Pedas, A., Tamme, E.: Smoothing transformation and spline collocation for linear fractional boundary value problems. App. Math. Comput. **283**, 234–250 (2016)
13. Lee, B.G., Lyche, T., Mørken, K.: Some examples of quasi-interpolants constructed from local spline projectors. In: Lyche, T., Schumaker, L.L. (eds.) Mathematical Methods for Curves and Surfaces (Oslo 2000), pp. 243–252. Vanderbilt University Press, Nashville (2001)
14. Li, C., Chen, A.: Numerical methods for fractional partial differential equations. Int. J. Comput. Math. **95**, 1048–1099 (2018)
15. Li, C., Zeng, F.: Numerical Methods for Fractional Calculus. A Chapman and Hall Book/CRC Press, London/Boca Raton (2015)
16. Lyche, T., Schumaker, L.L.: Local spline approximation methods. J. Approx. Theory **4**, 294–325 (1975)
17. Mainardi, F.: Fractional Calculus and Waves in Linear Viscoelasticity: An Introduction to Mathematical Models. World Scientific, Singapore (2010)

18. Oldham, K.B., Spanier, J.: The Fractional Calculus. Academic Press, New York (1974)
19. Pedas, A., Tamme, E.: On the convergence of spline collocation methods for solving fractional differential equations. J. Comput. Appl. Math. **235**, 3502–3514 (2011)
20. Pellegrino, E., Pezza, L., Pitolli, F.: A collocation method in spline spaces for the solution of linear fractional dynamical systems. Math. Comput. Simul. **176**, 266-278 (2020)
21. Pezza, L., Pitolli, F.: A multiscale collocation method for fractional differential problems. Math. Comput. Simul. **147**, 210–219 (2018)
22. Pezza, L., Pitolli, F.: A fractional spline collocation-Galerkin method for the fractional diffusion equation. Commun. Appl. Ind. Math. **9**, 104–120 (2018)
23. Pitolli, F.: Optimal B-spline bases for the numerical solution of fractional differential problems. Axioms **7**, 46 (2018)
24. Pitolli, F.: A fractional B-spline collocation method for the numerical solution of fractional predator-prey model. Fractal and Fractional **18**(2), 13 (2018)
25. Pitolli, F., Pezza, L.: A fractional spline collocation method for the fractional order logistic equation. In: Fasshauer, G., Schumaker, L. (eds.) Approximation Theory XV, San Antonio 2016. Proceedings in Mathematics and Statistics, vol. 201, pp. 307–318. Springer, Berlin (2018)
26. Podlubny, I.: Fractional Differential Equations: An Introduction to Fractional Derivatives, Fractional Differential Equations, to Methods of their Solution and Some of their Applications. Elsevier, Amsterdam (1999)
27. Sablonnière, P.: Bernstein type quasi-interpolants. In: Laurent, P.J., Le Méhauté, A., Schumaker L.L. (eds.) Curves and surfaces. Academic Press, Boston, pp. 421–426 (1991)
28. Sablonnière, P.: Recent progress on univariate and multivariate polynomial and spline quasi-interpolants. In: De Bruin, M.G., Mache, D.H., Szabados, J. (eds.) Trends and Applications in Constructive Approximation (ISNM), vol. 177, pp.229–245. Birkhäuser, Basel (2005)
29. Samko, S.G., Kilbas, A.A., Marichev, O.I.: Fractional Integrals and Derivatives: Theory and Applications. Gordon and Breach, London (1993).
30. Schoenberg, I.J.: On spline functions. In: Shisha, O. (ed.) Inequalities. Academic Press, New York, pp. 255–291 (1967)
31. Schumaker, L.L.: Spline Functions: Basic Theory. Cambridge University, Cambridge (2007)
32. Tarasov, V.E.: Fractional dynamics. In: Applications of fractional calculus to dynamics of particles, fields and media. In: Nonlinear Physical Science. Springer, New York (2010)
33. Unser, M., Blu, T.: Fractional splines and wavelets. SIAM Rev. **42**, 43–67 (2000)

Stochastic Collocation with Hierarchical Extended B-Splines on Sparse Grids

Michael F. Rehme and Dirk Pflüger

Abstract B-spline approximations with uniform isotropic tensor product grids soon reach computational limits, because the grid size increases exponentially with the dimensionality. Sparse grids are an established technique to mitigate this curse of dimensionality, and spatial adaptivity automatically selects only the most significant grid points. To compensate for missing boundary points of the sparse grids, the B-spline basis functions so far have been modified according to natural boundary conditions. However, modified B-splines do not span the polynomial space anymore and therefore lack a fundamental spline property. Recently we introduced hierarchical extended not-a-knot B-splines, which guarantee the polynomial basis property. Now we apply them to a subsurface flow uncertainty quantification benchmark, where we compare them to common spline bases on sparse grids, to Monte Carlo and to polynomial chaos expansion. The new basis improves the quality of quantities of interest, such as approximation error, mean and variance.

Keywords B-splines · Extension · Sparse grid · Uncertainty quantification · Stochastic expansion · Polynomial chaos

1 Introduction

Simulating real world processes through computer experiments [17] yields many benefits. Lower costs compared to real experiments, many executions in parallel and no risk to humans or the environment, just to name a few. However, computer experiments are never capable of simulating the real world comprehensively and always must be a compromise of precision and complexity.

The field of uncertainty quantification deals with the inevitably limited knowledge of the real world, and allows for more realistic assessments of computer

---

M. F. Rehme (✉) · D. Pflüger
University of Stuttgart, Stuttgart, Germany
e-mail: michael.rehme@ipvs.uni-stuttgart.de
https://www.ipvs.uni-stuttgart.de/abteilungen/sse

G. E. Fasshauer et al. (eds.), *Approximation Theory XVI*, Springer Proceedings in Mathematics & Statistics 336, https://doi.org/10.1007/978-3-030-57464-2_12

experiment results. This is done by introducing uncertainty to the input parameters and observing how the uncertainty propagates through the model and influences the results [29]. To increase the accuracy of the predictions for the underlying process more uncertain input parameters can be added, such that the computer experiment takes more aspects of the real world into account. However, the run-times and necessary computational resources increase with the complexity of the model.

This problem can be dealt with by creating a surrogate that is a sufficiently accurate approximation of the original model, but much faster to evaluate. For several years B-spline basis functions [13] have been used for the creation of surrogate models. However, the number of grid points of classical uniform isotropic tensor product grids increases exponentially with the number of input parameters. This is known as the curse of dimensionality [2]. Sparse Grids [3, 31] are an established technique to mitigate the curse, in particular when created spatially adaptive[19]. Sparse Grids have successfully been applied in combination with B-splines for interpolation, optimization, regression and uncertainty quantification [15, 19, 21, 28]. When further increasing the dimensionality of the parameter space, the boundary points of sparse grids again introduce exponential growth rates, and thus must be omitted. The B-spline basis must compensate for this to prevent a dramatic loss in approximation quality.

So far only a heuristical boundary treatment has been used [19, 28]. The left- and right-most splines were modified to enforce second zero derivatives at the boundary of the parameter domain. However, this can be disadvantageous in many cases, where the objective function does not meet this requirement. In particular, modified not-a-knot B-splines do not preserve the ability of the original not-a-knot B-spline basis, including the boundary, to represent polynomials exactly, and therefore lack one of the most important spline properties.

Recently we have introduced hierarchical extended not-a-knot B-splines for usage on spatially adaptive sparse grids [20] based on the extension concept [14, 18]. This extended basis follows the premise of preserving the polynomial representation property. In this work, we apply the new basis for the first time to a subsurface flow benchmark from the field of uncertainty quantification [12]. With this we are able to demonstrate that the new basis does not only represent polynomials exactly, but also improves the approximation of general objective functions and quantities of interest. We compare our results with a simple Monte Carlo approach and the widely used polynomial chaos expansion [9, 30].

2 Sparse Grids

Full uniform isotropic tensor product grids are one of the most widely used discretization approaches. However, their amount of grid points increases like $\mathcal{O}(h^{-D})$, where h is the grid width and D is the dimensionality of the underlying space. This exponential growth prevents calculations already for moderately high-dimensional applications.

Sparse Grids are a discretization scheme designed to mitigate this curse and enable higher-dimensional approximations. The amount of grid points of non-boundary regular sparse grids of level l with grid width h_l only increases like $\mathcal{O}(h_l^{-1}(\log_2 h_l^{-1})^{D-1})$. At the same time, the L^2-interpolation error of interpolations with B-splines of degree n on regular sparse grids of level l still decays asymptotically like $\mathcal{O}(h_l^{n+1}(\log_2 h_l^{-1})^{D-1})$ [26], if the objective function is sufficiently smooth. This is only slightly worse than the full grid error convergence rate of $\mathcal{O}(h_l^{n+1})$.

In contrast to the widely used combination technique, also known as Smolyak scheme [27], we use spatially adaptive sparse grids [19]. These can automatically be customized for the quantity of interest, resolving locally finer in more important regions and coarser in less important ones. By doing so the number of grid points is potentially reduced even further. This is important, because every grid point means an expensive evaluation of the original model.

The definition of sparse grids is based on arbitrary hierarchical basis functions $\varphi_{l,i}$ of level l and index i. We now introduce sparse grids in this general form, but later will only use hierarchical spline functions as bases.

2.1 Regular Sparse Grids

Without loss of generality, throughout this work, we restrict ourselves to parameters in the unit hypercube $[0, 1]^D$. Let I_l be the hierarchical index set of level $l \in \mathbb{N}_0$,

$$I_l := \begin{cases} \{0, 1\}, & l = 0, \\ \{0 < i < 2^l \mid i \text{ odd}\}, & \text{else.} \end{cases} \tag{1}$$

Given univariate hierarchical basis functions $\varphi_{l,i}$ of level $l \in \mathbb{N}_0$ and index $i \in \mathbb{N}_0$, we define multivariate basis functions $\boldsymbol{\varphi}_{\mathbf{l},\mathbf{i}}$ via tensor products,

$$\boldsymbol{\varphi}_{\mathbf{l},\mathbf{i}} = \prod_{d=1}^{D} \varphi_{l_d,i_d}, \mathbf{l} \in \mathbb{N}_0^D, \mathbf{i} \in I_{\mathbf{l}} := I_{l_1} \times \cdots \times I_{l_D}, \tag{2}$$

where $\mathbf{l}$ and $\mathbf{i}$ are multi-indices. Let now $\mathcal{H}_{\mathbf{l}} := \{\mathbf{x}_{\mathbf{l},\mathbf{i}} = (x_{l_1,i_1}, \ldots, x_{l_D,i_D}) \mid \mathbf{i} \in I_{\mathbf{l}}\}$ for $x_{l_d,i_d} := i_d h_{l_d}$ be the anisotropic grid of level $\mathbf{l}$ with grid widths $h_{l_d} := 2^{-l_d}$. We define the hierarchical subspaces $W_{\mathbf{l}}$ of level $\mathbf{l}$ through the basis functions corresponding to $\mathcal{H}_{\mathbf{l}}$,

$$W_{\mathbf{l}} := \text{span}\{\boldsymbol{\varphi}_{\mathbf{l},\mathbf{i}} \mid \mathbf{i} \in I_{\mathbf{l}}\}. \tag{3}$$

Regular boundary sparse grids V_l^b of level $l \in \mathbb{N}_0$ in D dimensions are defined as the direct sum of these hierarchical subspaces,

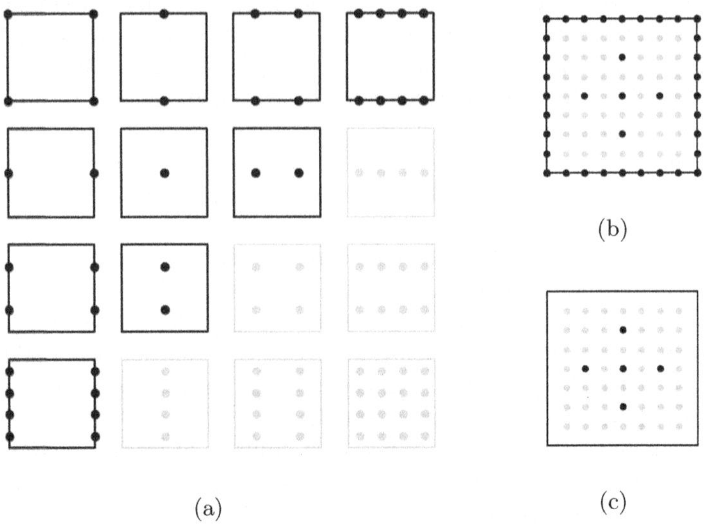

Fig. 1 (**a**) Hierarchical subspace scheme of level $l = 3$, (**b**) corresponding regular boundary sparse grid V_3^b and (**c**) corresponding regular nonboundary sparse grid V_3^s

$$V_l^b := \bigoplus_{|\mathbf{l}'|_1 \leq l} W_{\mathbf{l}'}, \tag{4}$$

where $|\mathbf{l}'|_1 := \sum_{d=1}^{D} l'_d$ is the discrete ℓ_1 norm of $\mathbf{l}'$. Unfortunately the number of boundary points of a boundary sparse grid grows like $\mathcal{O}(2^D)$. This growth is exponential, still preventing discretization for higher dimensional applications. Therefore the boundary points must be omitted. The D-dimensional nonboundary sparse grid V_l^s of level $l \in \mathbb{N}$ is defined as

$$V_l^s := \bigoplus_{|\mathbf{l}'|_1 \leq l,\ l'_d \geq 1 \forall d \in \{1,\ldots,D\}} W_{\mathbf{l}'}. \tag{5}$$

Figure 1 shows an illustration of the hierarchical subspace scheme, the corresponding regular boundary sparse grid and the corresponding regular nonboundary sparse grid.

2.2 *Spatial Adaptivity*

Regular sparse grids uniformly discretize the objective domain, spending too few grid points in regions of interest and too many grid points in regions of little significance. Spatially adaptive sparse grids [19] can automatically be adapted to the objective function. Given an initial sparse grid approximation, each basis function's benefit to the quantity of interest is estimated. Depending on this estimate, the grid points corresponding to the most significant basis functions are refined. This approach is more selective than classical dimensional adaptivity [11] and therefore allows the employment of even fewer grid points.

Let $\mathbf{x}_{\mathbf{l},\mathbf{i}}$ be a sparse grid point. We define its hierarchical children $C(\mathbf{l}, \mathbf{i})$ as all grid points $\mathbf{x}_{\mathbf{l}',\mathbf{i}'}$, for which there exists $r \in \{1, \dots, D\}$, s.t.

$$
\begin{aligned}
&l_d = l'_d, i_d = i'_d \quad \forall d \in \{1, \dots, D\} \setminus \{r\}, \\
&l'_r = l_r + 1, \\
&i'_r \in \{2i_r - 1, 2i_r + 1\}.
\end{aligned}
\tag{6}
$$

Let now $\mathcal{G}$ be a spatially refined grid,

$$
\mathcal{G} := \{\mathbf{x}_{\mathbf{l},\mathbf{i}} \mid (\mathbf{l}, \mathbf{i}) \in L\}, \tag{7}
$$

where $L \subset \{(\mathbf{l}, \mathbf{i}) \mid \mathbf{l} \in \mathbb{N}_0^D, \mathbf{i} \in I_\mathbf{l}\}$ is some finite level-index set. Note that this includes regular sparse grids as a special case. The set of all level-index pairs of refineable grid points, $L^{\text{ref}} \subseteq L$, is defined as

$$
L^{\text{ref}} := \{(\mathbf{l}, \mathbf{i}) \in L \mid C(\mathbf{l}, \mathbf{i}) \not\subset \mathcal{G}\}. \tag{8}
$$

The sparse grid $\mathcal{G}$ can now be refined, by iterating the following two steps until a given threshold for the total number of grid points is exceeded. First identify the level-index pair of the grid point $\mathbf{x}_{\mathbf{l}^*,\mathbf{i}^*} \in L^{\text{ref}}$ and corresponding basis function $\varphi_{\mathbf{l}^*,\mathbf{i}^*}$ with most influence on the quantity of interest. Second, add all its hierarchical children $C(\mathbf{l}^*, \mathbf{i}^*)$ to the grid.

Many criteria for the identification of $(\mathbf{l}^*, \mathbf{i}^*)$ exist. In this work we apply the standard surplus criterion [19]. It is based on the hierarchy of the basis, where larger interpolation coefficients $|\alpha_{\mathbf{l},\mathbf{i}}|$ imply a worse local approximation. Consequently we use

$$
(\mathbf{l}^*, \mathbf{i}^*) := \operatorname{argmax}_{(\mathbf{l},\mathbf{i}) \in L^{\text{ref}}} |\alpha_{\mathbf{l},\mathbf{i}}|. \tag{9}
$$

3 Basis Functions

Sparse grids are widely used in combination with the popular linear hat functions, i.e. B-splines of degree one. But if the objective function admits a certain smoothness, an approximation should preserve it or it would otherwise lose valuable information. Therefore in the last years B-splines have been used increasingly often on (spatially adaptive) sparse grids [15, 19, 28]. Their local support and arbitrary choosable degree result in their well-known approximation quality, while the underlying sparse grid keeps the number of necessary function evaluations small.

Before we define the new extended not-a-knot B-spline basis we must introduce the underlying classical not-a-knot B-splines. Furthermore we define modified not-a-knot B-splines to motivate the new basis. As is common, throughout this paper we only define and use splines of odd degrees.

3.1 B-Splines

Let $\xi := (\xi_0, \dots, \xi_{q+n})$ be a knot-sequence, i.e. a non-decreasing sequence of real numbers ξ_k for $k \in \{0, \dots, q+n\}$ and some $q \in \mathbb{N}_0$. The B-spline $b^n_{k,\xi}$ of index k and degree n is defined by the Cox-de-Boor recursion [4, 6],

$$b^n_{k,\xi}(x) = \begin{cases} \frac{x - \xi_k}{\xi_{k+n} - \xi_k} b^{n-1}_{k,\xi}(x) + \frac{\xi_{k+n+1} - x}{\xi_{k+n+1} - \xi_{k+1}} b^{n-1}_{k+1,\xi}(x) & n \geq 1, \\ \chi_{[\xi_k, \xi_{k+1}[}(x) & n = 0, \end{cases} \tag{10}$$

where $\chi_{[\xi_k, \xi_{k+1}[}(x)$ evaluates to one in the interval $[\xi_k, \xi_{k+1}[$ and zero elsewhere.

Originally, Schoenberg introduced B-splines with an infinite and uniform knot sequence $\xi^\infty_h = (\dots, \xi^\infty_{h,-1}, \xi^\infty_{h,0}, \xi^\infty_{h,1}, \dots)$, where $\xi^\infty_{h,k} = kh$ for grid width $h \in \mathbb{R}$ and index $k \in \mathbb{Z}$ [23]. The corresponding B-splines b^n_{k,ξ^∞_h} form a basis of $S^n_{\xi^\infty_h}$, the spline space of n times continuously differentiable piecewise polynomials on the knot intervals.

When using a finite knot sequence this desirable basis property is no longer valid, because the Schoenberg-Whitney conditions [13, 24] do not hold at the left-most and right-most knot intervals. A common approach to revalidate these conditions are not-a-knot B-splines [5, 28].

3.2 Not-a-Knot B-Splines

Not-a-knot B-splines are motivated by requiring continuity of the n-th derivatives at the $\frac{n-1}{2}$ left-most and $\frac{n-1}{2}$ right-most knots. This requirement is equivalent to excluding the according $n-1$ knots from the B-spline defining knot sequence ξ but keeping them in the set of interpolation nodes.

Without loss of generality we restrict ourselves to uniform B-splines of level $l \in \mathbb{N}_0$ on the unit interval $[0, 1]$ using the uniform knot sequence $\xi^{n,\mathrm{u}}_l := (\xi^{n,\mathrm{u}}_{l,0}, \dots, \xi^{n,\mathrm{u}}_{l,2^l+2n})$, where $\xi^{n,\mathrm{u}}_{l,k} := (k-n)h_l$ for grid width $h_l := 2^{-l}$. Consequently, we derive $\xi^{n,\mathrm{nak}}_l := (\xi^{n,\mathrm{nak}}_{l,0}, \dots, \xi^{n,\mathrm{nak}}_{l,2^l+n+1})$, the uniform not-a-knot sequence of level l and degree n as

$$\xi_{l,k}^{n,\mathrm{nak}} := \begin{cases} \xi_{l,k}^{n,\mathrm{u}}, & k = 0, \ldots, n, \\ \xi_{l,k+(n-1)/2}^{n,\mathrm{u}}, & k = n+1, \ldots, 2^l, \\ \xi_{l,k+n-1}^{n,\mathrm{u}}, & k = 2^l+1, \ldots, 2^l+n+1. \end{cases} \tag{11}$$

The definition of $\xi_{l,k}^{n,\mathrm{nak}}$ is only applicable if $l \geq \lceil \log_2(n+1) \rceil$. Otherwise we cannot exclude $n-1$ knots from the sequence. Therefore, if $l < \lceil \log_2(n+1) \rceil$, we use $\xi_{l,k}^{n,\mathrm{nak}} := \xi_{l,k}^{n,\mathrm{u}}$ and Lagrange polynomials

$$L_{l,k}(x) := \prod_{\substack{0 \leq m \leq 2^l, \\ m \neq k}} \frac{x - \xi_{l,m}^{n,\mathrm{u}}}{\xi_{l,k}^{n,\mathrm{u}} - \xi_{l,m}^{n,\mathrm{u}}}, \quad k = 0, \ldots, 2^l \tag{12}$$

as basis functions. This ensures a basis for the polynomial space on the first levels.

Finally, the not-a-knot B-spline basis $b_{l,k}^{n,\mathrm{nak}}$ of degree n, level l and index k is given by

$$b_{l,k}^{n,\mathrm{nak}}(x) := \begin{cases} b_{k,\xi_{l,k}^{n,\mathrm{nak}}}^{n}(x) & l \geq \lceil \log_2(n+1) \rceil, \\ L_{l,k}(x) & l < \lceil \log_2(n+1) \rceil. \end{cases} \tag{13}$$

The knot-sequence $\xi_{l,k}^{n,\mathrm{nak}}$ still includes the boundary points $\xi_{l,0}^{n,\mathrm{nak}} = 0$ and $\xi_{l,2^l+n+1}^{n,\mathrm{nak}} = 1$. Because the number of boundary points of higher-dimensional sparse grids dominates the total number of grid points, the boundary points must be omitted. However, simply excluding the boundary points, and thus the corresponding B-spline basis functions, impairs the approximation quality at the boundaries. Therefore an appropriate boundary treatment is necessary.

3.3 *Modified Not-a-Knot B-Splines*

So far modified not-a-knot B-splines [28] are used to compensate for the missing boundary points. Motivated by an application with natural boundary conditions, they were defined to enforce zero second derivatives at the domain's boundaries. The resulting basis functions extrapolate towards the boundaries, as can be seen in Fig. 3. Consequently the modified not-a-knot B-spline $b_{l,k}^{n,mod}$ of degree n, level l and index k is defined as,

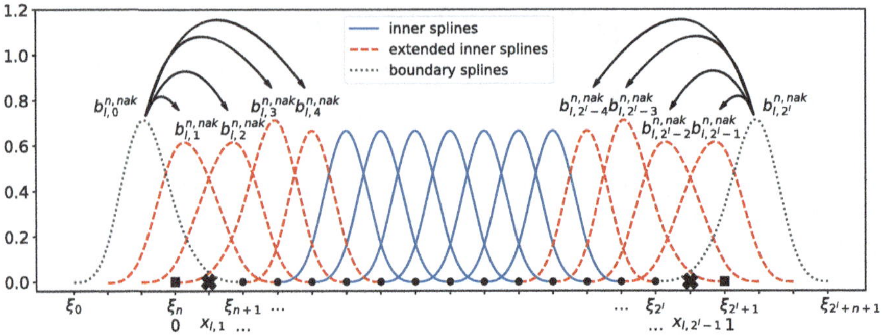

Fig. 2 Schematic visualization of the extension of not-a-knot B-splines of degree $n = 3$ on a one-dimensional regular grid of level $l = 4$. The boundary splines with indices $J_l = \{0, 16\}$ are added to the $n + 1$ next inner splines $I_l(0) = \{1, 2, 3, 4\}$ and $I_l(16) = \{12, 13, 14, 15\}$, indicated with arrows

$$b_{l,k}^{n,\text{mod}}(x) := \begin{cases} 1 & l = 1, k = 1, \\ b_{l,k}^{n,\text{nak}}(x) + b_{l,k-1}^{n,\text{nak}}(x) & l \geq 2, k = 1, n = 1, \\ b_{l,k}^{n,\text{nak}}(x) - \dfrac{\frac{d^2}{dx^2} b_{l,k}^{n,\text{nak}}(0)}{\frac{d^2}{dx^2} b_{l,k-1}^{n,\text{nak}}(0)} b_{l,k-1}^{n,\text{nak}}(x) & l \geq 2, k \in \{1, \ldots, \frac{n+1}{2}\}, n > 1, \\ b_{l,2^l-k}^{n,\text{mod}}(1-x) & l \geq 2, k \in \{2^l - \frac{n+1}{2}, \ldots, 2^l - 1\}, \\ b_{l,k}^{n,\text{nak}}(x) & \text{otherwise.} \end{cases} \tag{14}$$

Note, that for linear splines of degree $n = 1$ the second derivatives always vanish. Therefore the modification is defined as the linear continuation of the left-most and right-most inner splines.

Some applications require zero second derivatives, and thus are accurately representable by modified not-a-knot B-splines. However, this condition does not hold in general and modified not-a-knot B-splines are not capable of representing arbitrary functions. In particular the standard monomial basis $\{x^m \mid 0 \leq m \leq n\}$ for the polynomial space $\mathbb{P}^n$ has second derivatives unequal to zero for $n \geq 2$. The modified not-a-knot B-spline basis is thus not even capable of exactly representing polynomials, which is one of the most important properties for spline bases.

3.4 *Extended Not-a-Knot B-Splines*

The extension of B-splines was originally introduced in the context of WEB-splines [14] and later generalized for hierarchical subdivision schemes [18]. Recently we have introduced hierarchical extended not-a-knot B-splines for the usage on sparse grids [20].

The idea of the extension is to add the omitted splines $b_j, j \in J_l := \{0, 2^l\}$ to the remaining splines in such a way, that their contribution to the capability of representing polynomials is preserved. In a first step, we interpolate $\{P_m \mid m \in M := \{0, \ldots, n+1\}\}$ a basis for the polynomial space $\mathbb{P}^n$ with the regular not-a-knot B-spline basis including the boundary splines. Let $l \geq \lceil \log_2(n+2) \rceil$, then the polynomial basis is represented exactly by definition of the not-a-knot B-splines. This results in interpolation coefficients $\alpha_{m,i}$, such that

$$P_m = \sum_{k=0}^{2^l} \alpha_{m,k} b_{l,k}^{n,nak} \quad \forall m \in M. \tag{15}$$

In practice we use the monomials $P_m = x^m$, but the theory is independent of this particular choice.

In a next step, we identify the closest $n+1$ inner indices $I_l(j)$ for each index $j \in J_l$. Now the coefficients $\alpha_j,\ j \in J_l$ are represented as linear combinations of the coefficients $\alpha_i,\ i \in I_l(j)$, i.e.

$$\alpha_j = \sum_{i \in I_l(j)} e_{i,j} \alpha_i, \tag{16}$$

where $e_{i,j} \in \mathbb{R}$ are the extension coefficients. See Fig. 2 for an illustration.

Let $J_l(i) := \{j \in J_l \mid i \in I_l(j)\}$ be the dual of $I_l(j)$ and $P \in \mathbb{P}^n$ be an arbitrary polynomial. Following eq. (15) it holds

$$P = \sum_{m \in M} p_m P_m = \sum_{m \in M} \sum_{i \in I_l} p_m \alpha_{m,i} b_{l,i}^{n,\text{nak}} + \sum_{m \in M} \sum_{j \in J_l} p_m \alpha_{m,j} b_{l,j}^{n,\text{nak}} \tag{17}$$

for uniquely defined coefficients $p_m, \alpha_{m,i}, \alpha_{m,j} \in \mathbb{R}$. Exploiting the finiteness of the sets M, I_l and J_l, we interchange the sums,

$$P = \sum_{i \in I_l} \left(\sum_{m \in M} p_m \alpha_{m,i} \right) b_{l,i}^{n,\text{nak}} + \sum_{j \in J_l} \left(\sum_{m \in M} p_m \alpha_{m,j} \right) b_{l,j}^{n,\text{nak}}. \tag{18}$$

Because $J_l(i)$ is the dual of $I_l(j)$, and by the definition of the extension coefficients $e_{i,j}$ in eq. (16), it holds

$$P = \sum_{i \in I_l} \underbrace{\left(\sum_{m \in M} p_m \alpha_{m,i} \right)}_{=:\beta_i} \underbrace{\left(b_{l,i}^{n,\text{enak}} + \sum_{j \in J_l(i)} e_{i,j} b_{l,j}^{n,\text{nak}} \right)}_{=: b_{l,i}^{n,e}} \tag{19}$$

$$= \sum_{i \in I_l} \beta_i b_{l,i}^{n,e}. \tag{20}$$

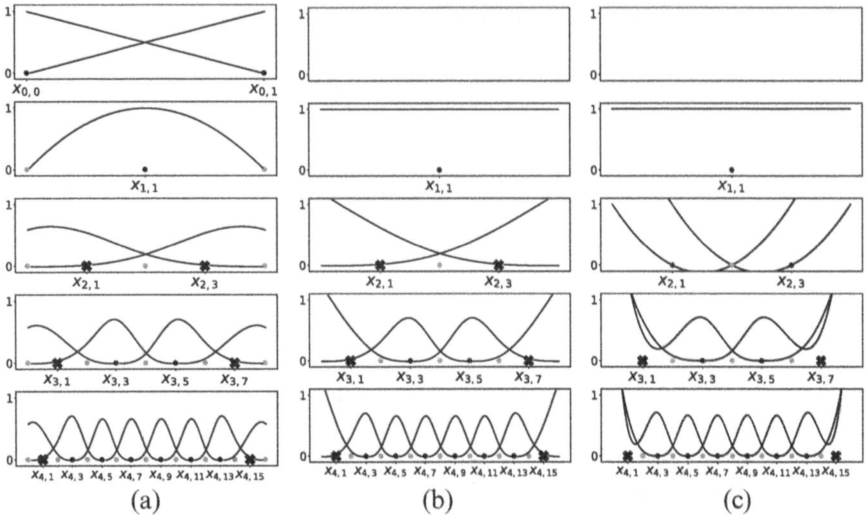

Fig. 3 (**a**) Hierarchical not-a-knot B-splines with boundary basis functions, (**b**) hierarchical modified not-a-knot B-splines and (**c**) hierarchical extended not-a-knot B-splines of degree 3 and levels 0, 1, 2, 3 and 4, respectively. The not-a-knot change in the knot sequence is illustrated with crosses at $x_{l,1}$ and $x_{l,2^l-1}$

Consequently the extended not-a-knot B-spline $b_{l,i}^{n,e}$ of degree n, level l and index i is defined through eq. (19)

$$b_{l,i}^{n,e} := \begin{cases} b_{l,i}^{n,\text{nak}} + \sum_{j \in J_l(i)} e_{i,j} b_{l,j}^{n,\text{nak}} & l \geq \lceil \log_2(n+2) \rceil, \\ L_{l,i}(x) & l < \lceil \log_2(n+2) \rceil, \end{cases} \tag{21}$$

where again Lagrange polynomials are employed on lower levels to ensure the polynomial basis property, as long as there are not enough inner knots for the extension.

For the usage on sparse grids, all presented B-spline basis functions are applied in the hierarchical manner introduced in Eq. (3). See Fig. 3 for an illustration. Recently we showed that the hierarchical extended not-a-knot B-spline basis fulfills the desired polynomial representation property [20].

4 Expansion Methods

The field of uncertainty quantification generalizes the concept of numerical modeling by introducing nondeterminism, thereby allowing more accurate simulations of the real world. Instead of real values, parameters are random variables obeying probability density functions. The uncertainty of the input parameters is then propagated through the model resulting in uncertain outputs. In order to estimate likely outcomes of the model, stochastic values such as mean and standard deviation can be calculated. Two of the most widely used techniques to calculate these values are stochastic collocation and polynomial chaos expansion (PCE).

Formally let $(\Omega, \mathcal{F}, P)$ be a complete probability space with $\Omega \subset \mathbb{R}^D$ being the D-dimensional sample space of all possible outcomes, $\mathcal{F}$ the σ-algebra of events and $P : \mathcal{F} \to [0, 1]$ the probability measure. Without loss of generality we assume $\Omega \subseteq [0, 1]^D$. Let $\mathbf{X} := (X_1, \ldots, X_D) \in \Omega$ be a random vector consisting of D random variables. We assume that the according random variables admit statistically independent probability density functions $\varrho_1, \ldots, \varrho_D$ and thus the random vector is distributed according to their product distribution $\boldsymbol{\varrho} := \prod_{d=1}^{D} \varrho_d$.

4.1 Stochastic Collocation

Stochastic collocation is based on the process of replacing the original objective function f by a surrogate $\tilde{f}$, and performing stochastic analysis on the surrogate. We create the surrogate as a linear combination of B-splines $b_{\mathbf{l},\mathbf{i}}$ on an adaptively created sparse grid $\mathcal{G}$ with level-index set L,

$$f \approx \tilde{f} := \sum_{(\mathbf{l},\mathbf{i}) \in L} \alpha_{\mathbf{l},\mathbf{i}} \mathbf{b}_{\mathbf{l},\mathbf{i}}, \tag{22}$$

where the coefficients $\boldsymbol{\alpha}_{\mathbf{l},\mathbf{i}}$ are computed via interpolation at the sparse grid points. From this we approximate the mean $\mathbb{E}(f)$ and variance $\mathbb{V}(f)$ of the objective function using Gauss-Legendre quadrature,

$$\mathbb{E}(f) \approx \mathbb{E}(\tilde{f}) = \int_{[0,1]^D} \tilde{f}(\mathbf{X}) \boldsymbol{\varrho}(\mathbf{X}) d\mathbf{X} \tag{23}$$

$$\approx \sum_{k} \tilde{f}(x_k) \boldsymbol{\varrho}(x_k) \omega_k, \tag{24}$$

$$\mathbb{V}(f) \approx \mathbb{V}(\tilde{f}) = \mathbb{E}(\tilde{f}^2) - \mathbb{E}(\tilde{f})^2, \tag{25}$$

where x_k are the points and ω_k the weights of the quadrature rule. The order of the quadrature rule is chosen depending on the distribution $\boldsymbol{\varrho}$. Being piecewise polynomials, splines are exactly integrated by the Gauss-Legendre quadrature rule of order $(n+1)/2$ with respect to a uniform probability density function. Therefore, if any of the density functions ϱ_d is uniform, the quality of the approximation $\tilde{f}$ directly propagates to the quality of the stochastic values.

4.2 Polynomial Chaos Expansion

Generalized polynomial chaos is based on the Wiener-Askey scheme [30], where Hermite, Legendre, Laguerre, Jacobi and generalized Laguerre polynomials are used to model the effects of uncertainties of normal, uniform, exponential, beta and gamma distributed random variables respectively. These polynomials are optimal for the according distribution in the sense, that they are orthogonal with respect to the according inner product [9].

If other distribution types are required, nonlinear variable transformations like Rosenblatt [22] and Nataf [7] can be applied, but convergence rates are typically decreased by this [9]. Alternatively orthogonal polynomials matching the given distribution can be numerically generated [8]. For a fair comparison, our numerical examples all obey the distributions from the Wiener-Askey scheme. Note however, that stochastic collocation with B-splines on sparse grids is not limited in the type of distribution and can be applied directly for any given distribution.

The actual chaos expansion takes the form

$$f(\mathbf{X}) = \gamma_0 \Phi_0 + \sum_{d=1}^{D} \gamma_d \Phi_1(X_d) + \sum_{d=1}^{D} \sum_{t=1}^{d} \gamma_{d,t} \Phi_2(X_d, X_t) + \dots, \tag{26}$$

where Φ_d are the basis functions from the Wiener-Askey scheme and each additional set of nested summation introduces an additional order of polynomials. Usually the order-based indexing is replaced by term-based indexing to simplify the representation. Consequently,

$$f(\mathbf{X}) = \sum_{\mathbf{k}=\mathbf{0}}^{\infty} \boldsymbol{\gamma}_{\mathbf{k}} \Psi_{\mathbf{k}}(\mathbf{X}), \tag{27}$$

where there is a direct correspondence between $\gamma_{d,t,\dots}$ and $\boldsymbol{\gamma}_{\mathbf{k}}$ and between $\Phi_t(X_d, X_t, \dots)$ and $\Psi_{\mathbf{k}}(\mathbf{X})$, which are multivariate polynomials.

The PCE coefficients $\boldsymbol{\gamma}_{\mathbf{k}}$ are calculated via spectral projection, taking advantage of the orthogonality of the polynomials to extract each coefficient,

$$\boldsymbol{\gamma}_{\mathbf{k}} = \frac{\langle f, \Psi_{\mathbf{k}} \rangle}{\langle \Psi_{\mathbf{k}}^2 \rangle} = \frac{1}{\langle \Psi_{\mathbf{k}}^2 \rangle} \int_{[0,1]^D} f(\mathbf{X}) \Psi_{\mathbf{k}} \varrho(\mathbf{X}) d\mathbf{X}. \tag{28}$$

The integral in eq. (28) must be numerically calculated. In high-dimensional settings usually regular sparse grids based on the combination technique are used [9] and we too use this approach.

Once the expansion coefficients have been calculated, the desired stochastic quantities follow directly, because of the orthogonality of the polynomials,

$$\mathbb{E}(f) = \boldsymbol{\gamma}_0, \tag{29}$$

$$\mathbb{V}(f) = \sum_{\mathbf{k}=0}^{\infty} \gamma_{\mathbf{k}^2} \langle \Psi_{\mathbf{k}}^2 \rangle_{\varrho}. \tag{30}$$

In practice the expansion representation of the variance must be truncated, thus PCE tends to underestimate the variance.

5 Numerical Results

We will measure the interpolation error between an objective function $f : \Omega \to \mathbb{R}$ and a surrogate $\tilde{f} : \Omega \to \mathbb{R}$ with the normalized root-mean-square error (NRMSE). For $R \in \mathbb{N}$ given samples $\{\mathbf{x}_r \in \Omega \mid r = 1, \dots, R\}$, the NRMSE is defined as

$$\frac{1}{f_{\max} - f_{\min}} \sqrt{\frac{\sum_{r=1}^{R} (f(\mathbf{x}_r) - \tilde{f}(\mathbf{x}_r))^2}{R}}, \tag{31}$$

where $f_{\max} := \max_{r=1,\dots,R} f(\mathbf{x}_r)$ and $f_{\min} := \min_{r=1,\dots,R} f(\mathbf{x}_r)$. In our examples we used $R = 100000$. Mean and variance errors are measured relatively,

$$\varepsilon_{\mathbb{E}} = \frac{|\mathbb{E}(f) - \mathbb{E}(\tilde{f})|}{\mathbb{E}(f)}, \qquad \varepsilon_{\mathbb{V}} = \frac{|\mathbb{V}(f) - \mathbb{V}(\tilde{f})|}{\mathbb{V}(f)}. \tag{32}$$

All results in this chapter, except for polynomial chaos expansion, were calculated with our software SG^{++} [19], a general toolbox for regular and spatially adaptive sparse grids. It is available open-source for usage and comparison [25]. Our spatial adaptivity algorithm was set up to refine up to 25 points in each refinement step, starting with a regular sparse grid of level 0 for not-a-knot B-splines on boundary sparse grids and level 1 otherwise.

In practice the extension coefficients must be calculated only once. This allows an efficient implementation of the new basis. The precalculated extension coefficients we used are listed in Table 1.

Table 1 Extension coefficients $e_{i,j}$ for the degrees $n \in \{1, 3, 5\}$ based on $P_m = x^m$, $m \in M$. Only the coefficients for the extension at the left boundary are shown, i.e. $j = 0$. The right boundary is treated symmetrically. For degree 5 and level 3, the left and right extensions overlap, resulting in a special case

n	$[e_{1,0}, \dots, e_{n+1,0}]$
1	$[2, -1]$
3	$[5, -10, 10, -4]$
5	$\begin{cases} [8, -28, 42, -35, 20, -6] & l = 3, \\ [8, -28, 56, -70, 56, -21] & l > 3 \end{cases}$

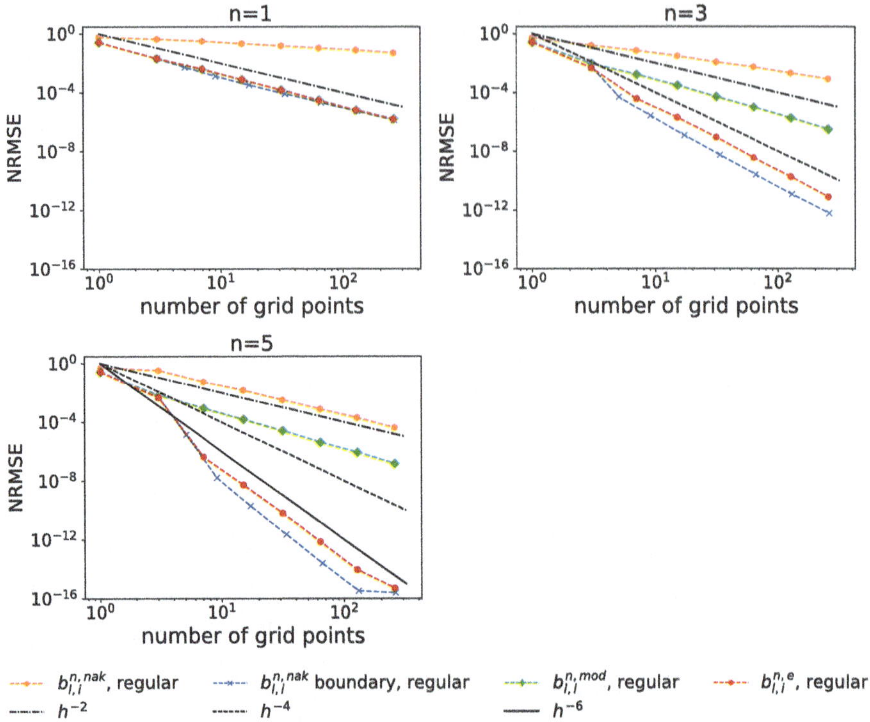

Fig. 4 Normalized root mean square error for the interpolation of $f(x) = \exp(x)$ with not-a-knot B-splines with and without boundary points, modified not-a-knot B-splines and extended not-a-knot B-splines on regular sparse grids for degrees $n \in \{1, 3, 5\}$

5.1 Exponential Objective Function

We first verify the improved convergence rates of extended not-a-knot B-splines in a simple setup, which illustrates why the new basis functions were necessary. We interpolate the one-dimensional exponential function with the common spline functions used in sparse grid context and measure the NRMSE, see Fig. 4.

The not-a-knot B-splines without boundary points or any boundary treatment converge very slowly, clearly showing the need for appropriate boundary treatment. The modified not-a-knot B-splines converge faster but are still far away from the optimal convergence rates of $\mathcal{O}(h^{-(n+1)})$. Only not-a-knot B-splines with boundary points and extended not-a-knot B-splines reach the optimal convergence rates.

In this one-dimensional example the additional costs of the two boundary points are negligible. However, in higher-dimensions 2^D boundary points of a level 0 sparse grid can already exceed the computational limits, leaving extended not-a-knot B-splines as the only viable alternative.

Table 2 The input variables and according distributions for the borehole model

Variable	Distribution	Description
r_w	$N(\mu = 0.1, \sigma = 0.0161812)$	Radius of borehole
r	Lognormal$(\mu = 7.71, \sigma = 1.0056)$	Radius of influence
T_u	Uniform[63070, 115600]	Upper aquifer transmissivity
H_u	Uniform[990, 1110]	Upper aquifer potentiometric head
T_l	Uniform[63.1, 116]	Lower aquifer transmissivity
H_l	Uniform[700, 820]	Lower aquifer potentiometric head
L	Uniform[1120, 1680]	Borehole length
K_w	Uniform[9855, 12045]	Borehole hydraulic conductivity

5.2 Borehole Model

The next example is a real world application, modeled in 1983 by Harper and Gupta for the office of nuclear waste isolation [12]. Since then, it has been used many times for testing new approximation methods, e.g. in [16, 32]. A borehole is drilled through an aquifer above a nuclear waste repository, through the repository, and to an aquifer below. The input parameter ranges are defined in Table 2, the response $Q \in \mathbb{R}$ is the flow in m^3/yr and is given by

$$Q = \frac{2\pi T_u (H_u - H_l)}{\ln(r/r_w)\left(1 + \frac{2LT_u}{\ln(r/r_w) r_w^2 K_w} + \frac{T_u}{T_l}\right)}. \tag{33}$$

In terms of calculating mean and variance we compare our method to the polynomial chaos expansion implementation of the DAKOTA library [1] and Monte Carlo. We compare calculated means and variances to a reference solution computed with extended not-a-knot B-splines of degree 5 on a spatially adaptive sparse grid with 35,000 grid points. We verified this reference solution by calculating another reference solution using DAKOTA's polynomial chaos expansion based on a sparse grid of level 5 with 34,290 grid points. The difference between both results for mean and variance is smaller than 10^{-11}.

Figure 5 shows the NRMSE, the relative mean error and relative variance errors for all introduced B-splines on regular and spatially adaptive sparse grids, simple Monte Carlo and polynomial chaos expansion. For this problem B-splines of degree $n = 5$ performed best and the plots show only these results. However, the free choice of the B-spline degree makes the approach very flexible and allows to react to local features of general objective functions. While higher degree approximations are in general better for smooth functions, they can start to oscillate, making lower degrees advantageous.

B-splines without boundary points or any boundary treatment barely converge, again demonstrating the urgent need for compensation, when omitting the boundary points. For B-splines with boundary points the errors do converge, but slower than for modified or extended not-a-knot B-splines, which can resolve the inner domain

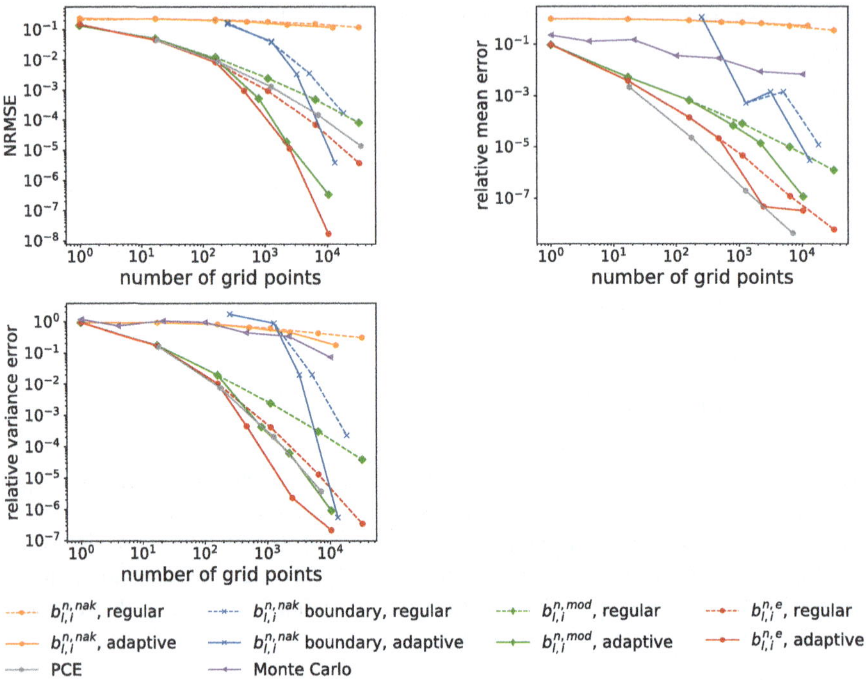

Fig. 5 Normalized root mean square error for the approximation of the borehole model and calculation of its mean and variance with not-a-knot B-splines with and without boundary points, modified not-a-knot B-splines and extended not-a-knot B-splines of degree 5 on regular and adaptive sparse grids, polynomial chaos expansion and Monte Carlo

much finer. Of these two, the extended not-a-knot B-splines perform significantly better. In all cases spatial adaptivity increases the convergence rate significantly over regular Sparse Grids.

The polynomial chaos expansion's NRMSE is worse than that of modified and extended not-a-knot B-splines. That is because the underlying global polynomials cannot react to local features, as the spline bases can. However, its approximation of the mean is best among all shown methods. This can be explained by Eq. (29). The mean of a polynomial chaos approximation is directly given by its first coefficient γ_0 and independent of all other terms. So the mean of a polynomial chaos approximation can be disproportionately better than its overall approximation quality. The variance approximation on the other hand, which is calculated according to Eq. (30), theoretically relies on all, infinitely many, coefficients. In practice the sum must be truncated. Consequently the polynomial chaos expansion tends to underestimate the variance and it can be seen, that the extended not-a-knot B-splines on spatially adaptive sparse grids approximate the variance better.

As expected the simple Monte Carlo approach is easily outperformed by almost all other techniques.

6 Conclusions and Outlook

In this article we have demonstrated the need for proper boundary treatment when creating surrogates with B-splines on sparse grids for moderately high-dimensional problems. We have shown that modified not-a-knot B-splines are not sufficient if the objective function does not have second zero derivatives at the boundary. Our recently introduced extended not-a-knot B-splines performed significantly better in a real world uncertainty quantification benchmark. Not only the overall approximation is improved but also the derived stochastic quantities of interest. The results of our new method are comparable to, and for some quantities of interest even outperform, widely used polynomial chaos expansion. This makes the technique an interesting alternative, in particular for objective functions with local features that often can hardly be resolved by global polynomial approaches.

For this work we used the standard surplus-based refinement criterion. However, other refinement criteria based on means or variances have successfully been used in the context of uncertainty quantification and sparse grids [10]. These might improve our techniques results even further.

Acknowledgments Funded by Deutsche Forschungsgemeinschaft (DFG, German Research Foundation) under Germany's Excellence Strategy EXC2075-390740016.

References

1. Adams, B.M., et al.: M.S.E.: Dakota, a multilevel parallel object-oriented framework for design optimization, parameter estimation, uncertainty quantification, and sensitivity analysis. Sandia Technical Report, SAND2014-4633 (2014). Updated May 2019
2. Bellman, R.: Adaptive Control Processes: A Guided Tour. Rand Corporation. Research Studies. Princeton University Press, New Jersey (1961)
3. Bungartz, H., Griebel, M.: Sparse grids. Acta Numer. **13**, 147–269 (2004)
4. Cox, M.G.: The numerical evaluation of B-splines. IMA J. Appl. Math. **10**(2), 134–149 (1972)
5. De Boor, C.: The method of projection as applied to the numerical solution of two point boundary value problems using cubic splines. Ph.D. Thesis, Citeseer (1966)
6. De Boor, C.: On calculating with B-splines. J. Approx. Theory **6**(1), 50–62 (1972)
7. Der Kiureghian, A., Liu, P.L.: Structural reliability under incomplete probability information. J. Eng. Mech. **112**(1), 85–104 (1986)
8. Eldred, M.: Recent advances in non-intrusive polynomial chaos and stochastic collocation methods for uncertainty analysis and design. In: 50th AIAA/ASME/ASCE/AHS/ASC Structures, Structural Dynamics, and Materials Conference 17th AIAA/ASME/AHS Adaptive Structures Conference 11th AIAA No, p. 2274 (2009)
9. Eldred, M., Burkardt, J.: Comparison of non-intrusive polynomial chaos and stochastic collocation methods for uncertainty quantification. In: 47th AIAA Aerospace Sciences Meeting Including the New Horizons Forum and Aerospace Exposition, p. 976 (2009)
10. Franzelin, F.: Data-Driven Uncertainty Quantification for Large-Scale Simulations. Verlag Dr. Hut, München (2018)
11. Gerstner, T., Griebel, M.: Dimension–adaptive tensor–product quadrature. Computing **71**(1), 65–87 (2003)

12. Harper, W.V., Gupta, S.K.: Sensitivity/uncertainty analysis of a borehole scenario comparing Latin hypercube sampling and deterministic sensitivity approaches. Technical Report, Battelle Memorial Institute (1983)
13. Höllig, K., Hörner, J.: Approximation and Modeling with B-Splines. SIAM, Philadelphia (2013)
14. Höllig, K., Reif, U., Wipper, J.: Weighted extended B-spline approximation of Dirichlet problems. SIAM J. Numer. Anal. **39**(2), 442–462 (2001)
15. Jiang, Y., Xu, Y.: B-spline quasi-interpolation on sparse grids. J. Complex. **27**(5), 466–488 (2011)
16. Kersaudy, P., Sudret, B., Varsier, N., Picon, O., Wiart, J.: A new surrogate modeling technique combining kriging and polynomial chaos expansions–application to uncertainty analysis in computational dosimetry. J. Comput. Phys. **286**, 103–117 (2015)
17. Koehler, J., Owen, A.: Computer experiments. Handb. Stat. **13**, 261–308 (1996)
18. Martin, F.: WEB-Spline Approximation and Collocation for Singular and Time-Dependent Problems. Shaker Verlag, Herzogenrath (2017)
19. Pflüger, D.: Spatially Adaptive Sparse Grids for High-Dimensional Problems. Verlag Dr. Hut, München (2010)
20. Rehme, M.F., Pflüger, D.: Hierarchical extended B-splines for approximations on sparse grids. In: Proceedings of the 5th Workshop on Sparse Grids and Applications (2018)
21. Rehme, M.F., Franzelin, F., Pflüger, D.: B-splines on sparse grids for surrogates in uncertainty quantification. J. Reliab. Eng. Syst. Saf. (2019). Submitted
22. Rosenblatt, M.: Remarks on a multivariate transformation. Ann. Math. Stat. **23**(3), 470–472 (1952)
23. Schoenberg, I.J.: Contributions to the problem of approximation of equidistant data by analytic functions. Q. Appl. Math. **4**, 45–99 and 112–141 (1946)
24. Schoenberg, I.J., Whitney, A.: On pólya frequency functions. III. The positivity of translation determinants with an application to the interpolation problem by spline curves. Trans. Am. Math. Soc. **74**(2), 246–259 (1953)
25. SG++ sparse grid library. https://github.com/SGpp/SGpp
26. Sickel, W., Ullrich, T.: Spline interpolation on sparse grids. Appl. Anal. **90**(3–4), 337–383 (2011)
27. Smolyak, S.A.: Quadrature and interpolation formulas for tensor products of certain classes of functions. In: Doklady Akademii Nauk, vol. 148, pp. 1042–1045. Russian Academy of Sciences, Moscow (1963)
28. Valentin, J.: B-splines on sparse grids. Algorithms and application to higher-dimensional optimization. Ph.D. Thesis. University of Stuttgart, IPVS (2019)
29. Walker, W.E., Harremoës, P., Rotmans, J., Van Der Sluijs, J.P., Van Asselt, M.B., Janssen, P., Krayer von Krauss, M.P.: Defining uncertainty: a conceptual basis for uncertainty management in model-based decision support. Integr. Assess. **4**(1), 5–17 (2003)
30. Xiu, D., Karniadakis, G.E.: The wiener–askey polynomial chaos for stochastic differential equations. SIAM J. Sci. Comput. **24**(2), 619–644 (2002)
31. Zenger, C.: Sparse grids. Notes Numer. Fluid Mech. **31**, 241–251 (1991)
32. Zhou, Q., Qian, P.Z., Zhou, S.: A simple approach to emulation for computer models with qualitative and quantitative factors. Technometrics **53**(3), 266–273 (2011)

Trivariate Interpolated Galerkin Finite Elements for the Poisson Equation

Tatyana Sorokina and Shangyou Zhang

Abstract When applying finite element method to the Poisson equation on a domain in $\mathbb{R}^3$, we replace some Lagrange nodal basis functions by bubble functions whose dual functionals are the values of the Laplacian. To compute the coefficients of these Laplacian basis functions instead of solving a large linear system, we interpolate the right hand side function in the Poisson equation. The finite element solution is then the Galerkin projection on a smaller vector space. We construct a qudratic and a cubic nonconforming interpolated finite elements, and quartic and higher degree conforming interpolated finite elements on arbitrary tetrahedral partitions. The main advantage of our method is that the number of unknowns that require solving a large system of equations on each element is reduced. We show that the interpolated Galerkin finite element method retains the optimal order of convergence. Numerical results confirming the theory are provided as well as comparisons with the standard finite elements.

Keywords Conforming finite element · Conforming finite element · Interpolated finite element · Tetrahedral grid · Poisson equation

1 Introduction

When solving partial differential equations using finite element method, the full space P_k of polynomials of degree $\leq k$ on each element is typically used in order to achieve the optimal order of approximation. Occasionally, the P_k polynomial space may be enriched by the so-called bubble functions. This is done for stability or continuity, while the order of approximation is not increased, cf. [2–4, 6–10, 15–

T. Sorokina (✉)
Department of Mathematics, Towson University, Towson, MD, USA
e-mail: tsorokina@towson.edu

S. Zhang
Department of Mathematical Sciences, University of Delaware, Newark, DE, USA
e-mail: szhang@udel.edu

G. E. Fasshauer et al. (eds.), *Approximation Theory XVI*, Springer Proceedings in Mathematics & Statistics 336, https://doi.org/10.1007/978-3-030-57464-2_13

18]. The only exception when a proper subspace of P_k is used while retaining the optimal order of convergence ($O(h^k)$ in the H^1-norm) can be found in [12, 13]. In these papers, we constructed a harmonic finite element method for solving the Laplace equation (1.1),

$$\begin{aligned} -\Delta u &= 0, \quad \text{in } \Omega, \\ u &= f, \quad \text{on } \partial\Omega, \end{aligned} \tag{1.1}$$

where Ω is a bounded polygonal domain in $\mathbb{R}^2$. In this method only harmonic polynomials are used in constructing the finite element space because the exact solution is harmonic. However, the harmonic finite element method of [12, 13] cannot be applied (directly) to the Poisson equation,

$$\begin{aligned} -\Delta u &= f, \quad \text{in } \Omega, \\ u &= 0, \quad \text{on } \partial\Omega, \end{aligned} \tag{1.2}$$

where Ω is a bounded polyhedral domain in $\mathbb{R}^3$.

Let $\mathcal{T}_h$ be a tetrahedral grid of size h on a polyhedral domain Ω in $\mathbb{R}^3$. Let $\partial\mathcal{K} = \cup_{K\in\mathcal{T}_h}\partial K$. When using the standard Lagrange finite elements to solve (1.2), the solution is given by

$$u_h = \sum_{\mathbf{x}_i\in\partial\mathcal{K}\backslash\partial\Omega} u_i\phi_i + \sum_{\mathbf{x}_j\in\Omega\backslash\partial\mathcal{K}} u_j\phi_j + \sum_{\mathbf{x}_k\in\partial\Omega} c_k\phi_k, \tag{1.3}$$

where $\{\phi_i, \phi_j, \phi_k\}$ are the nodal basis functions at element-boundary, element-interior and domain boundary, respectively, c_k are interpolated values on the boundary, and both u_i and u_j are obtained from the Galerkin projection by solving of a linear system of equations.

In [14] an interpolated Galerkin finite element method is proposed for the 2D Poisson equation. In this paper, we extend this idea to the trivariate setting. The main idea can be described as follows. We add non-harmonic polynomial basis functions to the harmonic finite element solution of [12, 13] to obtain a solution to (1.2). That is, the solution is obtained as

$$u_h = \sum_{\mathbf{x}_i\in\partial\mathcal{K}\backslash\partial\Omega} u_i\phi_i + \sum_{\mathbf{x}_j\in\Omega\backslash\partial\mathcal{K}} c_j\phi_j + \sum_{\mathbf{x}_k\in\partial\Omega} c_k\phi_k, \tag{1.4}$$

where both c_j and c_k are interpolated values (of the right hand side function f, or of the boundary condition), and only u_i are obtained from the Galerkin projection. In these constructions, the linear system of Galerkin projection equations involves only the unknowns on $\partial\mathcal{K} \setminus \partial\Omega$. The number of unknowns on each element is reduced from $\binom{k+3}{3}$ to $2k^2+2$, i.e., from $O(k^3)$ to $O(k^2)$. Compared to the standard finite element, the new linear system is smaller (good for a direct solver) and has a better condition number (by numerical examples in this paper.)

This method is similar to, but different from, the standard Lagrange finite element method with static condensation. In the latter, internal degrees of freedom on each element remain unknowns and are represented by the element-boundary unknowns. For example, the Jacobi iterative solutions of condensed equations are identical to those of original equations with a proper unknown ordering (internal unknowns first). That is, the static condensation is a method for solving linear systems of equations arising from the high order finite element discretization, which does not define a different system. In the new method the coefficients of some degrees of freedom are no longer unknowns but given directly by the data. For ease of analysis we use a local integral of the right hand side function f to determine these coefficients. We can simply use the pointwise values of f instead. The new method is like the standard Lagrange finite element method when some "boundary values" are given on every element.

The paper is organized as follows. In Sects. 2 and 3, for arbitrary tetrahedral partitions, we construct a P_2 and a P_3 nonconforming interpolated Galerkin finite elements with one internal Laplacian basis function for each tetrahedron. In Sect. 4, for arbitrary tetrahedral partitions, we construct quartic and higher degree conforming interpolated Galerkin finite elements with $\binom{k-1}{3}$ internal Laplacian basis functions for each tetrahedron. In Sect. 5, we show that the interpolated Galerkin finite element solution converges at the optimal order. In Sect. 6, numerical tests are provided to compare the interpolated Galerkin finite elements (P_2 to P_6) with the standard ones.

2 The P_2 Nonconforming Interpolated Galerkin Finite Element

Let $\mathcal{T}_h$ be a quasi-uniform tetrahedral grid of size h on a polyhedral domain Ω in $\mathbb{R}^3$. On all interior tetrahedra, a P_2 nonconforming finite element function must have continuous moments of degree one. Let $K := [\mathbf{x}_1, \mathbf{x}_2, \mathbf{x}_3, \mathbf{x}_4]$ be a tetrahedron in $\mathcal{T}_h$ with vertices v_i, and let $(\lambda_1, \lambda_2, \lambda_3, \lambda_4)$ be the barycentric coordinates associated with K. That is, λ_i is a linear function on K assuming value 1 at $\mathbf{x}_i$, and vanishing on the face F_i opposite vertex $\mathbf{x}_i$. From [1, 5], we know that there is only one nonconforming quadratic bubble function per K:

$$\phi_0 = c_0(2 - 4\sum_{i=1}^{4} \lambda_i^2), \tag{2.1}$$

where the constant c_0 is determined by (2.2) below, satisfying three vanishing 1-moment conditions on every face F_i of K, and one 0-moment of Laplacian condition on K:

$$\int_{F_i} \phi_0 \lambda_j^\alpha \lambda_k^\beta \lambda_l^\gamma = 0, \quad i \neq j \neq k \neq l, \ \alpha+\beta+\gamma = 1,$$

$$\int_K \Delta\phi_0 \phi_0 \, d\mathbf{x} = 1. \tag{2.2}$$

For a P_2 element on K, there are ten domain points located at the vertices and mid-edges of K. Let $\{\psi_i\}_{i=1}^{10}$ be the Lagrange basis functions of a conforming P_2 element on K, i.e., each ψ_i assumes value 1 at one domain point and vanishes at the remaining nine. We define the interpolated Galerkin finite element basis as follows:

$$\phi_i = \psi_i - \phi_0 \int_K \Delta\psi_i \phi_0 d\mathbf{x},$$

$$\int_K \Delta\phi_i \, d\mathbf{x} = 0, \quad i = 1, \ldots, 10.$$

We define the P_2 nonconforming interpolated Galerkin finite element space by

$$\begin{aligned} V_h = \{v_h \mid\ & v_h \text{ has continuous 1-moments on face triangle,} \\ & v_h \text{ has vanishing 1-moments on boundary triangle,} \\ & v_h|_K = \sum_{i=1}^{10} c_i\phi_i + u_0\phi_0 \ \text{ on each } K \in \mathcal{T}_h\}. \end{aligned} \tag{2.3}$$

The interpolated Galerkin finite element solution for the Poisson equation (1.2) is defined by

$$u_h = \sum_{K\in\mathcal{T}_h} \Big(\sum_{i=1}^{10} c_i\phi_i - \phi_0 \int_K f(\mathbf{x})\phi_0 d\mathbf{x} \Big) \in V_h \tag{2.4}$$

such that

$$(\nabla_h u_h, \nabla_h v_h) = (f, v_h) \quad \forall v_h = \sum_{K\in\mathcal{T}_h} (\sum_{i=1}^{10} v_i\phi_i) \in V_h, \tag{2.5}$$

where ∇_h denotes a picewise defined gradient, and the dependency of ϕ_i on K is omitted for brevity of notation.

3 The P_3 Nonconforming Interpolated Galerkin Finite Element

Let K be a tetrahedron with the associated barycentric coordinates $(\lambda_1, \lambda_2, \lambda_3, \lambda_4)$, as defined in Sect. 2. Then there is precisely one nonconforming cubic bubble function on K,

$$\phi_0 = c_0\Big(\sum_{i=1}^{4}(5\lambda_i^3 + 90\frac{\lambda_1\lambda_2\lambda_3\lambda_4}{\lambda_i}) - 3\Big), \tag{3.1}$$

satisfying vanishing 1-moment conditions on every face F_i of K, and one 0-moment of Laplacian condition on K:

$$\int_{F_i} \phi_0 \lambda_j^\alpha \lambda_k^\beta \lambda_l^\gamma = 0, \quad i \neq j \neq k \neq l,\ \alpha + \beta + \gamma = 2,$$
$$\int_K \Delta\phi_0\phi_0 \, d\mathbf{x} = 1.$$

For cubic finite elements, there are twenty domain points in each K, and twenty Lagrange basis functions $\{\psi_i\}_{i=1}^{20}$. We define the interpolated Galerkin finite element basis as follows

$$\phi_i = \psi_i - \phi_0 \int_K \Delta\psi_i\phi_0 d\mathbf{x}, \quad i = 1, .., 20.$$

The P_3 nonconforming interpolated Galerkin finite element space is defined by

$$\begin{aligned} V_h = \{v_h \mid\ & v_h \text{ has continuous 2-moments on face triangle,} \\ & v_h \text{ has vanishing 2-moments on boundary triangle,} \\ & v_h|_K = \sum_{i=1}^{20} c_i\phi_i + u_0\phi_0 \text{ on each } K \in \mathcal{T}_h\}. \end{aligned} \tag{3.2}$$

The P_3 interpolated Galerkin finite element solution for the Poisson equation (1.2) is defined by

$$u_h = \sum_{K\in\mathcal{T}_h} \Big(\sum_{i=1}^{20} c_i\phi_i - \phi_0 \int_K f(\mathbf{x})\phi_0 d\mathbf{x}\Big) \in V_h \tag{3.3}$$

such that

$$(\nabla_h u_h, \nabla_h v_h) = (f, v_h) \quad \forall v_h = \sum_{K \in \mathcal{T}_h} (\sum_{i=1}^{20} v_i \phi_i) \in V_h. \tag{3.4}$$

4 The P_k, $k \geq 4$, Conforming Interpolated Galerkin Finite Element

Let K be a tetrahedron with the associated barycentric coordinates $(\lambda_1, \lambda_2, \lambda_3, \lambda_4)$, as defined in Sect. 2. For $k \geq 4$, there are $\binom{k-1}{3}$ domain points strictly interior to K. We shall refer to them as internal degrees of freedom. In this section, we define a P_k interpolated Galerkin conforming finite element on general tetrahedral grids, where the internal degrees of freedom are determined by interpolating the values of the function f on the right hand side of (1.2).

We first deal with $\binom{k+3}{3} - \binom{k-1}{3} = 2k^2 + 2$ domain points on the boundary of K:

$$\mathcal{D} := \Big\{(i_1\mathbf{x}_1 + i_2\mathbf{x}_2 + i_3\mathbf{x}_3 + i_4 + \mathbf{x}_4)/k \mid 0 \leq i_j \leq k, \sum_{j=1}^{4} i_j = k, \prod_{j=1}^{4} i_j = 0\Big\}. \tag{4.1}$$

The first $(2k^2+2)$ linear functionals $F_l := u(\xi_l)$, $\xi_l \in \mathcal{D}, l = 1, \ldots, 2k^2+2$, (the dual basis of the finite element basis) are nodal values at these face Lagrange nodes. The remaining $\binom{k-1}{3}$ linear functionals are the weighted Laplacian $(k-4)$-moments corresponding to the strictly interior domain points. Let $\mathcal{B}$ be a basis for P_{k-4}, and let

$$\Big\{F_j(\Delta u) = \int_K p_j \prod_{i=1}^{4} \lambda_i \Delta u \, d\mathbf{x} \mid p_j \in \mathcal{B}, \;\; j = 2k^2+3, \ldots, \binom{k+3}{3}\Big\}. \tag{4.2}$$

Lemma 1 *The set of linear functionals in* (4.1) *and* (4.2) *uniquely determines a polynomial of degree* $\leq k$.

Proof We have a square linear system of equations. Thus, we only need to show the uniqueness of the solution. Let u_h have zero values for all these linear functionals. Therefore, u_h is identically zero on the boundary of K. Then

$$u_h = u_4 \prod_{i=1}^{4} \lambda_i \qquad \text{for some } u_4 \in P_{k-4}.$$

Letting $p = u_4$ in (4.2), we obtain

$$0 = \int_K u_4 \prod_{i=1}^{4} \lambda_i \Delta u \, d\mathbf{x} = -\int_K \nabla u_h \cdot \nabla u_h d\mathbf{x}$$

and, consequently, $\nabla u_h = 0$ on K. Thus, u_h is a constant on K. As $u_h = 0$ on ∂K, $u_h = 0$.

Let $\{\phi_i\}_{i=1}^{\dim P_k}$ be the basis of P_k dual to the set of linear functions defined by (4.1) and (4.2). In particular, the first $2k^2 + 2$ functions ϕ_i are dual to (4.1), and the remaining ones are dual to (4.2) Then, the P_k ($k \geq 4$) interpolated Galerkin finite element space is defined as follows:

$$V_h = \{v_h \in H_0^1(\Omega) \, : v_h|_K = \sum_{i=1}^{2k^2+2} c_i \phi_i + \sum_{j=2k^2+3}^{\dim P_k} v_j \phi_j \text{ on each } K \in \mathcal{T}_h\}, \tag{4.3}$$

where each ϕ_i and ϕ_j depend on K. The interpolated Galerkin finite element solution for the Poisson equation (1.2) is defined by

$$u_h = \sum_{K \in \mathcal{T}_h} \Big(\sum_{i=1}^{2k^2+2} c_i \phi_i - \sum_{j=2k^2+3}^{\dim P_k} F_j(f) \phi_j \Big) \in V_h \tag{4.4}$$

such that

$$(\nabla u_h, \nabla v_h) = (f, v_h) \quad \forall v_h = \sum_{K \in \mathcal{T}_h} (\sum_{i=1}^{2k^2+2} v_i \phi_i) \in V_h. \tag{4.5}$$

5 Convergence Theory

We prove convergence for conforming and nonconforming interpolated Galerkin finite elements separately. The conforming case is considered first.

Theorem 1 *Let u and u_h be the exact solution of* (1.2) *and the finite element solution of* (4.5)*, respectively. Then*

$$\|u - u_h\|_1 \leq C h^k \|u\|_{k+1}, \tag{5.1}$$

where $\| \cdot \|_i$ is the standard Sobolev $H^i(\Omega)$ norm

Proof Testing (1.2) by $v_h = \sum_{K \in \mathcal{T}_h} \sum_{i=1}^{2k^2+2} v_i \phi_i \in H_0^1(\Omega)$, we have

$$(\nabla u, \nabla v_h) = (f, v_h). \tag{5.2}$$

Subtracting (4.5) from (5.2),

$$(\nabla(u - u_h), \nabla v_h) = 0. \tag{5.3}$$

On one element K, testing (1.2) by $v_h = \phi_j \in H_0^1(K)$ for $j > 2k^2 + 2$, using (4.2) we obtain

$$\begin{aligned}(\nabla(u - u_h), \nabla\phi_j) &= -\int_K \Delta u\phi_j d\mathbf{x} + \int_K \Delta u_h\phi_j d\mathbf{x} \\ &= \int_K f\phi_j d\mathbf{x} - F_j(f) = 0.\end{aligned} \tag{5.4}$$

Combining (5.3) and (5.4) implies

$$\begin{aligned}|u - u_h|_1^2 &= (\nabla(u - u_h), \nabla(u - I_h u)) \\ &\le |u - u_h|_1 |u - I_h u|_1,\end{aligned}$$

where I_h is the interpolation operator to V_h. The following inequalities complete the proof:

$$\|u - u_h\|_1 \le C|u - u_h|_1 \le C|u - I_h u|_1 \le Ch^k \|u\|_{k+1}.$$

Next we consider the two nonconforming cases.

Theorem 2 *Let u and u_h be the exact solution of* (1.2) *and either the finite element solution of* (2.5) *or of* (3.4)*, respectively. Then*

$$|u - u_h|_{1,h} \le Ch^k \|u\|_{k+1}, \tag{5.5}$$

where $|\cdot|_{1,h}^2 = (\nabla_h\cdot, \nabla_h\cdot)$, $k = 2$ and 3 for (2.5) *and* (3.4)*, respectively, and $\|\cdot\|_{k+1}$ is the standard Sobolev $H^{k+1}(\Omega)$ norm.*

Proof We shall prove the case of $k = 2$. The proof of the cubic case is similar. Let $\tilde{u}_h = \sum_{K\in\mathcal{T}_h} \left(\sum_{i=1}^{10} \tilde{u}_i\phi_i + \tilde{u}_0\phi_0\right) \in V_h$ be the Galerkin finite element solution, i.e.,

$$(\nabla_h \tilde{u}_h, \nabla_h v_h) = (f, v_h) \quad \forall v_h \in V_h. \tag{5.6}$$

Testing (5.6) by $v_h = \phi_0$ on some $K \in \mathcal{T}_h$, we get

$$(\nabla_h \tilde{u}_h, \nabla_h \phi_0) = (f, \phi_0)_K = \int_K f(\mathbf{x})\phi_0 \, d\mathbf{x},$$

$$(\nabla_h \tilde{u}_h, \nabla_h \phi_0) = -\int_K \Delta \tilde{u}_h \phi_0 \, d\mathbf{x}$$
$$= -\tilde{u}_0 \int_K \Delta \phi_0 \phi_0 \, d\mathbf{x} = -\tilde{u}_0,$$

where in the integration by parts, we use the fact $\nabla \tilde{u}_h \cdot \underline{\mathrm{n}}$ is a polynomial of a smaller (in fact, one less) degree on the boundary of K. That is, $u_h = \tilde{u}_h$, i.e., u_h satisfies (5.6).

Let $w_h \in V_h$. Then

$$|u - u_h|_{1,h} \leq |u - w_h|_{1,h} + |u_h - w_h|_{1,h} = |u - w_h|_{1,h}$$
$$+ \sup_{v_h \in V_h} \frac{(\nabla_h(u_h - w_h), \nabla_h v_h)}{|v_h|_{1,h}} \leq |u - w_h|_{1,h}$$
$$+ \sup_{v_h \in V_h} \frac{(\nabla_h(u - u_h), \nabla_h v_h)}{|v_h|_{1,h}} + \sup_{v_h \in V_h} \frac{(\nabla_h(u - w_h), \nabla_h v_h)}{|v_h|_{1,h}}$$
$$\leq 2|u - w_h|_{1,h} + \sup_{v_h \in V_h} \frac{(\nabla_h(u - u_h), \nabla_h v_h)}{|v_h|_{1,h}}.$$

The first term is the bounded by the interpolation error, i.e., the right hand side of (5.5). We estimate the second term. Let $[v_h]$ denote the jump on an (internal) triangle e of $\mathcal{T}_h$, after choosing an orientation for e. Then

$$(\nabla_h(u - u_h), \nabla_h v_h) = \sum_{K \in \mathcal{T}_h} \int_\partial K \frac{\partial u}{\partial \underline{\mathrm{n}}} v_h dS = \sum_{e \in \partial \mathcal{T}_h} \int_e \frac{\partial u}{\partial \underline{\mathrm{n}}} [v_h] dS$$
$$= \sum_{e \in \partial \mathcal{T}_h} \int_e \left(\frac{\partial u}{\partial \underline{\mathrm{n}}} - \Pi_e \frac{\partial u}{\partial \underline{\mathrm{n}}} \right) (v_h|_{e^+} - \Pi_e v_h|_{e^+} - v_h|_{e^-} + \Pi_e v_h|_{e^-}) dS$$
$$= \left(\sum_{e \in \partial \mathcal{T}_h} \int_e \left(\frac{\partial u}{\partial \underline{\mathrm{n}}} - \Pi_e \frac{\partial u}{\partial \underline{\mathrm{n}}} \right)^2 dS \right)^{1/2} \left(\sum_{e \in \partial \mathcal{T}_h} \int_{e^\pm} (v_h - \Pi_e v_h)^2 dS \right)^{1/2},$$

where Π_e is the L^2 projection onto the space of bivariate linear polynomials $P_1(e)$. By the trace inequality, we continue above estimation,

$$(\nabla_h(u - u_h), \nabla_h v_h) \leq C \left(\sum_{e \in \partial \mathcal{T}_h} \int_e \left(\frac{\partial u}{\partial \underline{\mathrm{n}}} - \frac{\partial I_h u}{\partial \underline{\mathrm{n}}} \right)^2 dS \right)^{1/2}$$
$$\cdot \left(\sum_{K \in \mathcal{T}_h} (\frac{1}{h} \|v_h - E_h v_h\|^2_{L^2(K)} + h \|\nabla(v_h - E_h v_h)\|^2_{L^2(K)}) \right)^{1/2}$$

$$\leq C\left(\sum_{K\in\mathcal{T}_h}(\frac{1}{h}\|\nabla(u-I_hu)\|^2_{L^2(K)}+h\|D^2(u-I_hu)\|^2_{L^2(K)})\right)^{1/2} h^{1/2}|v_h|_{1,h}$$

$$\leq C\left(\sum_{K\in\mathcal{T}_h}(h^{2k-1}|u|^2_{H^{k+1}(K)}\right)^{1/2} h^{1/2}|v_h|_{1,h}\leq Ch^k|u|_{k+1}|v_h|_{1,h},$$

where I_h is the standard interpolation operator to V_h, see [11], and $E_hv_h \in P_k(K)$ is a stable extension of (moments of) $\Pi_e v_h$ inside K. The proof is complete.

6 Numerical Tests

Let the domain of the boundary value problem (1.2) be $\Omega = [0, 1]^3$, and let $f(x) = 3\pi^2 \sin\pi x \sin\pi y \sin\pi z$. The exact solution is $u(x, y) = \sin\pi x \sin\pi y \sin\pi z$. In all numerical tests on P_k interpolated Galerkin finite element methods in this section, we choose a family of uniform grids shown in Fig. 1.

We solve problem (1.2) first by the P_2 interpolated Galerkin conforming finite element method defined in (2.3), and by the P_2 nonconforming finite element method, on same grids. The errors and the orders of convergence are listed in Table 1. We have one order of superconvergence for the interpolated Galerkin finite element method (2.5), in both H^1 semi-norm and L^2 norm. We note that the standard P_2 conforming finite element method has one order of superconvergence in both H^1 semi-norm and L^2 norm. But the nonconforming P_2 element has the optimal order of convergence only.

Next we solve the same problem by the interpolated Galerkin P_3 finite element method (3.4) and by the P_3 nonconforming finite element method. The errors and the orders of convergence are listed in Table 2. Both methods converge in the optimal order.

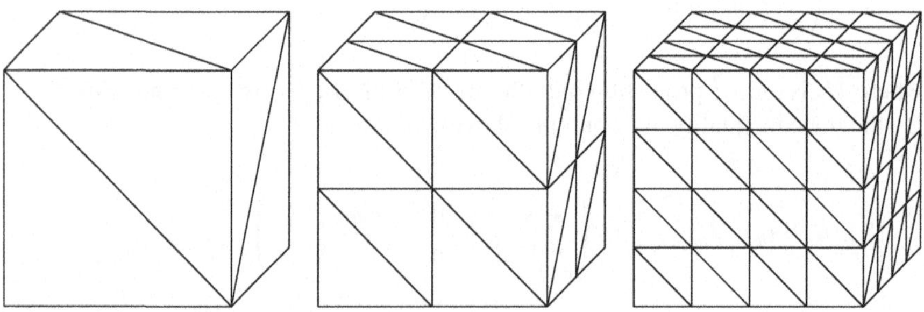

Fig. 1 The first three levels of grids

Table 1 The error $e_h = I_h u - u_h$ and the order of convergence, by the P_2 interpolated Galerkin finite element (2.3) and by the P_2 nonconforming finite element

	$\|e_h\|_0$	h^n	$\|e_h\|_1$	h^n	$\|e_h\|_0$	h^n	$\|e_h\|_1$	h^n
Grid	P_2 interpolated element				P_2 nonconforming element			
3	0.3298E−02	3.2	0.5367E−01	2.3	0.3769E−02	2.9	0.1079E+00	1.8
4	0.2538E−03	3.7	0.8380E−02	2.7	0.4439E−03	3.1	0.2751E−01	2.0
5	0.1704E−04	3.9	0.1150E−02	2.9	0.5315E−04	3.1	0.6861E−02	2.0
6	0.1089E−05	4.0	0.1493E−03	2.9	0.6542E−05	3.0	0.1712E−02	2.0
7	0.6859E−07	4.0	0.1898E−04	3.0	0.8141E−06	3.0	0.4276E−03	2.0

Table 2 The error $e_h = I_h u - u_h$ and the order of convergence, by the P_3 interpolated Galerkin finite element (3.2) and by the P_3 nonconforming finite element

	$\|e_h\|_0$	h^n	$\|e_h\|_1$	h^n	$\|e_h\|_0$	h^n	$\|e_h\|_1$	h^n
Grid	P_3 interpolated element				P_3 nonconforming element			
3	0.3943E−03	4.0	0.1453E−01	2.7	0.3957E−03	4.0	0.1464E−01	2.7
4	0.2320E−04	4.1	0.1995E−02	2.9	0.2341E−04	4.1	0.2004E−02	2.9
5	0.1423E−05	4.0	0.2589E−03	2.9	0.1439E−05	4.0	0.2598E−03	2.9
6	0.8854E−07	4.0	0.3283E−04	3.0	0.8964E−07	4.0	0.3293E−04	3.0

Table 3 Comparison of P_4 interpolated Galerkin and conforming Lagrange finite elements

	P_4 interpolated element				P_4 Lagrange element			
Grid	$\|e_h\|_0$	h^n	$\|e_h\|_1$	h^n	$\|e_h\|_0$	h^n	$\|e_h\|_1$	h^n
3	0.5624E−04	4.8	0.2450E−02	3.8	0.5577E−04	4.8	0.2471E−02	3.8
4	0.1794E−05	5.0	0.1587E−03	3.9	0.1789E−05	5.0	0.1592E−03	4.0
5	0.5592E−07	5.0	0.1000E−04	4.0	0.5587E−07	5.0	0.1002E−04	4.0
# unknowns	225471				250047			
# iterations	927				3050			
CPU	95.5				308.6			

Finally, we solve the problem by the interpolated Galerkin P_4, P_5, and P_6 finite element methods, (4.3) with $k = 4, 5, 6$, and by the P_4, P_5, and P_6 conforming finite element methods. The errors and the orders of convergence are listed in Tables 3, 4, and 5. The optimal order of convergence is achieved in every case. Also in the table, we list the number of unknowns, the number of conjugate iterations used in solving the resulting linear system of equations, and the computing time, on the last level computation. The number of unknowns for the P_6 element is only about 2/3 of that of the P_6 Lagrange element. The number of iterations for the P_6 interpolated element is less than 1/16 of that of the Lagrange element. The conditioning of the system of the new element is much better while giving also a slightly better solution. For the P_6 elements, the new method uses less than 1/10 of the computer time than that of the standard finite element.

Table 4 Comparison of P_5 interpolated Galerkin and conforming Lagrange finite elements

	P_5 interpolated element				P_5 Lagrange element			
Grid	$\|e_h\|_0$	h^n	$\|e_h\|_1$	h^n	$\|e_h\|_0$	h^n	$\|e_h\|_1$	h^n
2	0.3343E−03	5.7	0.9230E−02	4.7	0.3402E−03	5.7	0.9354E−02	4.7
3	0.5719E−05	5.9	0.3282E−03	4.8	0.5739E−05	5.9	0.3295E−03	4.8
4	0.8999E−07	6.0	0.1065E−04	4.9	0.8967E−07	6.0	0.1068E−04	4.9
# unknowns	47031				59319			
# iterations	877				7080			
CPU	31.1				267.0			

Table 5 Comparison of P_6 interpolated Galerkin and conforming Lagrange finite elements

	P_6 interpolated element				P_6 Lagrange element			
Grid	$\|e_h\|_0$	h^n	$\|e_h\|_1$	h^n	$\|e_h\|_0$	h^n	$\|e_h\|_1$	h^n
1	0.2234E−02	0.0	0.5041E−01		0.3098E−02	0.0	0.6026E−01	
2	0.5798E−04	5.3	0.2134E−02	4.6	0.5866E−04	5.7	0.2153E−02	4.8
3	0.5037E−06	6.8	0.3700E−04	5.8	0.5046E−06	6.9	0.3713E−04	5.9
# unknowns	8327				12167			
# iterations	876				14335			
CPU	18.0				181.5			

Acknowledgement The first author is partially supported by a grant from the Simons Foundation #235411 to Tatyana Sorokina.

References

1. Alfeld, P., Sorokina T.: Linear differential operators on bivariate spline spaces and spline vector fields. BIT Numer. Math. **56**(1), 15–32 (2016)
2. Arnold, D.N., Boffi, D., Falk, R.S.: Approximation by quadrilateral finite elements. Math. Comput. **71**(239), 909–922 (2002)
3. Brenner, S.C., Scott, L.R.: The mathematical theory of finite element methods. In: Texts in Applied Mathematics, vol. 15. Springer, New York (2008)
4. Falk, R.S., Gatto P., Monk, P.: Hexahedral H(div) and H(curl) finite elements. ESAIM Math. Model. Numer. Anal. **45**(1), 115–143 (2011)
5. Fortin, M.: A three-dimensional quadratic nonconforming element. Numer. Math. **46**, 269–279 (1985)
6. Hu, J., Huang Y., Zhang S.: The lowest order differentiable finite element on rectangular grids. SIAM Numer. Anal. **49**(4), 1350–1368 (2011)
7. Hu, J., Zhang, S.: The minimal conforming H^k finite element spaces on R^n rectangular grids. Math. Comput. **84**(292), 563–579 (2015)
8. Hu, J., Zhang, S.: Finite element approximations of symmetric tensors on simplicial grids in R^n: the lower order case. Math. Models Methods Appl. Sci. **26**(9), 1649–1669 (2016)
9. Huang, Y., Zhang, S.: Supercloseness of the divergence-free finite element solutions on rectangular grids. Commun. Math. Stat. **1**(2), 143–162 (2013)
10. Schumaker, L.L., Sorokina, T., Worsey, A.J.: A C1 quadratic trivariate macro-element space defined over arbitrary tetrahedral partitions. J. Approx. Theory **158**(1), 126–142 (2009)

11. Scott, L.R., Zhang, S.: Finite element interpolation of nonsmooth functions satisfying boundary conditions. Math. Comput. **54**, 483–493 (1990)
12. Sorokina, T., Zhang, S.: Conforming harmonic finite elements on the Hsieh-Clough-Tocher split of a triangle. Int. J. Numer. Anal. Model. **17**(1), 54–67 (2020)
13. Sorokina, T., Zhang, S.: Conforming and nonconforming harmonic finite elements. Appl. Anal. https://doi.org/10.1080/00036811.2018.1504031
14. Sorokina, T., Zhang, S.: An interpolated Galerkin finite element method for the Poisson equation (preprint)
15. Zhang, S.: A C1-P2 finite element without nodal basis. Math. Model. Numer. Anal. **42**, 175–192 (2008)
16. Zhang, S.: A family of 3D continuously differentiable finite elements on tetrahedral grids. Appl. Numer. Math. **59**(1), 219–233 (2009)
17. Zhang, S.: A family of differentiable finite elements on simplicial grids in four space dimensions (Chinese). Math. Numer. Sin. **38**(3), 309–324 (2016)
18. Zhang, S.: A P4 bubble enriched P3 divergence-free finite element on triangular grids. Comput. Math. Appl. **74**(11), 2710–2722 (2017)

Index

G. E. Fasshauer et al. (eds.), *Approximation Theory XVI*, Springer Proceedings in Mathematics & Statistics 336, https://doi.org/10.1007/978-3-030-57464-2

S

T

V

W

Z